Radio Astronomy

Other books by John D. Kraus

Antennas
Electromagnetics

Radio Astronomy

John D. Kraus, Ph.D.
*Professor of Astronomy and Electrical Engineering,
and Director of the Radio Observatory,
The Ohio State University*

With a chapter on Radio-Telescope Receivers by
Martti E. Tiuri
*Doctor of Technology
Professor of Radio Engineering,
Institute of Technology, Helsinki, Finland*

McGraw-Hill Book Company
New York   St. Louis   San Francisco   Toronto   London   Sydney

Radio Astronomy

Preface

Radio astronomy embraces a wide range of topics from astrophysical phenomena to receiver and antenna design. The aim of this book is to bring together a balanced selection and treatment of these topics that is elementary enough to serve as an introduction to radio astronomy yet is sufficiently detailed to be useful as a teaching text and reference work.

Chapters 3, 4, and 5 cover topics, such as radiation laws and wave polarization and propagation, that are fundamental to radio astronomy. These are followed by Chapters 6 and 7 on antennas and receivers. Chapter 8, which is the final and longest chapter, deals with radio sources and our present state of knowledge concerning them as deduced from both radio and optical observations.

The relation of radio astronomy to other techniques for the exploration of space is discussed in Chapter 1, which also includes a brief history of the early years of radio astronomy. Chapter 2 brings together a variety of astronomy fundamentals that may be useful to those who do not have an astronomical background. There are extensive tables throughout the book of important parameters, formulas, and objects. The appendix includes several lists of radio sources.

Except for Chapter 7, the book is an outgrowth of lecture material presented by the author in courses on radio astronomy and related topics at the Ohio State University over a period of years. Much of the material is suitable for use at the last years of undergraduate or first year of graduate level. As preparation it is desirable that a student have a knowledge of vector analysis and of elementary electromagnetic and circuit theory. Numerous references to the literature and very extensive bibliographies are given to facilitate further study of a particular subject. There are many worked examples. These are supplemented by extensive problem sets having a considerable range in difficulty. Some problems serve to illustrate points not covered in the text. Answers to most of the problems are included.

For the most part the rationalized mks system of units is employed, but in a few cases other systems are used where they are more convenient or appropriate. The meaning and units attached to symbols are explicitly detailed following key equations throughout the book.

The book has dozens of tables, over 100 problems, over 300 figures, and over 500 references. In the appendix there are lists of over 1,000 radio sources.

I wish to express my appreciation to my colleagues and students for helpful comments and suggestions. In particular I should like to thank Prof. Robert G. Kouyoumjian of the Ohio State University, who gave very valuable assistance on Chapter 5. Useful suggestions on individual chapters or parts of chapters were made by Profs. W. K. Bonsack, G. W. Collins II, H. C. Ko, D. S. Mathewson, and T. K. Menon of the Ohio State University Departments of Astronomy and Electrical Engineering and P. N. Myers and S. R. O'Donnell of the Ohio State University Radio Observatory staff. Also Professors Ko and Menon have supplied the basis for several problems. The official U.S. Navy photographs of several astronomical objects, kindly supplied by Dr. K. Aa. Strand, were taken with the 61-inch astrometric reflector at the U.S. Naval Observatory, Flagstaff, Arizona. The photographs showing Cygnus A and Centaurus A were kindly provided by Dr. Thomas A. Matthews of the California Institute of Technology.

Professor Martti E. Tiuri of the Finland Institute of Technology, who wrote Chapter 7 on radio-telescope receivers, was a visiting professor at the Ohio State University Radio Observatory during 1961–1962. I have edited Dr. Tiuri's manuscript to make the symbols and terminology consistent with those used in the rest of the book, and any errors in his chapter are my responsibility.

Although great care has been exercised, some errors in the text, tables, lists, or figures will inevitably occur. Anyone finding them will do me a great service by writing me about them so that they can be corrected in subsequent printings.

John D. Kraus

Contents

1

Introduction

1-1 Radio Astronomy and the Exploration of the Universe Until a few decades ago man's knowledge of the universe outside the earth came almost entirely from optical-astronomy observations. Beginning millenniums ago with purely visual techniques, astronomy made rapid advances after the invention of the optical telescope in the early seventeenth century and the application of photographic methods in the last century. All observations were in the visible part of the electromagnetic spectrum in a band about one octave wide. During the last three decades astronomical observations at radio wavelengths have created a new branch of astronomy called *radio astronomy*. The older astronomy in the visible spectrum is now often called *optical astronomy* to distinguish it from the newer branch.

The positions of optical and radio astronomy in the electromagnetic spectrum coincide with the two principal transparent bands of the earth's atmosphere and ionosphere. These transparent bands are commonly referred to as the optical and radio windows. A graph of the relative transparency of the earth's atmosphere is presented in Fig. 1-1, with transparency as ordinate ranging from 0 (opaque) to 1 (perfectly transparent) as a function of wavelength on a logarithmic scale. The optical window extends from about 0.4 to 0.8 micron (1 octave) while the much broader radio window extends from about 1 cm to 10 m (about 10 octaves). The values of 1 cm and 10 m are nominal and arbitrary. Because of some relatively transparent bands in the millimeter region and occasional ionospheric "holes" at decameter wavelengths, more extreme limits of the radio window may be placed at 1 mm and 150 m. The short-wavelength limit is a function of the atmospheric composition, cloud cover, etc., while the long-wavelength limit depends on the electron density in the ionosphere. This, in turn, is a function of the time of day, solar activity, etc.

The division of astronomy into two branches, optical and radio, is a consequence of the fact that almost all observations to date have been ground-based, with the earth's atmosphere interposed between the celestial source and the observer. With the advent of artificial earth satellites, from simple instrumented types to complex, manned space stations, it is becoming possible to conduct astronomical observations above the atmosphere (Liller, 1961; Kraushaar and Clark, 1962). The National Aeronautics and Space Administration has extensive programs under way for orbiting solar observatories (OSO) and orbiting astronomical observatories (OAO). In the near future, astronomical observations may be made over the complete range of the electromagnetic spectrum from the shortest gamma rays to the longest radio waves. The different techniques needed to cover this broad spectrum will result in the development of special orbiting telescopes for gamma rays, X rays, ultraviolet rays, infrared rays, and radio waves (particularly at the lower radio frequencies cut off by the ionosphere). Current observations from rockets and high-altitude balloons form an important prelude to this bridging of the gaps in the spectrum (Friedman, 1959; Liller, 1961).

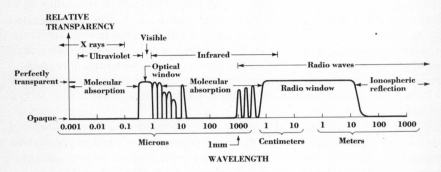

Fig. 1-1. Electromagnetic spectrum showing relative transparency of the earth's atmosphere and ionosphere.

Since all electromagnetic-wave energy is conveyed by photons, the new astronomy embracing the full electromagnetic spectrum may be described collectively as *photon astronomy.* The highest-frequency gamma rays have photon energies of a billion electron volts. On the other hand, a 1-kc radio wave has a photon energy of only a few micromicro electron volts. The range in energy, or frequency, between these extremes is about 10^{20}. This range is so large that widely different techniques will be required in different parts of the spectrum, and it is likely that the techniques in the radio part of the spectrum will continue to be distinctive, with such specialized components as antennas, wave guides, and superheterodyne receivers. Although the various parts of the spectrum may have very dissimilar techniques, the observational data gathered from photons of all energies will form a unified result.

Even at the radio frequencies for which the atmosphere and ionosphere are essentially transparent, there may be important advantages for a space observatory since absorption, refraction, and noise effects will be less.

Another reason is that interference from terrestrial radio transmitters may be reduced. At present, the invasion of more and more channels by larger numbers of higher-powered transmitters seriously threatens the future of ground-based radio astronomy. Although some ground-based radio astronomy may continue to be possible in spite of such interference, it may be necessary for ultimate sensitivity to place radio telescopes in space stations orbiting the earth or to seek even quieter locations, such as may be provided by a lunar-based observatory on the back side of the moon. In choosing an interference-free site, consideration must also be given to the powerful radio emissions from the sun. To radio astronomers not interested in studying these emissions, the sun can be a troublesome source of interference, making nighttime observations necessary or desirable. To reduce such interference a radio telescope in space would require special shielding and other precautions, and even so, the ultimate sensitivity might be achieved only when the telescope was in the shadow of the earth or the moon. The best prospects for avoiding both terrestrial and solar interference would appear to be provided by a lunar-based radio telescope situated on the back side of the moon and operating during the lunar night. But at kilometer wavelengths even such a location may be unsuitable and a site on one of the asteroids may be indicated (Reber, 1964).

So far only a *passive* astronomy, in which celestial objects are studied by means of their natural radiation, has been considered. In recent years, an *active* phase of astronomy has been developed using radar techniques. Here the radiations are man-generated. Radio echoes of signals from powerful terrestrial transmitters have been bounced off the moon, the sun, and several planets. Laser-generated infrared rays have also been reflected back to earth from the moon. However, the time interval required for the waves to make the round-trip journey and the fact that the echo power decreases as the fourth power of the distance impose limitations on the practical range of the radar method. Nevertheless, it is now a powerful technique for studies within the solar system, and its capabilities and importance in this realm will undoubtedly grow (Goldstein, 1964; James, 1964).

Besides emitting photons of electromagnetic energy, celestial objects may radiate in other ways. For example, a massive double-star system sends out gravity waves. It is presumed that gravitational energy, like electromagnetic energy, is quantized and that the carriers, called gravitons, travel with the speed of light. Devices for detecting gravitons are under development (Weber, 1960, 1961). If gravity-wave detection proves feasible, a new branch of astronomy may develop. Since gravitons and photons are both bosons, gravity-wave (or graviton) astronomy and electromagnetic-wave (or photon) astronomy may both be grouped as branches of boson astronomy, as suggested in Fig. 1-2.

This chart also lists other present and future techniques for space exploration. Thus, acoustic waves or magnetohydrodynamic waves generated near the sun's surface can propagate to the earth through the sun's outer corona. This tenuous medium is a gaseous envelope which extends beyond the earth, and observations of mechanical vibrations or motions of the acoustic, magnetohydrodynamic, or other type traveling through it may be regarded as constituting a *pneumatic astronomy*, as suggested in Fig. 1-2.

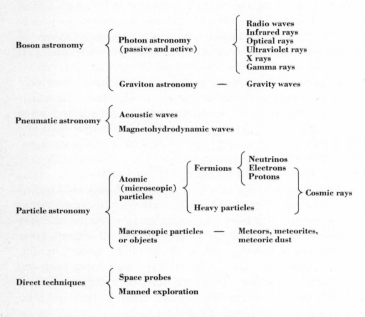

Fig. 1-2. Present and future techniques of space exploration classified into several "astronomies."

Another technique of space exploration or astronomical investigation relies on observations of atomic particles which stream toward the earth in huge numbers from different directions. They include elementary particles such as electrons and protons and heavier, more complex atomic particles. The more energetic of these particles are frequently referred to as *cosmic rays*. Being charged particles, they are deflected by the earth's magnetic field and interstellar fields, so that their direction of origin is difficult to determine. However, some appear to originate on the sun and others from objects, like the Crab nebula, which are also strong radio sources. By studying only cosmic rays of very high energy (about 10^{18} ev) the effect of deflection by the magnetic field is minimized (Rossi, 1959).

Yet another elementary particle with potential for studies of our universe is the neutrino (Chiu and Stabler, 1961). Having no charge, this particle is not deflected by magnetic fields, so that it may travel directly from its point of origin. But the detection of such a neutral virtually massless particle is a formidable problem. Nevertheless, pioneering attempts are in progress to map the neutrino radiation from space (Morrison, 1962).

Neutrinos, electrons, and protons, or collectively fermions, and the heavier atomic particles may be regarded as the carriers for a branch of space exploration called *particle astronomy*. Macroscopic particles, such as meteors, could also be included as a subdivision of this branch, as suggested in Fig. 1-2. Although the groupings in Fig. 1-2 may be convenient, the techniques for detecting the different carriers will vary greatly.

After a meteor falls to the earth's surface it is termed a meteorite. Such objects can be handled and subjected to laboratory tests and analyses. Meteoric dust brought to earth by rain or snow can be studied in the same way. The difference here is that these particles or objects can be examined and reexamined in a leisurely way. They are not so transient as individual photons or fermions.

With the development of artificial satellites and space probes the way has been opened for a more direct approach to many phases of extraterrestrial exploration. Such probes can transport sensing devices for on-the-spot measurements at distant locations and telemeter their findings back to earth. But a human being's capacity for observation and interpretation of diverse and unexpected phenomena will make manned exploration the ultimate in space exploration. As suggested in Fig. 1-2, probes and manned exploration can be grouped together under the heading of *direct techniques*, in contrast to the indirect methods involving observations at a distance by boson or particle astronomy.

The classification of techniques in Fig. 1-2 is not intended to be complete in all respects. Thus, only a few fundamental particles have been listed, and no attempt has been made to include any antiparticles. The classification, like any such scheme, is also quite arbitrary and groupings could be made into different "astronomies." The principal purpose of Fig. 1-2 is to place radio astronomy in perspective in relation to other present and future techniques of space exploration.

The difficulties and high cost of exploration by probes and man tend to limit the range of such sorties. Ultimately the solar system may be spanned but the jump of several light years to the nearest stars puts the use of direct techniques for exploration beyond the planets into the distant future, if, indeed, trips to the stars are ever possible at all (Purcell, 1963).

As in the past, the indirect, at-a-distance techniques will continue to be the only ones available for studying the more remote parts of our universe.   The very small prospective range of probes and manned space travel compared to radio and optical astronomy is illustrated by the six sketches of Fig. 2-8.   Whereas radio and optical telescopes can penetrate far out in the sixth sketch to 5,000 mega-light-years or more, man has not yet covered the interval in the first sketch (1 light-second), and probes have functioned only to the inner parts of the second sketch (1 light-hour).

The objects at the greatest measured distances (5×10^9 light-years) are relatively strong when radio techniques are used but faint optically. It is likely that many of the weaker radio sources are still further away, perhaps near the limits of the universe and beyond the range of optical telescopes. Radio photons are of much lower energy than optical photons and hence are emitted in larger numbers for the same energy radiated. Accordingly, radio astronomy has a fundamental advantage over optical astronomy, or any other photon astronomy, in the detection of distant bjects because it has the least tendency to be photon-limited, i.e., have insufficient photons for significant measurements (von Hoerner, 1964). Thus, radio astronomy is not only the most promising technique currently available for studying the most distant parts of the universe but it may be the only one.   The potential range of neutrino and graviton techniques is still a question for the future.

Observations at the greatest possible distance are essential to an understanding of both the origin and type of universe in which we live. And this knowledge, in turn, may give some clue as to its ultimate destiny.

1-2   A Short History of the Early Years of Radio Astronomy†

The science of radio astronomy had its beginnings in the experiments of Karl G. Jansky in 1931.   Jansky was a radio engineer at the Holmdel, New Jersey, field site of the Bell Telephone Laboratories.   He had been assigned the problem of studying the direction of arrival of thunderstorm static.   This information would be useful since, if a predominant direction was found, beam antennas for transoceanic radio-telephone circuits might be designed to give a minimum response in this direction, thereby improving the signal-to-noise ratio for the circuit.   To study the problem Jansky built a vertically polarized unidirectional beam antenna of the Bruce curtain type, as shown in Fig. 1-3.   The antenna, about 100 ft long by 12 ft high, was mounted on four wheels running on a circular horizontal track so that the antenna could rotate in azimuth.   A synchronous motor turned the structure one revolution every 20 min.   Operating at a wavelength of 14.6 m

† A more complete history of radio astronomy replete with personal anecdotes will be found in Pfeiffer (1956); see also Southworth (1956).

Fig. 1-3. Karl G. Jansky and the rotating antenna array with which he made the first observations of radio waves of extraterrestrial origin. (*Photograph courtesy of Bell Telephone Laboratories.*)

(20.5 Mc), the antenna was connected to a sensitive receiver with the receiver output connected in turn to a pen-on-paper recorder with long time constant.

In a paper published in the December, 1932, issue of the *Proceedings of the Institute of Radio Engineers*, Jansky (1932) reported the first results with his equipment. He was able to identify three groups of static as follows: (1) static from local thunderstorms, (2) static from distant thunderstorms found to come principally from southerly directions, and (3) ". . . a steady hiss type static of unknown origin." A sample record of this third type of hiss static observed by Jansky on February 24, 1932, is shown in Fig. 1-4. Continuing his discussion of the third type, Jansky stated that it produced "a hiss in the phones that can hardly be distinguished from the hiss caused by set noise. The direction of arrival of this static changes

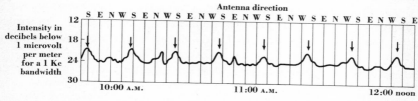

Fig. 1-4. Record obtained by Jansky on Feb. 24, 1932. Peaks (indicated by arrows) occurred at 20-min intervals as the antenna beam swept through the plane of our galaxy. Note that the direction of the peak shifted from nearly south to southwest in about 2 hr. (*After Jansky, 1932.*)

gradually throughout the day going almost completely around the compass in twenty-four hours." Further he said, "This type of static was first definitely recognized only last January [1932]. Previous to this time it had been considered merely as interference from an unmodulated carrier." Thus, if a date of birth is to be given for the science of radio astronomy, January, 1932, would appear to be appropriate.

Because of the 24-hr effect, Jansky speculated that the new static might in some way be associated with the sun. However, in a paper published in October, 1933, entitled "Electrical Disturbances Apparently of Extraterrestrial Origin" (Jansky, 1933), he indicated that further observations "lead to the conclusion that the direction of arrival of these waves is fixed in space, i.e., that the waves come from some source outside the solar system." He gave the approximate coordinates of the region from which the waves appeared to come as 18 hr right ascension and 10° south declination, with an error of less than 30 min in right ascension and 30° in declination. This position with the indicated tolerance embraces the position of the center of our galaxy.

In a third paper† which appeared in October, 1935, Jansky (1935a) was able to demonstrate further that the "radiations are received any time the antenna is directed towards some part of the Milky Way system, the greatest response being obtained when the antenna points towards the center of the system. This fact leads to the conclusion that the source of these radiations is located in the stars themselves or in the interstellar matter distributed throughout the Milky Way." Jansky noted that, if stars were the source, strong radiation should be observed from the sun, whereas at no time had he detected any solar radio radiation. His conclusion that stars are not an important source of the galactic radiation was correct. However, it so happened that Jansky had made his observations during a sunspot minimum. Had he continued his observations a few years more, he would undoubtedly have detected solar radiation during a period of high sunspot activity. But Jansky was soon transferred to other research activities, and years were to pass before further advances were made in the still unrecognized new science of radio astronomy.

Jansky was well aware not only of the astronomical importance of his discovery but also of its practical implications. For example, he wrote in 1935 that "this star static, as I have always contended, puts a definite

---

† I was privileged to hear this paper, which was presented by Jansky on July 3, 1935, at the Summer National Convention of the Institute of Radio Engineers meeting in Detroit, Michigan. In this paper Jansky correlated the radio radiations with various parts of the Milky Way structure. In retrospect, it is probably correct to say that this was the first observational paper on radio astronomy. Although historically this was a very significant occasion, it attracted but little attention. There were scarcely two dozen persons in Jansky's audience.

limit upon the signal strength that can be received from a given direction at a given time and when a receiver is good enough to receive that minimum signal it is a waste of money to spend any more on improving the receiver" (Jansky, 1935*b*). Thus, by 1935 Jansky had identified the origin of the radio radiation with the structure of our galaxy. He had detected the radiation at 14.6 m and also at 10 m. He also understood how this background radiation set a limit to useful receiver sensitivity. He realized that progress in radio astronomy would require larger antennas with sharper beams which could be pointed easily in different directions. In fact, he proposed the construction of a parabolic-mirror antenna 100 ft in diameter for use at meter wavelengths. However, he obtained no support for his proposal, and radio astronomy languished.

In 1937, Grote Reber, a radio engineer living at Wheaton, Illinois, became interested in Jansky's work and constructed a parabolic-reflector antenna 31 ft in diameter in the backyard of his home. This antenna, shown in Fig. 1-5, was mounted as a meridian-transit instrument, being steerable only in declination and relying on the earth's rotation to sweep the antenna beam in right ascension.

The most promising assumption at the time was that the radiation found by Jansky would obey Planck's blackbody radiation law and therefore be stronger at shorter wavelengths. Accordingly, Reber made his first observations at 9.1 cm (3,300 Mc) but with negative results. He then modified his receiver to operate at a longer wavelength of 33 cm (910 Mc). Once more he was unable to detect any celestial radiation, and he again rebuilt his receiver to work at the still longer wavelength of 1.87 m (160 Mc). By the spring of 1939 Reber obtained definite indications of radiation at this wavelength, which showed a marked concentration in the plane of the galaxy, and he published some preliminary results (Reber, 1940*a*, 1940*b*). Although Reber's antenna was smaller than Jansky's, he used a much shorter wavelength, with the result that his antenna beam was considerably sharper. Reber's antenna had a conical beam about 12° between half-power points, while Jansky's beam was fan-shaped with a half-power width of about 30° in the narrower direction.

Reber devoted considerable effort to an understanding of the limitations of his receiving equipment. He recognized that the antenna-receiver combination, which is now usually called a radio telescope, acts like a bolometer, or heat-measuring device, in which the radiation resistance of the antenna measures the equivalent temperature of distant parts of space to which it is projected by the antenna response pattern (Reber, 1942).

While continually improving his equipment, Reber undertook a systematic survey of the sky and in 1944 published the first maps of the radio sky (Reber, 1944). These maps show the background radiation at 1.87 m wavelength (160 Mc) in units of watts per square meter per circular degree per megacycle of bandwidth. The maps constitute the first extensive quantitative measurements of radio radiation from the sky and are remarkably good even compared to present-day maps. Considering the primitive equipment Reber had available, they are a great tribute to his ability. In Reber's words the maps show the maximum radiation "to be

Fig. 1-5. Grote Reber's meridian-transit radio telescope. Many modern radio telescopes bear a striking resemblance to this early instrument.

in the constellation of Sagittarius. Minor maxima appear in Cygnus, Cassiopeia, Canis Major, and Puppis. The lowest minimum is in Perseus." Reber also reported detecting emission from the sun.

There is an interesting story concerning Reber's 1940 paper in the *Astrophysical Journal*. When Reber submitted this paper, Dr. Otto Struve, editor of the *Astrophysical Journal*, was undecided about the wisdom of publishing it. It was the first time a paper reporting radio emission from the sky had been submitted to an astronomical publication, and most persons were either skeptical or puzzled about the results. Dr. Struve did not wish to publish an article which might later prove to be incorrect or inaccurate, and he could find no reviewer willing to defend the paper.

Thus he had ample grounds for rejecting Reber's contribution; but reasoning that a good paper rejected was a greater evil than a poor paper accepted, he approved the article for publication. It appeared during World War II, and although Holland was then under German occupation, a copy of the *Astrophysical Journal* carrying Reber's article eventually reached the observatory at Leiden.

Professor Jan H. Oort, director of the Leiden Observatory, was very much interested in the article and discussed Reber's findings with some of his associates. Oort was quick to perceive that the radiation Reber and Jansky had observed must be a continuum, i.e., radiation extending over a broad spectrum region from wavelengths of, perhaps, less than 1 m to many meters in length. Oort suggested that if some monochromatic-line radiation existed in the radio spectrum, significant advances could result. He referred the matter to Dr. Hendrick van de Hulst, a young astronomer at Leiden, who considered various possible mechanisms which might yield a line or single-frequency type of radiation in the radio spectrum. In 1944 van de Hulst reported that the neutral hydrogen of interstellar space was a likely source. It had a natural frequency of 21.1 cm (1,420 Mc), corresponding to a transition between two closely spaced energy levels of the ground state related to the electron spin or magnetic-dipole orientation. The low probability of the transition (see Sec. 8-9*c*) and lack of knowledge of the density of neutral hydrogen in space made it uncertain whether the line could be detected. However, van de Hulst (1945) suggested that a search be made.

On March 25, 1951, Ewen and Purcell (1951) at Harvard University detected the line in emission. Within a few weeks it was also detected by Muller and Oort (1951) at Leiden and by Christiansen at Sydney, Australia. Subsequently Hagen and McClain (1954) at the U.S. Naval Research Laboratory detected the line in absorption.

Observations of the hydrogen line have already been of immense value, and research at this wavelength has become one of the most important and active phases of radio astronomy. One of the most spectacular results has been the mapping of the structure of our galaxy (see Sec. 8-9*c*). The Dutch group at Leiden has been a leader in this work.

In retrospect it is interesting to note that Reber's paper,† which might not have been published at all, turned out to be one of the classic papers of radio astronomy and also an important catalytic agent in the formation of a new field of research dealing with the hydrogen line.

I became acquainted with Reber in 1941, when we were both employed at the U.S. Naval Ordnance Laboratory, Washington, D.C. We were

---

† The above story concerning Reber's paper has been assembled from conversations I have had with Dr. Otto Struve, Dr. H. C. van de Hulst, and Dr. Grote Reber.

working on top-secret projects having nothing to do with radio. Away from the laboratory conversation turned to radio astronomy, a completely unclassified subject. Reber described his equipment and observations of the Milky Way structure with contagious enthusiasm that furthered my interest. But it was not until about 10 years later, at the Ohio State University, that I had the opportunity to begin a program of radio-astronomy research. As noted in Sec. 8-7, my first radio-astronomy experiment, made with Arthur Adel in 1934, to detect solar radio emission was unsuccessful.

Reber subsequently continued his survey of the sky at shorter wavelengths with his 31-ft backyard telescope and in 1948 published radio maps of the sky at 62.5 cm (480 Mc) (Reber, 1948). He recognized that progress in radio astronomy required the availability of much larger radio telescopes, and in 1948 he proposed the construction of a steerable 220-ft-diameter radio telescope of excellent design. However, he was unable to elicit any interest in its construction.

Following World War II vigorous programs in radio astronomy were begun in England and Australia. In 1946 construction was started on a 218-ft-diameter fixed parabolic radio telescope at the University of Manchester, England. This was completed in 1947, and in 1948 plans were formulated for the construction of a 250-ft-diameter steerable radio telescope at the same institution. Construction began in 1953, with completion in 1957. For many years this telescope has been the largest fully steerable radio telescope in the world (see Fig. 6-38 for photograph).

At Malvern, England, Hey, Parsons, and Phillips (1946) began a survey of the radio sky at a wavelength of 1.7 m (64 Mc). During their survey they noted that the radiation from a region in Cygnus appeared to fluctuate appreciably in intensity about once per minute or so. Bolton and Stanley (1948) observed this Cygnus region with a sea interferometer made by mounting an antenna on a cliff overlooking the ocean near Sydney, Australia. With the high resolution of their interferometer they found that the Cygnus source had an angular diameter of less than 8 min of arc. This was the first intimation that at least some of the celestial radio radiation might come from sources of very small angular extent. The results of Bolton and Stanley on the Cygnus source were confirmed by Ryle and Smith (1948) at Cambridge University, England, using a two-element type of interferometer. Ryle and Smith also discovered the even more intense radio source in Cassiopeia, whose declination is too far north to be observed at Sydney, Australia.

The fluctuations in intensity of the Cygnus source (now called Cygnus A), first noted by Hey, Parsons, and Phillips, were thought for several years to be fluctuations inherent in the source. However, later studies demonstrated that the fluctuations are caused by inhomogeneities in the earth's ionosphere. The fluctuations, now usually called scintillations, are analogous to the twinkling of visual stars as caused by irregularities in the earth's atmosphere.

The position of Cygnus A reported by Bolton and Stanley was remote from any likely visual object which might be the source. Only faint visual stars were in the neighborhood. Cygnus A, however, was such a strong source that it seemed worthwhile to make a diligent search for an optically observable object which might be identified with it. This required more accurate information regarding its position. In 1951, Graham Smith, at Cambridge University, England, obtained a very precise position for the radio source Cygnus A (Smith, 1951), which he communicated in August of that year to Walter Baade of the Mount Palomar Observatory staff. During the next month, with the aid of the 200-in. Mount Palomar telescope, Dr. Baade made long-exposure photographs centered on the Cygnus A position. At Graham Smith's coordinates for Cygnus A, Baade noted a faint smudge like a distant galaxy, which, however, differed from the usual appearance of such an object. The galaxy appeared to be double (see photograph in Fig. 8-60), and Baade speculated that it might represent two distant galaxies in collision. Although the collision hypothesis is no longer generally held, there has been no new satisfactory explanation for the unusual appearance. A spectrum of Cygnus A obtained subsequently by Minkowski made it possible to estimate the distance as about 200 million light-years. Revisions of the distance scale now make the distance estimate for Cygnus A more like 600 million light-years.

Cygnus A has a radio power output of 10^{38} watts. If it were 10 times as far away, it would still be a relatively strong radio source but near the limit of detection of the largest optical telescopes. This spectacular result suggested that radio astronomy might play a leading role in the exploration of the most distant parts of the universe and gave fresh impetus to the subject.

About three years prior to Baade's identification, John G. Bolton (1948) at Sydney, Australia, had identified a strong radio source in Taurus with the Crab nebula. This was the first celestial object emitting radio waves, other than the sun, to be identified with a visual object. In 1950 Brown and Hazard (1951), using the fixed 218-ft-diameter parabola at the University of Manchester, England, detected radio emission from the great Andromeda nebula at 1.9 m wavelength (158 Mc). Although it is one of the nearest of external galaxies, it is not a very strong radio source. It has a radio power output of 10^{32} watts, as compared to 10^{38} watts for Cygnus A. This large difference led to the division of external galaxies into two types, the so-called *normal galaxies*, like the Andromeda nebula, and the *radio galaxies*, like Cygnus A, with much greater (10^3 to 10^8 more) power output.

This was the status of radio astronomy 10 to 15 years ago. Since that time new findings have come with such increasing rapidity that it is difficult to recount them in sequence. A discussion of much of our present state of knowledge is given in Chap. 8.

1-3   A Modern Radio Telescope   The antennas and receivers used in radio astronomy vary so widely that it is difficult to designate any par-

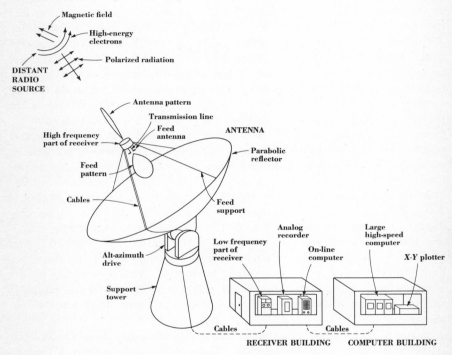

Fig. 1-6.   A modern radio telescope.

ticular arrangement as typical. However, to show the interrelation of some of the principal components, a hypothetical radio telescope is illustrated in Fig. 1-6.

The antenna with its feed collects the radio power from the distant celestial source. After preamplification at the feed point, the signal power is conveyed by cable to the main receiver in a building adjacent to the antenna tower. Here it is amplified further, detected, and integrated, and the output is displayed on an analog recorder (pen on moving paper chart). It may also be recorded in digital form on paper or magnetic tapes for further processing by high-speed computers. Or, as suggested in Fig. 1-6, the output is processed initially by an on-line computer and then fed to a large high-speed computer, which ultimately displays the final result on a large XY plotting machine.

General References

BROWN, R. H., and A. C. B. LOVELL: "The Exploration of Space by Radio," John Wiley & Sons, Inc., New York, 1958.

KRAUS, J. D.: Recent Advances in Radio Astronomy, *IEEE Spectrum*, vol. 1, pp. 78–95, September, 1964.

PAWSEY, J. L., and R. N. BRACEWELL: "Radio Astronomy," Oxford University Press, Fair Lawn, N. J., 1955.

PFEIFFER, J.: "The Changing Universe," Random House, Inc., New York, 1956.

REBER, G.: Early Radio Astronomy at Wheaton, Illinois, *Proc. IRE*, vol. 46, pp. 15–23, January, 1958.

SMITH, F. G.: "Radio Astronomy," Penguin Books, Inc., Baltimore, 1960.

SOUTHWORTH, G. C.: Early History of Radio Astronomy, *Sci. Monthly*, vol. 82, pp. 55–66, February, 1956.

STEINBERG, J. L., and J. LEQUEUX: "Radio Astronomy," translated by R. N. Bracewell, McGraw-Hill Book Company, New York, 1963.

Special References

BAADE, W., and R. MINKOWSKI: Identification of the Radio Sources in Cassiopeia, Cygnus A, and Puppis A, *Astrophys. J.*, vol. 119, pp. 206–214, January, 1954

BOLTON, J. G.: Discrete Sources of Galactic Radio-frequency Noise, *Nature*, vol. 162, p. 141, 1948.

BOLTON, J. G., and G. J. STANLEY: Observations on the Variable Source Radio-frequency Radiation in the Constellation Cygnus, *Australian J. Sci. Res. Ser. A*, vol. 1, p. 58, 1948.

BOLTON, J. G., and O. B. SLEE: Positions of Three Discrete Sources of Galactic Radio-frequency Radiation, *Nature*, vol. 164, p. 101, 1949.

BROWN, R. H., and C. HAZARD: Radio Emission from the Andromeda Nebula, *Monthly Notices Roy. Astron. Soc.*, vol. 3, pp. 357–367, 1951.

CHIU, H. Y., and R. C. STABLER: Emission of Photoneutrinos and Pair Annihilation Neutrinos from Stars, *Phys. Rev.*, vol. 122, pp. 1317–1322, May 15, 1961.

ESHLEMAN, V. R., and A. M. PETERSON: Radar Astronomy, *Sci. Am.*, pp. 50–59, August, 1960.

EWEN, H. I., and E. M. PURCELL: Radiation from Galactic Hydrogen at 1420 Mc/s, *Nature*, vol. 168, pp. 356–357, Sept. 1, 1951.

FISHER, P. C., and A. J. MEYEROTT: Stellar X-ray Emission, *Astrophys. J.*, vol. 139, pp. 123–142, Jan. 1, 1964.

FORWARD, R. L.: General Relativity for the Experimentalist, *Proc. IRE*, vol. 49, pp. 892–904, May, 1961.

FRIEDMAN, H.: Rocket Astronomy, *Sci. Am.*, vol. 200, pp. 52–59, June, 1959.

GOLDSTEIN, R. M.: Radar Investigations of the Planets, *IEEE Trans. Military Electron.*, vol. MIL-8, pp. 199–206, July–October, 1964.

GORDON, W. E.: Radar Back Scatter from the Earth's Ionosphere, *IEEE Trans. Military Electron.*, vol. MIL-8, pp. 206–210, July–October, 1964.

GREENSTEIN, J. L., L. G. HENYEY, and P. C. KEENAN: Interstellar Origin of Cosmic Radiation at Radio Frequencies, *Nature*, vol. 157, p. 805, June 15, 1946.

HAGEN, J. P., and E. F. McCLAIN: Galactic Absorption of Radio Waves, *Astrophys. J.*, vol. 120, pp. 368–370, September, 1954.

HENYEY, L. G., and P. C. KEENAN: Interstellar Radiation from Free Electrons and Hydrogen Atoms, *Astrophys. J.*, vol. 91, pp. 625–630, June, 1940.

HEY, J. S., S. J. PARSONS, and J. W. PHILLIPS: Fluctuations in Cosmic Radiation at Radio Frequencies, *Nature*, vol. 158, p. 234, 1946.

JAMES, J. C.: Radar Echoes from the Sun, *IEEE Trans. Military Electron.*, vol. MIL-8, pp. 210–225, July–October, 1964.

JANSKY, K. G.: Directional Studies of Atmospherics at High Frequencies, *Proc. IRE*, vol. 20, pp. 1920–1932, December, 1932.

JANSKY, K. G.: Electrical Disturbances Apparently of Extraterrestrial Origin, *Proc. IRE*, vol. 21, pp. 1387–1398, October, 1933.

JANSKY, K. G.: A Note on the Source of Interstellar Interference, *Proc. IRE*, vol. 23, pp. 1158–1163, October, 1935a.

JANSKY, K. G.: Personal letter dated Sept. 20, 1935b.

KRAUSHAAR, W. L., and G. W. CLARK: Gamma Ray Astronomy, *Sci. Am.*, vol. 206, pp. 52–61, May, 1962.

LILLER, W. (ed.): "Space Astrophysics," McGraw-Hill Book Company, New York, 1961.

MORRISON, P. On Gamma-ray Astronomy, *Nuovo Cimento*, vol. 7, pp. 858–865, Mar. 16, 1955.

MORRISON, P.: Neutrino Astronomy, *Sci. Am.*, vol. 207, pp. 90–98, August, 1962.

MULLER, C. A., and J. H. OORT: The Interstellar Hydrogen Line at 1,420 Mc/sec and an Estimate of Galactic Rotation, *Nature*, vol. 168, pp. 357–358, 1951.

PURCELL, E: in A. G. W. Cameron (ed.), "Interstellar Communication," W. A. Benjamin, Inc., New York, 1963.

REBER, G.: Cosmic Static, *Astrophys. J.*, vol. 91, pp. 621–624, June 1940a.

REBER, G.: Cosmic Static, *Proc. IRE*, vol. 28, pp. 68–70, February, 1940b.

REBER, G.: Cosmic Static, *Proc. IRE*, vol. 30, pp. 367–378, August, 1942.

REBER, G.: Cosmic Static, *Astrophys. J.*, vol. 100, pp. 279–287, November, 1944.

REBER, G.: Cosmic Static, *Proc. IRE*, vol. 36, pp. 1215–1218, October, 1948.

REBER, G.: Hectometer Cosmic Static, *IEEE Trans. Military Electron.*, vol. MIL-8, pp. 257–263, July–October, 1964.

ROSSI, B.: High Energy Cosmic Rays, *Sci. Am.*, vol. 201, pp. 134–146, November, 1959.

RYLE, M., and F. G. SMITH: A New Intense Source of Radio-frequency Radiation in the Constellation of Cassiopeia, *Nature*, vol. 162, p. 462, 1948.

SMITH, F. G.: An Accurate Determination of the Position of Four Radio Stars, *Nature*, vol. 168, p. 555, 1951.

VAN DE HULST, H. C.: Radio Waves from Space, *Ned. Tijdschr. Natuurk.*, vol. 11, pp. 201–221, 1945.

VON HOERNER, S.: Requirements for Cosmological Studies in Radio Astronomy, *IEEE Trans. Military Electron.*, vol. MIL-8, pp. 282–288, July–October, 1964.

WEBER, J.: Detection and Generation of Gravitational Waves, *Phys. Rev.*, vol. 117, pp. 306–313, Jan. 1, 1960.

WEBER, J.: "General Relativity and Gravitational Waves," Interscience Publishers, Inc., New York, 1961.

YANG, C. N.: "Elementary Particles," Princeton University Press, Princeton, N. J., 1961.

Special Journal Issues on Radio Astronomy

IEEE Trans. Military Electron., vol. MIL-8, July–October, 1964.
IEEE Trans. Antennas Propagation, vol. AP-12, December, 1964.
Proc. IRE Australia, vol. 24, February, 1963.
IRE Trans. Antennas Propagation, vol. AP-9, January, 1961.
Proc. IRE, vol. 46, January, 1958.

2
General Astronomy Fundamentals

2-1 Introduction In this chapter a number of general astronomy topics forming an essential background to a study of radio astronomy are discussed. The topics include brief descriptions of the solar system, our galaxy, and the universe. These are followed by discussions of coordinate systems, measures of time, distance, magnitude, and luminosity. A number of other astronomical topics are treated in later chapters in connection with the interpretation of radio data.

The main purpose of the chapter is to provide reference information useful in connection with other chapters of the book. It is not intended that more than a very brief treatment of these background topics be given, and the reader is referred to the many excellent astronomy texts for more detailed information and for discussions of topics not included here. A number of such texts are listed in the bibliography at the end of this chapter.

2-2 The Solar System Man's home in space is the *earth*, a planet circling the sun (symbol ☉) at a mean distance of about 150×10^6 km. The mean earth-sun distance is taken as one astronomical unit (AU) (see Sec. 2-14).

The plane of the earth's orbit around the sun is called the *plane of the ecliptic*. As shown in Fig. 2-1, the ecliptic is the circle on the celestial sphere† at the intersection of the sphere and the plane of the earth's orbit. The earth's axis of rotation is not perpendicular to the plane of the ecliptic but inclined about 23°.5 to the perpendicular.

There are nine planets, including the earth, revolving around the sun. The closest to the sun is Mercury, at a mean distance of about 58×10^6 km, and the most distant is Pluto, at a mean distance of 6×10^9 km. The orbits of all the planets are relatively close to the plane of the ecliptic, except that of Pluto, which is inclined about 17°.

Six of the nine planets are known to have satellites, as follows: the

† An imaginary surface at a large distance with the earth at its center.

earth, 1 (the moon); Mars and Neptune, 2; Uranus, 5; Saturn, 9; and Jupiter, 12.

In addition to the nine planets there are thousands of small bodies revolving around the sun between the orbits of Mars and Jupiter. These objects, called *asteroids*, may be regarded as minor planets. Most of them are probably less than 50 km in diameter. Only a few are over 150 km in diameter. The largest, Ceres, is 770 km in diameter. Although most

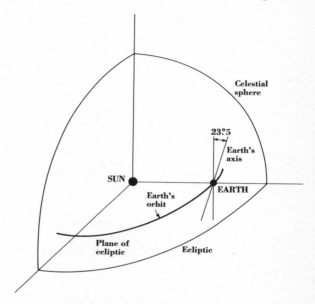

Fig. 2-1.   Earth's orbit and plane of the ecliptic.

asteroids have orbits lying between those of Mars and Jupiter, a few have orbits which extend considerably inside the orbit of Mars or outside that of Jupiter.

The system including the sun, its planets, and asteroids is collectively referred to as the *solar system*. Important characteristics of the planets and the four largest asteroids are summarized in Table 2-1. In this table the planets and asteroids are arranged in order of increasing orbital distance from the sun. This distance is given in millions of kilometers, astronomical units, and light time, i.e., the time for light to travel the distance. More information on the sun and the solar system is given in Chapter 8.

In addition to the sun, planets, and asteroids, the solar system contains significant amounts of gas, dust, and small solids, including meteors and comets, which may be collectively referred to as *interplanetary matter*.

Table 2-1
Planets and asteroids

Planet or asteroid	Symbol	Mean distance from the sun			Diameter (mean), 10^3 km	Inclination		Sidereal period	Length of day	Orbital velocity (km sec^{-1})
		10^6 km	AU	Light time		Orbital	Axial			
Mercury	☿	57.9	0.39	193^s	5	7°	?	88^d	88^d(?)	48
Venus	♀	108.3	0.72	362^s	12.4	3°24'	?	224^d7	250^d(?)	35
Earth	⊕	149.7	1.00†	499^s	12.742	0°†	23°26'	365^d26	1^d†	29.8
Mars	♂	228.1	1.52	761^s	6.9	1°51'	25°12'	687^d	1^d37^m	24
Vesta		354	2.36	20^m	0.4	7°		3^y6	5^h20^m	
Juno		391	2.61	22^m	0.2	13°		4^y4	7^h13^m	
Pallas		410	2.74	23^m	0.5	35°		4^y6		
Ceres		415	2.77	23^m	0.8	11°		4^y6	9^h05^m	
Jupiter	♃	778.7	5.20	43^m	139.8	1°18'	3°7'	11^y9	9^h55^m	14
Saturn	♄	1,427.7	9.54	79^m	115.1	2°29'	26°45	29^y5	10^h38^m	10
Uranus	♅	2,872.4	19.19	2^h7	51	0°46'	98°	84^y0	10^h42^m	7
Neptune	♆	4,500.8	30.07	4^h2	50	1°47'	29°	164^y8	15^h48^m	5
Pluto	♇	5,914.8	39.46	5^h5	13(?)	17°19'	?	247^y7	?	5

† By definition.

2-3   Our Galaxy   The sun is but one of many millions of stars which form a huge flattened system turning like a great disk or wheel in space. An idealized sketch of this system, called our *galaxy* or *Milky Way* system, is presented in Fig. 2-2a. There is a dense central region, or *nucleus*. Outside the nucleus are spiral arms extending out to a large distance. The spiral arms and space between them occupy a space having the shape of a flat *disk*. The arms thin out near the edge of the disk. This makes it difficult to assign a definite diameter to the disk, but it is at least 100,000 light-years.

Since we are situated inside our galaxy, we cannot observe it from an external vantage point like that assumed in the sketch of Fig. 2-2a. With optical telescopes it is possible, however, to observe other galactic systems at great distances from our own. Some of these are presumably rather similar to our galaxy. An example of such a galaxy is M 81 in Ursa Major,† a photograph of which is shown in Fig. 2-2b. Although our galaxy

† The designation M 81 signifies object 81 in Messier's catalog of 103 objects. This catalog was compiled by Charles Messier (1730–1817), the French astronomer who had an interest in comets. External galaxies and other objects of nebulous appearance in low-power telescopes were listed by Messier to prevent loss of observing time by comet searchers who might mistake these objects for comets and spend time observing them until they found that they did not move as comets do. A list of Messier's objects is given in Appendix 4.

may differ from M 81 in many details, the photograph may be regarded as suggesting how our galaxy might appear if viewed from a great distance. In the photograph of Fig. 2-2*b* the central part of the galaxy is overexposed, giving the appearance of a larger nucleus than is actually present.

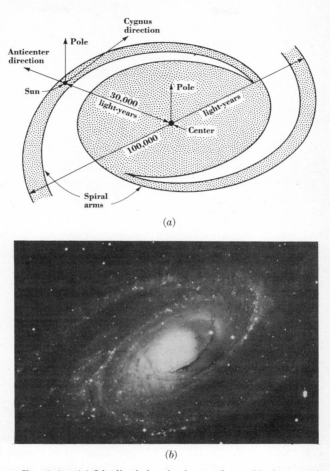

(*a*)

(*b*)

Fig. 2-2.   (*a*) Idealized sketch of our galaxy; (*b*) photograph of nearby galaxy M 81 in Ursa Major. (*Official U.S. Navy photograph, courtesy of Dr. K. Aa. Strand.*)

The solar system is situated in one of the spiral arms of our galaxy at a distance of about 30,000 light-years from the center. The sun and the stars in its neighborhood revolve slowly around the nucleus, completing one revolution in about 300 million years. The intersection of the central plane of the disk with the celestial sphere forms the *galactic equator*.

The total mass of our galaxy is estimated to be about 10^{10} to 10^{11} solar masses.

An idealized cross section through our galaxy is presented in Fig. 2-3a. A spherical region, or halo, surrounding the disk is also shown. It is of lower density than the disk or nucleus, but its actual composition,

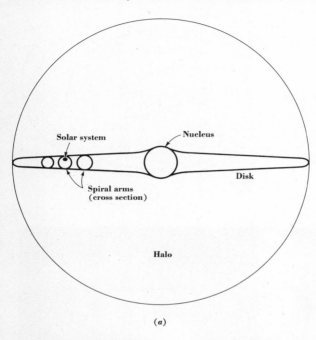

(*a*)

Fig. 2-3. (*a*) Idealized cross section through our galaxy; (*b*) photograph of galaxy NGC 4565, which is seen edge on (facing page). (*Official U.S. Navy photograph, courtesy of Dr. K. Aa. Strand.*)

shape, and extent are uncertain. By way of comparison, a photograph of a distant galaxy (NGC 4565)† seen edge on is shown in Fig. 2-3b. The flat disk and bright nucleus are striking features of this object. There is also a dark band of obscuring matter along the central plane. No trace

† The designation NGC 4565 refers to object 4565 in the New General Catalog, published by J. L. E. Dreyer in 1888. This catalog of nebulas and star clusters superseded the General Catalog, published in 1864 by Sir John Herschel. The Andromeda nebula, which is object 31 in Messier's catalog, is object 224 in the New General Catalog. Thus, it is referred to as M 31 or NGC 224. The New General Catalog has two later supplements, known as the Index Catalogs (IC). The New General Catalog and two Index Catalogs contain a total of over 13,000 objects.

appears in the photograph, however, of a halo, although one might be present.   This very tenuous region is best observed by radio techniques. The structure of our galaxy is discussed in more detail in Chap. 8.

Before the nature of the external galaxies was understood, they were classed as nebulas because of their diffuse, nebulous appearance.   Later, when it was found that they were outside our galaxy, they were designated as *extragalactic nebulas* to distinguish them from diffuse, nebulous objects in our own galaxy.   The trend at present is to refer to the extragalactic nebulas as *external galaxies*.

(*b*)

The stars nearest to the sun are at a distance of about 4 light-years, and within a radius of 10 light-years from the sun there are only about a dozen stellar objects.   This star density of approximately 1 star per 350 cubic light-years is probably typical of spiral-arm regions in our galaxy at our distance from the nucleus.   However, between the arms or farther out in the galaxy the density must be less, while in the arms nearer the nucleus or in the nucleus itself the density may be much higher.

Within 16 light-years of the sun 55 stellar objects have been found. Of these 31 are isolated stars (like the sun), 18 are paired in 9 binary-star systems (such as Sirius), and 6 form two triple-star systems.   Table 2-2 lists the 10 known stellar objects nearest to the sun with their apparent magnitude and luminosity.   Only two of these objects, Alpha Centauri and Sirius, are visible with the unaided eye.   From a star's apparent magnitude and its distance its luminosity can be calculated (see Sec. 2-15). The sun's luminosity is taken as unity.   All the objects have a much lower luminosity than the sun except for Alpha Centauri A and B, which are of the same order as the sun, and Sirius A, which has 23 times the luminosity of the sun.

Table 2-2
The ten stellar objects nearest to the sun
with the sun included for reference†

Name	Distance, Light-years	Apparent magnitude	Luminosity
Sun	0.0	−26.9	1‡
Alpha Centauri A ⎱ Triple	4.3	0.3	1.0
Alpha Centauri B ⎰ system	4.3	1.7	0.28
Alpha Centauri C ⎰	4.3	11	0.00005
Barnard's star	6.0	9.5	0.0004
Wolf 395	7.7	13.5	0.00002
Luyten 726-8 A ⎱ Binary	7.9	12.5	0.00004
Luyten 726-8 B ⎰ system	7.9	13.0	0.00003
Lalande 21185	8.2	7.5	0.005
Sirius A ⎱ Binary	8.7	−1.6	23
Sirius B ⎰ system	8.7	7.1	0.008

† After van de Kamp, 1954.
‡ By definition.

2-4 Extragalactic Systems and the Universe Our galaxy, or Milky Way system, is but one of billions of similar systems which fill space in all directions as far as the largest telescopes can penetrate. Although these external galaxies are of various types, many are flattened systems like our own, which suggests that they are also rotating objects. The spacing between galaxies is on the average about 3 million light-years or some 30 times the diameter of our own galaxy.

The nearest galaxy similar to our own is the Andromeda nebula (M 31), shown in the photograph of Fig. 2-4. It is believed to be at a distance of about 2 million light-years and to have a diameter of about 150,000 light-years. It is thus approximately the same size as our own galaxy. Since it subtends an angle of several degrees, it has been possible to study M 31 in much greater detail than the more distant galaxies (Hubble, 1936; Baade, 1951; Mayall, 1951).

Associated with the Andromeda nebula are two subsystems, or "satellites," M 32 and NGC 205. These appear as the small bright regions near the center of the system. M 32 is below the nucleus and NGC 205 above and to the right of the nucleus in the figure. The object with four spikes and a halo is the overexposed image of a star in our galaxy which lies in the direction of M 31. The spikes are caused by diffraction from internal supports in the telescope tube (Danjon, 1960). The halo is a halation ring due to internal reflections in the photographic plate. Such aberrations in an optical telescope are present on all star images but are noticeable only on the brighter objects. They are analogous, but not necessarily similar, to the minor lobes in the antenna pattern of a radio telescope (see Chap. 6).

The nearest external galaxies are the Magellanic clouds, which form a prominent feature of the southern sky. Known as the large Magellanic cloud (LMC) and small Magellanic cloud (SMC), they have diameters of about 40,000 and 15,000 light-years, respectively, and are both situated nearly 200,000 light-years from the center of our galaxy. Although a

Fig. 2-4. Nearby galaxy M 31 in Andromeda, a spiral galaxy of the Sb type. (*Perkins Observatory photograph.*)

slight tendency toward spiral structure is apparent in the larger cloud, both clouds are classified as galaxies of irregular form. Figure 2-5 is a photograph of the larger cloud.

Our own galaxy and the two Magellanic clouds form a small cluster or relatively compact group of galaxies, the distance between the three being small compared to the distances to the next nearest galaxies, such as the Andromeda nebula. Since our galaxy is considerably larger than the Magellanic clouds, one may regard the latter as satellites of our galaxy, analogous to the subsystems M 32 and NGC 205 of the Andromeda nebula (M 31).

Galaxies, in general, can be classified on a morphological basis into three main types, *irregular* (I), *spiral* (S), and *elliptical* (E). Hubble further classified the spiral galaxies into three types, Sa, Sb, and Sc. These subdivisions refer to the tightness of the spiral structure, Sa indicating a tightly wound spiral, Sc an open spiral, and Sb an intermediate form.

The spiral galaxies M 81 (Fig. 2-2*b*) and M 31 (Fig. 2-4) are classified as Sb types. Our own galaxy is also presumed to be an Sb type. In some spiral galaxies first distinguished by H. D. Curtis (1936) the nucleus is not circular or spherical but is elongated and has the appearance of a bright bar. Hubble classified such galaxies as *barred spirals* (SB type) with three subdivisions (SBa, SBb, and SBc) depending on the tightness of the spiral

Fig. 2-5. The large Magellanic cloud, an irregular galaxy. (*Mount Stromlo Observatory photograph.*)

structure. The galaxy NGC 1300, shown in Fig. 2-6, is an example of a barred spiral classified as an SBb type. In a barred spiral the arms attach to the ends of the bar, whereas in the ordinary spiral the arms emerge from a central nucleus.

Elliptical galaxies are ones of symmetrical shape but without apparent internal structure. They have been subdivided by Hubble according to their ellipticity from E0 for a circular shape to E7 for a highly flattened type. The classification is based on the appearance of the galaxy to us and not on the true form of the galaxy. Thus, a galaxy classified as E0 might actually be an E7 type but seen broadside. A truly spherical galaxy would, of course, be classified as an E0 regardless of its orientation with respect to us.

Irregular galaxies, designated I, are ones which cannot be readily classified as spiral or elliptical. As the name implies, they are of irregular or unsymmetrical form. These galaxies may or may not show perceptible structure. The Magellanic clouds are classified as irregular galaxies. The larger cloud (Fig. 2-5) has a structure suggestive of a barred spiral (SB), but it is hardly distinct enough to classify it as such.

Fig. 2-6. Galaxy NGC 1300, a galaxy of the barred spiral type SBb. (*Mount Wilson and Palomer Observatories photograph.*)

Hubble combined the classification of spiral and elliptical galaxies in a diagram having the shape of a tuning fork, as illustrated in Fig. 2-7. He inserted the S0 type as a transitional form between the spirals and the ellipticals.

On the basis of further studies, Hubble and, later, Sandage developed the original classification in more detail. Using these and other studies, de Vaucouleurs (1959a) made a further revision of the classification system with four principal classes, ellipticals (E), lenticulars (S0), spirals (S), and irregulars (I).

Many galaxies occur in double systems. In some of these, distortions of the spiral arms and bridges or links between them can be detected. Such systems have been studied by Zwicky (1959a).

Galaxies are not uniformly distributed in space but apparently tend to form groupings. These range in number from double systems, through small groups of a few galaxies, to large clusters that may contain a thousand galaxies. Our own galaxy with the Magellanic clouds forms a triple system of galaxies. The Andromeda nebula M 31 with its satellites M 32 and NGC 205 also forms a triple system. These two triple systems and several other nearby galaxies form a *local group* of galaxies about 2 million light-years across.

The nearest large cluster of galaxies is the one in Virgo. It is perhaps nearer than 20 million light-years and contains some 500 galaxies. There is evidence (de Vaucouleurs, 1958) that our local group and the Virgo cluster form part of a local *super galaxy* with its center in the Virgo cluster (see Sec. 2-12).

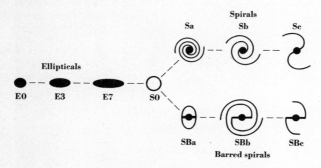

Fig. 2-7. Hubble's tuning-fork diagram of galaxy types.

One of the largest clusters of galaxies is the one in the constellation Coma Berenices, known as the Coma cluster. It is some 100 million light-years distant and contains nearly 1,000 galaxies. As far as the largest telescopes can penetrate, space appears to be filled with galaxies, most of which are in clusters. The diameter of the clusters appears to be of the order of 130 million light-years (Abell and Seligman, 1965).

The volume of space which telescopes can reach constitutes the observable part of our universe. The composition of our universe beyond and its ultimate extent and evolution are problems of *cosmology*, a subject discussed in Sec. 8-11.

2-5 Size Relationships Units of distance such as the astronomical unit or the light-year are so large compared to the distances of our ordinary experience that it is difficult to grasp their magnitude. For this reason, it is sometimes instructive to scale down astronomical dimensions to ones of more familiar size. In this way we may obtain a better appreciation of the relative dimensions involved. In the case of the solar system, for example, let the sun be scaled down from its diameter of 1,400,000 km (864,000 miles) to about 50 cm (20 in.).† When we apply this scale reduction to the rest of the solar system, the diameter of the earth is about 4 mm ($\frac{3}{16}$ in.) and its distance from the sun about 54 m (175 ft). Mars is about half the earth's diameter, and its distance from the sun about 82 m (270 ft). The giant planet of Jupiter is reduced on this scale to a diameter of 5 cm (2 in.), or slightly larger than a golf ball, moving in an orbit around the sun at a radius of about 280 m (920 ft). The most distant planet of the solar system, Pluto, is believed to be about the same size as the earth and would be at a distance of about 2,100 m (1.3 miles). Thus, this scaled-down solar system, with a sun of 50 cm in diameter at its center, has an overall diameter of about 4 km (2.5 miles). The large distance between objects and the relative emptiness of the solar system are apparent from this example.

If we extend the same scale reduction beyond the solar system, the distance of the next nearest star to the sun (about 4 light-years) is about 14,500 km (9,000 miles). Applying the same scale reduction to the distance from the solar system to the center of our galaxy (30,000 light-years) yields about 100 million km (62 million miles). This distance, even though it applies to a scaled-down galaxy, is so large as to be difficult to appreciate.

Another instructive way of comparing the dimensions of our solar system with those of our galaxy is to put all distances in terms of the time required for light, radio, or other electromagnetic waves to travel the distance. These waves all have the same velocity in vacuum of 300,000 km sec^{-1}. On the basis of travel time for these waves the circumference of the earth is 0.13 light-second, the distance of the earth from the sun nearly 500 light-seconds, and the diameter of the solar system about 11 light-hours. The distance from the sun to the next nearest star is about 4 light-years, and the distance of the sun from the center of the galaxy about 30,000 light-years. The diameter of our galaxy is more than 3 times this, or about 100,000 light-years. Table 2-3 summarizes these distances, along with other distances of interest.

A diagrammatic representation of the universe is presented in Fig. 2-8 by means of a series of sketches, with a large scale reduction between each sketch and the previous one (Kraus, 1964). The first diagram (Fig. 2-8*a*) is of the earth-moon system, with the earth at the center. The next diagram (Fig. 2-8*b*) is of the solar system, with the sun at the center. This diagram is reduced in scale from the earth-moon system diagram by a factor of 10,000. The third diagram (Fig. 2-8*c*) shows the solar neighborhood in an idealized manner. It is assumed that the stars have a uniform spacing of 5 light-years in a cubical lattice. The stars are, of course, not arranged in such a uniform manner, the actual distances of the stars

† This scale reduction is about 3 billion (3 × 10⁹) to 1.

Table 2-3
*Distances in travel time
of light and in parsecs*

	Light time	Parsecs†
New York to London	0.02 sec	
Circumference of earth	0.13 sec	
Earth to moon	1.3 sec	
Earth to sun (1 AU)	500 sec	
Earth to Venus (min)	2.3 min	
Earth to Venus (max)	14 min	
Earth to Mars (min)	2.8 min	
Earth to Mars (max)	21 min	
Sun to Pluto	5.5 hr	
Diameter of solar system	11 hr	
Sun to next nearest star	4 years	1
Sun to center of galaxy	30,000 years	9,000
Diameter of galaxy	100,000 years	30,000
Distance to nearby external galaxy (M 31)	2×10^6 years	600,000
Distance to Cygnus A	600×10^6 years	200×10^6
Distance to celestial horizon‡	$10,000 \times 10^6$ years	$3,000 \times 10^6$

† 1 parsec = 3.26 light-years.
‡ Assumes a uniformly expanding universe and a Hubble constant of 100 km sec^{-1} Mpc^{-1} (see Sec. 8–11).

from the sun being as listed in Table 2-2. Figure 2-8c is reduced in scale from the solar-system diagram by a factor of 10,000. Figure 2-8d is a diagrammatic representation of our galaxy, with the solar neighborhood at a distance of 30,000 light-years from the center of the galaxy. Again there is a scale reduction of 10,000 in this sketch as compared to the solar-neighborhood diagram. Figure 2-8e is an idealized representation of our galactic neighborhood, or metagalaxy. It is assumed that the galaxies have a uniform spacing of 2 million light-years in a cubical lattice. The galaxies are, of course, not arranged in such a uniform manner. Our galaxy is at the center of Fig. 2-8e. The scale reduction of this figure as compared to Fig. 2-8d is 40. The final diagram of the series (Fig. 2-8f) represents the observable universe bounded by the celestial horizon at a radial distance of 10,000 million light-years with our galactic neighborhood, or metagalaxy, at the center. A uniformly expanding universe is assumed, with the velocity of expansion equal to the velocity of light and the red shift infinite at the horizon. The space inside the celestial horizon contains, perhaps, 100 billion (10^{11}) galaxies† with an average intergalactic spacing of several million light-years. This diagram is reduced with respect to Fig. 2-8e by a factor of 5,000. The overall scale reduction of Fig. 2-8f with respect to Fig. 2-8a is 2×10^{17} to 1.

† Since a typical galaxy contains perhaps 10^{11} stars, the universe may contain some 10^{22} stars.

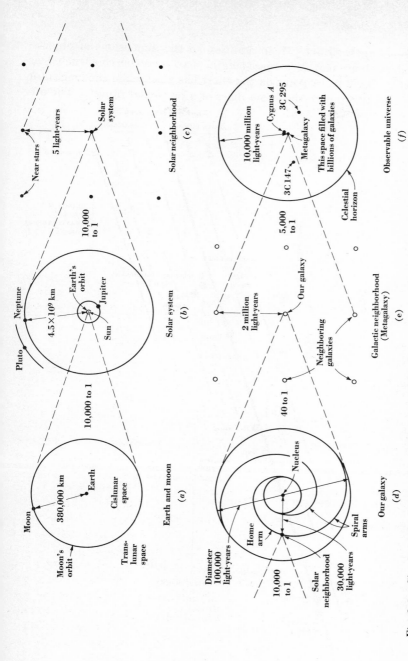

Fig. 2-8. Six sketches of the universe, progressing from the earth-moon system to the entire universe. Each sketch is reduced from the previous one by the scale factor indicated. The overall scale reduction from the first to last sketch is 2×10^{17}. The positions of a few of the more powerful and distant radio sources are shown in the last sketch.

31

2-6 Mass-length Chart If, in addition to the dimensions of objects, one considers their masses, a mass-length chart may be constructed as illustrated in Fig. 2-9. A point on the chart has as abscissa the length (or diameter) and as ordinate the (rest) mass of a particular object. The ob-

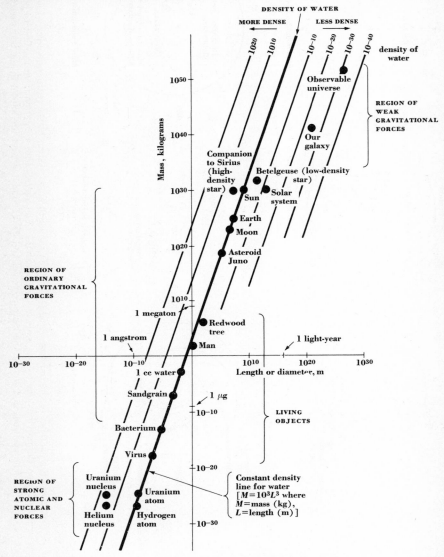

Fig. 2-9.   Mass-length chart giving mass and length or diameter of objects from atomic particles to the observable universe. The heavy slant line is for objects with the density of water and the lighter slant lines for densities greater or less in steps of 10 billion. The densities assume a cubical shape.

jects range from atomic particles to the observable universe. The heavy straight line gives the mass-to-length ratio for objects with the average density of water.‡ The lighter parallel lines indicate densities greater than and less than that of water in steps of 10 billion (10^{10}). Over the central region of the chart all the objects are close to the constant-density line for water. However, at the large end the objects tend to be less dense, while at the small end they tend to be more dense.

Both very small and very large dimensions involve uncertainties. For measuring a very small distance a wave of comparable length would be needed. One quantum of energy is given by

$$ h\nu = h\frac{c}{\lambda} \tag{2-1} $$

where h = Planck's constant ($= 6.63 \times 10^{-34}$ joule sec)
  c = velocity of light ($= 3 \times 10^8$ m sec^{-1})
  ν = frequency, cps
  λ = wavelength, m

For a wavelength of 10^{-40} m the quantum energy is of the order of 10^{15} joules, which is in the energy range of an atomic bomb. Clearly, it is impractical to measure such small dimensions and hence to speak meaningfully of them. A length of about 10^{-17} m is near the practical limit (Brillouin, 1964). There are also uncertainties involved in very large dimensions. Thus, distances in excess of a few billion (10^9) light-years need to be associated with a particular world model and Hubble constant to be significant (see Sec. 8-11).

2-7 Coordinate Systems To an observer on the earth there are several coordinate systems which are useful in dealing with the positions of celestial objects. These coordinate systems and their basic planes of reference are

(1) *Horizon system,* based on a plane parallel to the horizon
(2) *Equatorial coordinates,* based on a plane through the earth's equator
(3) *Ecliptic coordinates,* based on a plane through the earth's orbit
(4) *Galactic coordinates,* based on a plane parallel to the plane of our galaxy

---

‡ Although the line applies exactly only to objects of cubical shape, it gives an order of magnitude for most objects of other shape. For example, a spherical object of the same density as water has a mass-to-diameter ratio one-half that for a point on the water-density line in the chart. On the logarithmic scale of Fig. 2-9 a factor of 2 is a barely discernible difference.

(5) *Supergalactic* (*or metagalactic*) *coordinates,* based on a plane parallel to a plane in which there is an apparent local concentration of external galaxies

Each of these systems is described briefly in the following subsections.

2-8 Horizon System of Coordinates   In this system a plane through the observing point parallel to the horizon is the plane of reference.   In Fig. 2-10 the poles are the *zenith* (point overhead) and *nadir* (point underfoot).   The vertical circle through a celestial object and the zenith is the *object circle.*   The coordinates of the object are then given by the *azimuth,* or horizontal angle measured from an arbitrary reference direction (usually north) clockwise to the object circle, and the *altitude,* or elevation angle measured upward from the horizon to the object.   The great circle through the north and south points and the zenith is the *meridian.*   The circle through the east and west points and the zenith is the *prime vertical.*

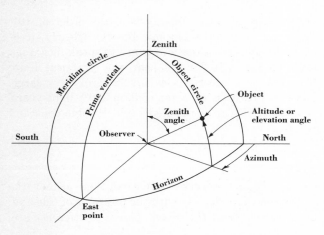

Fig. 2-10.   The horizon system of coordinates.

The coordinates of a celestial object in the horizon system change continuously during the day because of the earth's rotation.   Thus, although it may be convenient to use the altitude and azimuth angles in setting a telescope which is steerable around vertical and horizontal axes (altazimuth mounting), the horizon system is less convenient for specifying the position of a celestial object than a system in which the position is relatively fixed, as in the equatorial-coordinate system, described in the next section.

2-9 Equatorial Coordinates In this system the *earth's equator* is the plane of reference. The poles are at the intersection of the earth's axis with the celestial sphere, an imaginary surface at a large distance with the earth at its center. The poles are the *north celestial pole* (NCP) and the

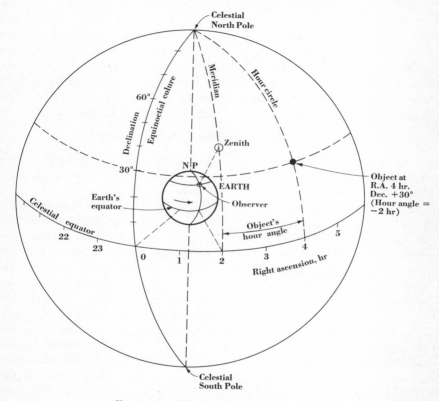

Fig. 2-11. The equatorial (or celestial) system of coordinates.

south celestial pole (SCP). The circle at the intersection of the plane of the earth's equator and the celestial sphere is the *celestial equator*. The great circle (Fig. 2-11) through the celestial poles and the object is the object's *hour circle,* and the great circle which passes through the celestial poles and the zenith is the *meridian circle.* The coordinates of a celestial object are given by the *declination,* or angle between the celestial equator and the object, and the *right ascension,* or angle measured from an arbitrary reference direction (the *vernal equinox*) to the object's hour circle. The *declination* (dec or δ) is expressed in degrees and is positive if the object is north of the equator and negative if south ($-90° \leq \delta \leq +90°$). In Fig. 2-11 an object is shown 30° north of the celestial equator (dec

+30°). For an observer at 30° north latitude this object would pass through the zenith when it transits the meridian.

Right ascension (RA or α) is measured eastward from the vernal equinox and is expressed either in degrees ($0° \leq \alpha < 360°$) or, more commonly, in hours, minutes, and seconds of time ($0^h \leq \alpha < 24^h$). The arc of the celestial equator between the object's hour circle and the meridian is the object's *hour angle* (HA). It increases with time; i.e., it is negative before the object transits the meridian and is positive after. It is measured either in degrees ($-180° \leq HA \leq +180°$) or, more commonly, in hours, minutes, and seconds of sidereal time. In Fig. 2-11 the object has a right ascension of 4 hr and at the time shown has an hour angle of -2 hr. The relation between the hour angle and right ascension of an object is given by

$$HA = RA \text{ of meridian} - RA \text{ of object}$$

LST − RA

The right ascension of the meridian is the same as the local *sidereal time* (see Sec. 2-14). A celestial object is on the meridian (hour angle zero) when the sidereal time is equal to the object's right ascension.

The reference point on the celestial equator from which right ascension is measured is at the intersection of the plane of the earth's equator and the plane of the earth's orbit, or the ecliptic (see Fig. 2-12). This intersection defines the equinoctial line. Opposite directions along this line are the *equinoxes*. The reference direction for the coordinate of right ascension is the *vernal equinox* ($RA = 0^h$). The sun passes this equinox in the spring (about March 21). The opposite equinox is the *autumnal equinox* ($RA = 12^h$). The sun passes the autumnal equinox in the autumn (about September 22). The great circle through the celestial poles and the equinoxes is the *equinoctial colure*. The points on the celestial equator 6 hr from the equinoxes are the *summer solstice* ($RA = 6^h$) and the *winter solstice* ($RA = 18^h$), the sun passing these solstices about June 22 and December 22, respectively. The great circle through the celestial poles and the solstices is the *solsticial colure*.

The *vernal equinox* γ is in the direction of the constellation Pisces. Years ago, however, because of precession of the earth's axis, it was in the direction of the constellation Aries, and for this reason the vernal equinox is still sometimes referred to as the *first point of Aries*.

The right ascension α and declination δ of a celestial object define its position in the sky in a relatively fixed manner, which is independent of the earth's diurnal rotation. However, because of a gradual *precession* of the earth's axis around the pole of the ecliptic there is a slow change in the equatorial coordinates for a fixed object in the sky. This change makes one cycle in about 26,000 years. Hence, to be explicit it is necessary to specify the date to which the right ascension and declination refer. This date is called the *epoch*. The epochs 1900.0 and 1950.0 have been used extensively and refer to the right ascension and declination on or about the dates of January 1, 1900, and January 1, 1950. Referring the coordinates to a standard epoch, such as 1950.0, greatly facilitates comparisons of celestial positions. The differences between the right ascension and declination for a given date or epoch and for epoch 1950.0 are given by†

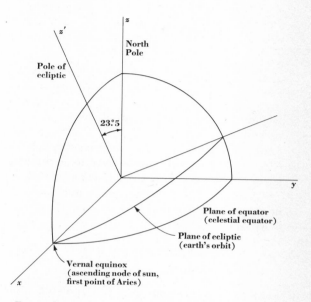

North Pole

Pole of ecliptic

23°.5

z'

z

y

Plane of equator (celestial equator)

Plane of ecliptic (earth's orbit)

Vernal equinox (ascending node of sun, first point of Aries)

x

Fig. 2-12. The plane of the earth's equator and the plane of the earth's orbit intersect along the equinoctial line.

$$\text{RA change} = \Delta\alpha = m + n \sin\alpha \tan\delta \qquad \text{per year} \qquad (2\text{-}2)$$
$$\text{Dec change} = \Delta\delta = n \cos\alpha \qquad \text{per year} \qquad (2\text{-}3)$$

where α = right ascension for 1950.0

δ = declination for 1950.0

$m = 3^s07327 \quad = 8^h5369 \; E-4 \; (time) = 1.28 E-2$

$n = 1^s33617$ in (2-2) or 20.0426 sec of arc in (2-3) $= 5.567 \times 10^{-3}$

The coordinates on the given date are then $\alpha + \Delta\alpha$ and $\delta + \Delta\delta$, with years reckoned as positive after 1950.0 and negative before. The quantities m an n in (2-2) and (2-3) vary slowly, resulting in a *secular variation* which needs to be taken into account in precise calculations over periods of many years (several decades). The variation can be judged from Table 2-4,

† See Appendix 7 for graphical presentation.

Table 2-4
Precession quantities for several epochs

Epoch	m Sec of time	n Sec of time	n Sec of arc
1800.0	3.07048	1.33703	20.0554
1850.0	3.07141	1.33674	20.0511
1900.0	3.07234	1.33646	20.0468
1950.0	3.07327	1.33617	20.0426
2000.0	3.07420	1.33589	20.0383
Change per century	+0.00186	−0.00057	−0.0085

which gives values for m and n for several epochs over two centuries (Smart, 1944).

Although (2-2) and (2-3) may be sufficiently accurate for many cases, greater accuracy, particularly over long intervals of time, is sometimes required. For such cases the following procedure can be used.

Suppose that the coordinates α_1 and δ_1 at epoch 1 are to be precessed to new coordinates α_2 and δ_2 N years later at epoch 2. First calculate $\Delta\alpha$ and $\Delta\delta$ from

$$\Delta\alpha = (m_1 + n_1 \sin \alpha_1 \tan \delta_1)N \tag{2-4}$$

$$\Delta\delta = (n_1 \cos \alpha_1)N \tag{2-5}$$

where m_1 and n_1 are constants (interpolated from Table 2-4) for epoch 1. Next calculate new changes $\Delta\alpha'$ and $\Delta\delta'$ from

$$\Delta\alpha' = [m_2 + n_2 \sin (\alpha_1 + \Delta\alpha) \tan (\delta_1 + \Delta\delta)]N \tag{2-6}$$

$$\Delta\delta' = [n_2 \cos (\alpha_1 + \Delta\alpha)]N \tag{2-6a}$$

where m_2 and n_2 are constants (interpolated from Table 2-4) for epoch 2. Finally, the coordinates at epoch 2 are given by

$$\alpha_2 = \alpha_1 + \frac{\Delta\alpha + \Delta\alpha'}{2} \tag{2-7}$$

$$\delta_2 = \delta_1 + \frac{\Delta\delta + \Delta\delta'}{2} \tag{2-7a}$$

When a celestial object crosses the meridian, it is said to *transit* the meridian or to be at *meridian transit*. Two such transits occur per day, the one nearer the zenith being called *upper culmination* and the one farther from the zenith *lower culmination*. If both culminations are above the horizon, the object is circumpolar. If both culminations are below the horizon, the object is not observable.

The declination of a celestial object on the meridian (see Fig. 2-13) is related to its zenith angle by

$$\delta = \phi + z_m \qquad (2\text{-}8)$$

where δ = declination, deg
   ϕ = latitude, deg
   z_m = zenith angle of object at upper culmination (positive when object is north of zenith and negative when south), deg

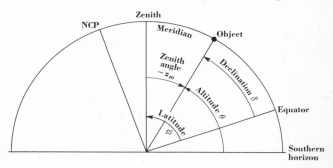

Fig. 2-13. Relation of declination, altitude, latitude, and zenith angle.

Since the zenith angle is equal to the altitude minus 90°, the declination is given by

$$\delta = \phi + \theta - 90° \qquad (2\text{-}9)$$

where δ = declination $(-90° \leq \delta \leq +90°)$, deg
   ϕ = latitude (positive north of equator and negative south of equator) $(-90° \leq \phi \leq +90°)$, deg
   θ = altitude angle above southern horizon at upper culmination $(0° \leq \theta \leq 180°)$, deg

A convenient way of obtaining the right ascension of a celestial object is by observing its time of meridian transit. The right ascension (see Fig. 2-14) is given by

$$\alpha = ST + UT\,(1 + C) - \lambda \qquad (2\text{-}10)$$

where α = right ascension of object
   ST = mean sidereal time (hour angle of the first point of Aries) at 0^h UT (obtained from "The American Ephemeris and Nautical Almanac" or equivalent ephemeris)
   UT = universal time of meridian transit of object at upper culmination
   C = correction to convert universal time into sidereal units (= 9.857 sec hr^{-1} after 0^hUT)

λ = longitude of observer (positive if west and negative if east of Greenwich)

Thus, (2-10) can be rewritten

$$\alpha = \mathrm{UT} + 9.857\mathrm{UT} - \lambda + \mathrm{ST} \tag{2-11}$$

The use of this relation may be demonstrated with the aid of the following illustrative example. Universal and sidereal times are discussed further in Sec. 2-14.

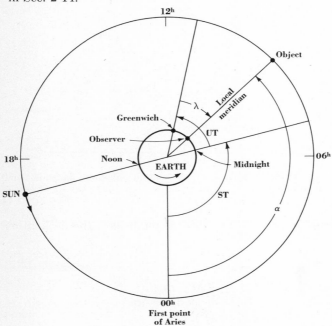

Fig. 2-14. Geometry for calculating local sidereal time or right ascension of the meridian.

Example. If the observed meridian transit of a radio source at upper culmination is 0600 (6:00 A.M. EST) on August 2, 1962, find the right ascension of the source.† The longitude at which the observation was made is 05ʰ32ᵐ00ˢ west (or positive). The declination of the object was observed to be 15° north.

Solution. First convert the Eastern standard time (EST) to universal time (UT) by adding 5 hr. To this add the correction for the 11 hr after 0ʰ UT. Next, subtract

---

† It is assumed that necessary corrections have been made for instrumental errors, such as misalignment of the feed with the telescope axis, so that 0600 is the actual time the source transits the meridian. It is assumed that refraction by the atmosphere and/or ionosphere is negligible.

the observer's longitude. Then add the mean sidereal time (hour angle of the first point in Aries) at 0^h UT on August 2, 1962, as obtained from the ephemeris. The total exceeds 24^h so this is subtracted to give the right ascension Thus,

$$EST = 06^h00^m00^s$$
$$Add \quad 05\ 00\ 00$$
$$UT = 11\ 00\ 00$$

$$Add\ correction = 9.857 \times 11\ sec = 00\ 01\ 48.5$$
$$11\ 01\ 48.5$$

$$Subtract\ longitude,\ \lambda = 05\ 32\ 00$$
$$05\ 29\ 48.5$$

$$Add\ mean\ sidereal\ time\ (ST) = 20\ 40\ 26.6$$
$$25\ 69\ 75.1$$

$$Subtract \quad 24\ 00\ 00$$
$$01\ 69\ 75.1$$

$$Right\ ascension,\ \alpha = 02\ 10\ 15.1$$

This is the mean right ascension at the time or epoch of the observation.

The effects of nutation (see Sec. 2-14) are accounted for in this result, but the effects of aberration (Sec. 2-14) are not. However, aberration will be near its minimum value for observations on the meridian near 6 A.M. and 6 P.M. local time at low declinations, conditions which are met in the above example. But at other times and/or at high declinations a correction for aberration may be required if a right-ascension accuracy of 1 sec of time or better is desired. Under these conditions the apparent (instead of mean) sidereal time should be used in the calculations. This then yields the apparent right ascension at the time of the observation. Then by using reduction formulas given in the "American Ephemeris" or Smart (1944) the apparent position is converted to the mean position at the nearest year epoch with the aid of the Besselian or independent day numbers. In the above example this epoch would be 1963.0. This conversion takes precession, nutation, and aberration into account. The mean position at the year epoch can then be precessed to other years by (2-2).

To convert the mean right ascension of the above example from the epoch of the date (August 2, 1962) to epoch 1950.0 the change $\Delta\alpha$ is subtracted from α, where

$$\Delta\alpha = (3^s07 + 1^s34 \sin \alpha \tan \delta)12.6$$
$$= (3^s07 + 1^s34 \sin 32°.6 \tan 15°)12.6 = 41^s1$$

Thus,

$$\alpha(1950.0) = 02^h09^m34^s0$$

This result is accurate to about 0^s1. For greater accuracy (2-7) should be used.

If, as mentioned above, the apparent instead of the mean sidereal time had been used in the example and the resulting apparent position converted to epoch 1963.0 using the Besselian or independent day numbers, a precession change of 13.0 years would be required to convert to epoch 1950.0.

2-10 Ecliptic System of Coordinates   In this system the ecliptic, or plane through the earth's orbit, is taken as the reference.   The orbits of most of the planets lie close to this plane (within 7° except for Pluto).   The coordinates are the *celestial longitude,* measured eastward along the ecliptic from the vernal equinox, and the *celestial latitude,* measured north (+) or south (−) from the ecliptic. This system is useful mainly in studies of the solar system.

2-11 Galactic Coordinates   In this system a plane through the sun parallel to the plane of the galaxy is taken as the plane of reference.   More precisely the plane is through the earth.   However, compared to galactic dimensions, the earth-sun distance is sufficiently small, so that this difference can usually be neglected.

The orientation of the galactic coordinates can be conveniently specified by the equatorial coordinates of one of the galactic poles.   Based on extensive optical measurements, a pole was adopted at $\alpha = 12^h\ 40^m$, $\delta = +28°$ (1900.0), for which conversion tables have been prepared by Ohlsson (1932) and Ohlsson et al, (1956).   The set published in 1932 gives the conversion from equatorial (1900.0) coordinates into galactic coordinates, and the more recent set, published in 1956, gives the conversion from galactic into equatorial (1958.0) coordinates.   The zero point of longitude for these coordinates is taken as the intersection of the galactic equator and the celestial equator (1900.0).

On the basis of numerous radio surveys of the galaxy it became apparent that the plane of the above system deviated by at least 1° from the principal plane of the galaxy, and a new system of galactic coordinates has been adopted by the International Astronomical Union (Blaauw et al., 1960; Gum, Kerr, and Westerhout, 1960; Gum and Pawsey, 1960; Blaauw, 1960; Oort and Rougoor, 1960).   The former system is now referred to as the *old system of galactic coordinates* and the recently adopted one as the 1958 revision or the *new system of galactic coordinates.*   The new pole is about 1°.5 from the old pole. A graphical conversion chart from equatorial (1950.0) coordinates to the new galactic coordinates and vice versa is given in Appendix 8.

The equatorial coordinates of the poles of the two systems are compared for three epochs in Table 2-5.

Table 2-5
Equatorial coordinates of north pole of
old and new galactic coordinates

	1900.0		1950.0		2000.0	
	α	δ	α	δ	α	δ
Old	12ʰ40ᵐ	$+28°$	12ʰ42ᵐ4	$+27°7$	12ʰ44ᵐ8	$+27°4$
New	12 46.6	$+27.7$	12 49	$+27.4$	12 51.4	$+27.1$

In the new system of galactic coordinates the reference point (of zero ongitude) has also been changed from the intersection of the galactic plane with the celestial equator to the direction of the center of the galaxy. The relation of the galactic plane to the equatorial plane and the ecliptic is llustrated by Fig. 2-15.

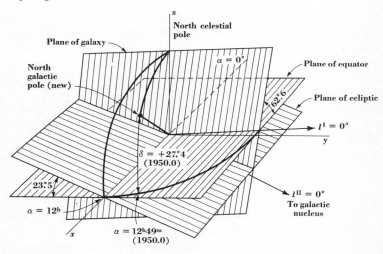

Fig. 2-15. Relation of plane of the galaxy to the plane of the earth's orbit (ecliptic) and plane of the earth's equator ($\delta = 0$).

The galactic coordinates of longitude l and latitude b are normally ꞵoth expressed in degrees. To distinguish the two systems the notation l^I, b^I is used for the old coordinates and l^{II}, b^{II} for the new coordinates. The ꞵole given in Table 2-5 is referred to as the *north galactic pole*. (It is closer ꞁo the north than to the south celestial pole.) Galactic latitude ranges from $-90°$ at the south galactic pole through $0°$ at the galactic equator to $+90°$ at the north galactic pole. Galactic longitude increases from 0 to 360°, with the direction of increasing longitude in the same sense as increasing right ascension at the galactic equator.[†]

The coordinates of the position $l^{II} = b^{II} = 0$ for three epochs of equatorial coordinates are listed in Table 2-6. The position of this point in the old system of galactic coordinates is $l^{I} = 327°69$, $b^{I} = -1°40$.

Table 2-6
Position of zero longitude and latitude of new
system of galactic coordinates ($l^{II} = b^{II} = 0$)
for three epochs of equatorial coordinates

	1900.0	1950.0	2000.0
α	$17^{h}39^{m}3$	$17^{h}42^{m}4$	$17^{h}45^{m}5$
δ	$-28°54'$	$-28°55'$	$-28°56'$

According to Gum, Kerr, and Westerhout (1960) the neutral-hydrogen layer in our galaxy is exceedingly flat out to a radius of 7 kpc from the galactic nucleus. They state that "the mean plane of the hydrogen in this region must have an important dynamical significance and may be regarded as the principal plane of the galaxy." Deviations from the mean plane in this region amount to less than 20 pc. The height of the sun above the principal plane of neutral hydrogen is about 4 pc but with a probable error 3 times as great (or ±12 pc). With the sun as center of the galactic coordinate system, the reference plane of the system through the sun and parallel to the galactic plane is thus very close to the actual principal plane of the galaxy. The geometry of our galaxy is discussed further in Chap. 8.

2-12 Supergalactic Coordinates The distribution of the brighter external galaxies in the Shapley-Ames catalog (1932) gives evidence of the existence of a metagalaxy, or supergalaxy, within which our own galaxy and local group of galaxies are situated. The galaxies in this supergalactic cluster appear to be concentrated close to a plane nearly perpendicular to the equatorial plane. The pole of this supersystem is placed by de Vaucouleurs (1953, 1958) at the coordinates $l^{I} = 15°$ and $b^{I} = +5°$ with the center of the system in the direction of the Virgo cluster of galaxies ($l^{I} = 255°$, $b^{I} = +75°$). A coordinate system based on the above pole is useful in studies of the distribution of external galaxies. The principal plane of this coordinate system may be taken through our position and at right angles to the above pole.

[†] Intersection of the galactic plane and celestial sphere.

The supergalactic system of coordinates is in an early stage of use, and no formal adoption of the system has yet been made by astronomical organizations. The system is mentioned here mainly to call attention to the fact that realms or hierarchies may exist which are of greater size than our own galaxy and for which appropriate coordinate systems may eventually be useful.

2-13 Coordinate-system Centers and Standards of Rest The position of a celestial object may be expressed conveniently in one or more of the coordinate systems described in the preceding sections. Although the origin of the coordinate system is the observer's position on the earth, the parallax (see Fig. 2-18) can be neglected in most cases if the center of the earth or even of the sun is taken as the origin. However, where motions or celestial mechanics are considered, the origin chosen would most conveniently correspond to the center of gravity (barycenter) of the system being considered. Thus, for studies of the motions of satellites around the earth, a *geocentric* equatorial system, or one with the center of the earth as origin, would be appropriate. For studies of the motion of the bodies of the solar system, a *heliocentric* ecliptic system, or one with an origin at the sun, would be appropriate. In a like manner, *galactocentric* or *supergalactocentric* coordinates would be appropriate for dynamical studies of the galaxy or supergalaxy, in galactic or supergalactic coordinates respectively. If the systems are all geocentric, as for position measurements only, a simple rotation suffices to go from one system to another. However, if the origin is also changed, as in dynamical considerations, both a rotation and a translation are required to transform from one system to another.

All celestial objects are in relative motion, and for some purposes a mixed arrangement may be useful. For example, in studies of galactic and extragalactic objects it may be convenient to take the earth or sun as origin for positional purposes but to refer motions to a *local standard of rest*, defined as the centroid of motion of the stars near the sun or local region of our home area of the galaxy (Chandrasekhar, 1942; Trumpler and Weaver, 1953; Edmondson, 1959). In this way the earth's orbital motion around the sun (about 30 km sec^{-1}) and the sun's drift of about 15 km sec^{-1} with respect to the centroid of motion of the local group of stars may be eliminated from the observed motions of galactic and extragalactic objects. The local standard of rest is used extensively in hydrogen-line studies of the galaxy, as discussed further in Chap. 8.

From an analysis of the radial velocities of 2,148 stars Campbell and Moore (1928) gave the *solar apex* (direction of sun's drift with respect to the centroid of motion of the local group) as $\alpha = 270°.6$, $\delta = +29°.2$ (1900.0) with a drift velocity of 19.7 km sec^{-1}. Vyssotsky and Janssen

(1951) derived a new position from analyses of over 400 A stars and over 400 K giants within 100 pc of the sun. The position they give for the apex is $\alpha = 265° \pm 1°2$, $\delta = +20°7 \pm 1°4$ (1900.0) with a drift velocity of 15.5 ± 0.4 km sec^{-1}. The circular velocity of the local standard of rest with respect to the galactic nucleus is believed to be about 275 km sec^{-1} (Godfredsen, 1961).[†] The velocity relations are shown diagrammatically in Fig. 2-16.

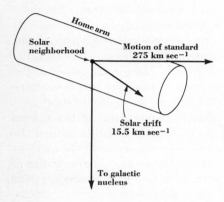

Fig. 2-16.   Local standard of rest.

The Campbell-Moore apex is now usually referred to as the *old standard of rest* and the Vyssotsky-Janssen apex as the *new standard of rest.* The velocities and positions are compared in Table 2-7 for three epochs.

Table 2-7
Solar motion and position of
old and new standards of rest

	Velocity km sec^{-1}	1900.0		1950.0		2000.0	
		α	δ	α	δ	α	δ
Old	19.7	270°6	+29°2	271°1	+29°2	271°6	+29°2
New	15.5	265.0	+20.7	265.5	+20.7	266	+20.7

A "Table for the Reduction of Velocities to the Local Standard of Rest" by MacRae and Westerhout (1956) has been published by the Lund Observatory. It gives the components of the earth's velocity and hydrogen-line frequency shifts with respect to the old standard of rest.

[†] A value of 250 km sec^{-1} is given by the IAU, 1964.

2-14 Measures of Time, Distance, and Motion The mks unit of
time is the *second*. The *civil second* is 1/86,400 part of a *civil or mean solar
day*. The civil day is an average (mean) day. It is very nearly 1/365.2422
part of a *tropical year*, defined as the interval between two successive passages of the sun through the vernal equinox. Civil time at the meridian
of Greenwich (0° longitude) is called *universal time*, or UT (also formerly
Greenwich mean time, or GMT). This is the time in which most observations are made and recorded, together with a calendar date and year.
Time signals are broadcast by radio in universal time (or times equal to
UT plus or minus an integral number of hours). Since 1925 the mean
solar day has been reckoned by astronomers from midnight, the same as
the ordinary civil day, but prior to 1925 it was reckoned by astronomers
from noon.

The number of days in a tropical year (and therefore the number of
civil days) is subject to slight variations of an irregular and unpredictable
kind (Clemence, 1960; Newton, 1960). It turns out that the tropical year
provides a more uniform time interval than the day. That is to say, time
can be kept more accurately by using the earth's orbital motion around
the sun than by using the earth's rotation (universal time). Hence, for
more precise timekeeping the tropical year has been taken as the standard,
and time based on this interval is called *ephemeris time* (ET). A second
of ephemeris time is defined as 1/31,556,925.9747 part of the tropical year
1900. It is not exactly equal to a second of universal time (= 1/86,400
of a mean solar day) but differs from it by the order of 1 part in 10^7 or less.
Although the tropical year is not constant, it changes in a uniform and
predictable fashion. Thus, the number of ephemeris seconds in any year
can be predicted with precision. Since the number of universal seconds in
a year cannot be predicted with equal precision, a correction of universal
time to give ephemeris time for a given date cannot be made until sometime after the date, months or years later. To reduce universal to ephemeris time the following relation is used:

$$ET = UT + \Delta T \tag{2-12}$$

where ET = ephemeris time
UT = universal time
ΔT = correction

Tables giving the correction ΔT are published in the "American Ephemeris
and Nautical Almanac" and "The Astronomical Ephemeris." Table 2-8
gives corrections for a few dates since 1910.

The interval between two successive upper culminations or transits of
the vernal equinox is defined as a mean *sidereal day*. As shown in Fig. 2-17,
the earth must rotate about 361° for a mean solar day and 1° less, or 360°,
for a sidereal day. As a result a sidereal day is about 4 min shorter than

a mean solar day. Since, however, a sidereal day is defined with respect to the vernal equinox, a point which precesses, a sidereal day is about 0.01 sec less than if it were defined with respect to a fixed point in space. At present

$$1 \text{ mean sidereal day} = 23^h56^m04\overset{s}{.}09054 \text{ universal time}$$

and

$$1 \text{ mean solar day} = 24^h03^m56\overset{s}{.}55536 \text{ mean sidereal time}$$

Thus,

$$86,400 \text{ sidereal seconds} = 86,164.09054 \text{ civil (universal) seconds}$$

It follows that

$$1 \text{ civil (universal) second} = 1.00273791 \text{ sidereal seconds†}$$

Table 2-8
Conversion from universal to
ephemeris time

Date	Correction ΔT	Difference
1910.5	$+10\overset{s}{.}28$	
1920.5	$+20.48$	$10\overset{s}{.}20$
1930.5	$+23.18$	2.70
1940.5	$+24.30$	1.12
1950.5	$+29.42$	5.12
1960.5	$+34$ (approx.)	6

and that there are 9.857 more sidereal seconds than civil (universal) seconds per civil (universal) hour. This is the correction factor used in the problem in Sec. 2-9. The local sidereal time is 00^h00^m when the vernal equinox is on the meridian. Hence, the right ascension of an object on the meridian is equal to the sidereal time, or, what is the same thing, the hour angle of the vernal equinox.

Timekeeping for long time intervals is now most accurately done with atomic clocks, which have a precision of 1 part in 10^9, or 1 sec in 30 years.

Where long time intervals are involved, it is often convenient to reckon time entirely in days instead of days, months, and years. The number of days reckoned from noon (1200 UT) of January 1, 4713 B.C., is called the *Julian date*. Julian days begin at noon civil time. Thus, midnight of January 1, 1961   (0000 UT, January 1, 1961) is Julian date 2,437,300.5.

---

† The value given in the "American Ephemeris and Nautical Almanac" is 1.0027379093. It also follows that 1 cycle per sidereal second equals 1.00273791 cycles per universal second.

The Julian day number at midnight (0000 UT)† of January 1 (midnight of New Year's Eve, or start of new Gregorian calendar year) is given for the twentieth century in Table 2-9 at 5-year intervals. Tables for the Julian day number at monthly intervals in the twentieth century and also other conversion tables are given in the "American Ephemeris and Nautical Almanac."

The time during the day is expressed as a decimal fraction. Thus, 2100 UT January 1, 1960, is Julian day number 2,436,935.375 since this time is 9 hr, or 9/24 = 0.375 part of one day, after the beginning of Julian day 2,436,935.0.

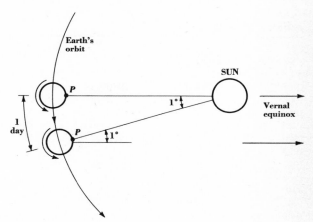

Fig. 2-17. Motion of earth in 1 sidereal day and in 1 solar day. (Drawing not to scale.)

For other purposes it may be more convenient to express time in terms of years and decimal fractions thereof, which can be done in terms of Besselian years. One *Besselian year* is defined as the period of a complete circuit of the mean sun in right ascension beginning at the instant when its right ascension is 18ʰ 40ᵐ. Since a calendar year is an integral number of days (365 or 366) while a Besselian year is not, an integral Besselian year number will ordinarily not coincide with the beginning of a calendar year. The difference may be about a day or less. Thus, the beginning of the Besselian year 1961 is designated 1961.0 and corresponds to the calendar date January 1, 1961, at 0207 UT. This is also equal to Julian date 2,437,300.588.

The instants of time or *epochs*, such as 1900.0 or 1950.0, to which the coordinates of right ascension and declination are referred, are in Besselian

† If the day begins at 12ʰ ephemeris time (1200 ET) instead of 12ʰ universal time (1200 UT), one may refer to the date as the *Julian ephemeris day number*.

year numbers. Thus, the epoch 1950.0 signifies the beginning of the Besselian year 1950, which started December 31, 1949, at 2209 UT.

In addition to the precession of the earth's axis around a perpendicular to the ecliptic with a period of about 26,000 years, there are small oscillations of the earth's axis around its mean position as it precesses. This nodding motion is called *nutation*. The principal nutation effect has a period of 19 years and causes a maximum displacement in the apparent position of a star of about 9 sec of arc. There are other nutations of shorter period but of smaller displacement (less than 1 sec of arc).

Table 2-9
Julian day numbers

Year	Julian day number		Year	Julian day number	
1900 A.D.	241	5020.5	1970 A.D.	244	0587.5
1905		6846.5	1975		2413.5
1910		8672.5	1980		4239.5
1915	242	0498.5	1985		6066.5
1920		2324.5	1990		7892.5
1925		4151.5	1995		9718.5
1930		5977.5	2000	245	1544.5
1935		7803.5			
1940		9629.5			
1945	243	1456.5			
1950		3282.5			
1955		5108.5			
1960		6934.5			
1965		8761.5			

The orbital motion of the earth produces an apparent change in the direction of a celestial object called *aberration*. The aberration for an object follows a yearly cycle. When the earth's motion is at right angles to the object's direction the aberration amounts to about 20 sec of arc.

Nutation and aberration corrections are never more than about 9 and 20 sec of arc, respectively. For an accuracy to the nearest minute of arc, no corrections need be made except for precession. However, for more precise position calculations it is necessary to correct for nutation and aberration as well as precession. This is discussed briefly in connection with the worked example in Sec. 2-9. For a more complete discussion, reference should be made to the "American Ephemeris" and texts such as Smart (1944).

For the nearer stars the change in the earth's position, due to its orbital motion, causes an appreciable apparent annual shift in position relative to the distant stars called the *parallactic displacement*. As shown in Fig. 2-18, the maximum value of this *parallax* is the angle *SPE* subtended at the star by the semidiameter of the earth's orbit. This maximum value is called the *heliocentric parallax*.

The distance of a star, for a heliocentric parallax of one second of arc, is called one parsec (pc). Thus,

$$d = \frac{1}{p}$$

where d = distance, pc

p = parallax, sec of arc

The *parsec* is used in expressing the largest astronomical distances. The *light-year* is another unit used for this purpose. It is equal to the distance

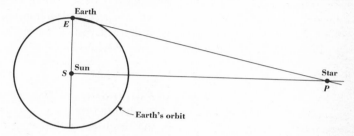

Fig. 2-18. Parallax angle *SPE*. *SP* equals 1 pc if *SPE* equals 1 sec of arc.

traveled in vacuum by electromagnetic waves in one year. For smaller distances, such as solar-system dimensions, the mean earth-sun distance, or *astronomical unit*,† is commonly employed. The values for the astronomical unit, light-year, and parsec are listed in Table 2-10.

Table 2-10

Unit	Kilometers	Miles	Other
Astronomical unit (AU)	1.496×10^8	9.3×10^7	
Light-year (LY)	9.460×10^{12}	5.9×10^{12}	0.307 pc
			6.324×10^4 AU
Parsec (pc)	3.086×10^{13}	1.9×10^{13}	3.262 LY
			2.063×10^5 AU

Precise position measurements of a star at different dates show changes in apparent position. Such changes may be due largely to precession, nutation, aberration, and parallax, but even when corrections are made for

† According to the "American Ephemeris and Nautical Almanac for 1961," p. 488, the astronomical unit is defined in such a way that the mean earth-sun distance is 1.00000003 AU. Recent radar measurements give 1 AU = 149,598,000 km (Goldstein, 1964).

all these effects, a small residual change, called *proper motion*, may remain. This proper motion is the resultant of the motion of the star in space and the solar motion. The largest known proper motion is about 10 sec of arc per year (Barnard's proper-motion star). The proper motion of most stars is, however, very much less, and only about 300 stars have proper motions of more than 1 sec of arc per year (Edmondson, 1959).

The proper motion of a star is at right angles to the sun-star line. If the sun-star distance is known, the *tangential velocity* can be obtained. The velocity along the line of sight, or *radial velocity*, may be obtained by measurements of Doppler displacements. The total velocity is equal to the square root of the sum of the squares of the tangential and radial velocities.

The motion of a star with respect to the centroid of motion of the stars in its neighborhood is called its *peculiar motion*. Thus, the sun's drift with respect to the local standard of rest is a peculiar motion which is directed toward the solar apex (see Sec. 2-13).

2-15 Stellar Magnitude and Luminosity Hipparchus (about 130 B.C.) and Ptolemy (about A.D. 150) divided the stars visible to the unaided eye into six magnitude groups, the brightest stars being classified as first magnitude and those just visible as the sixth magnitude. In 1827 Sir John Herschel found that the ratio of the intensity or apparent luminosity[†] of a first- to a sixth-magnitude star (difference of 5 magnitudes) was about 100 to 1. In 1854, Pogson concluded that the ratio of one magnitude to the next was about 2.5 to 1.

Now if Herschel's luminosity ratio of 100 for a 5-magnitude difference is adopted, it follows from Fechner's law that the luminosity ratio from one magnitude to the next should be $100^{1/5} = 2.512$. Or, in general, the ratio of the apparent luminosity l of a star with respect to a reference star of apparent luminosity l_0 is given by

$$\frac{l_0}{l} = 2.512^{m-m_0} \tag{2-13}$$

where m = magnitude of star of apparent luminosity l
       m_0 = magnitude of reference star of apparent luminosity l_0

It is to be noted that as the apparent luminosity of a star *decreases*, its magnitude *increases*.

In radio work, power ratios are commonly expressed in decibels (dB). Thus, we have

[†] The physical quantity implied by the intensity or luminosity is the light-power flux density (watts m^{-2}).

Luminosity ratio per magnitude $= 10 \log 100^{1/5} = 4 \text{ dB}†$

or

1 magnitude $= 4$ dB

The following table compares magnitude and decibel ratios:

Magnitudes	Decibels
0.25	1
1	4
2	8
5	20
10	40
20	80
30	120

From (2-13)

$$m - m_0 = 2.500 \log \frac{l_0}{l} \tag{2-14}$$

Since the power received from a star varies inversely as the square of the distance, we have that

$$\frac{l_0}{l} = \left(\frac{r}{r_0}\right)^2 \tag{2-15}$$

where $r =$ distance of star of apparent luminosity l
$r_0 =$ distance of star of apparent luminosity l_0
Substituting (2-15) in (2-14) yields

$$m - m_0 = 5 \log \frac{r}{r_0} \tag{2-16}$$

The observed magnitude of a star is called its *apparent magnitude*. However, to compare one star with another we need to know their intrinsic magnitudes, such as would be obtained if the stars were placed at a standard distance. The distance chosen as standard is 10 pc and (32.6 light-years), and the magnitude at this distance is termed the *absolute magnitude M*. If the reference star is placed at a distance of 10 pc, $m_0 = M$ and $r_0 = 10$, and (2-16) becomes

$$m - M = 5 \log \frac{r}{10} = 5 \log r - 5 \tag{2-17}$$

Now if both m and M refer to the same star, i.e., if m is its apparent (or observed) magnitude and M its absolute magnitude (or apparent magnitude at standard distance), the difference $m - M$ is a measure of the

† Ratio in bels $= \log (\text{power } A/\text{power } B)$. Ratio in decibels $= 10 \log (\text{power } A/\text{power } B)$.

star's distance and is called the *modulus of distance*. Thus, from (2-17) we obtain

$$\log r = \frac{m - M}{5} + 1 \tag{2-18}$$

where r = distance to star of apparent magnitude m, pc
   m = apparent magnitude of star
   M = absolute magnitude of star
$m - M$ = modulus of distance, magnitudes
Accordingly, if the modulus of distance (difference of the apparent and absolute magnitude) of a star is known, its distance follows directly from (2-18). Conversely, if the distance and apparent magnitude are known, the absolute magnitude can be obtained from (2-18).

2-16 Visual, Photographic, and Radio Magnitudes The magnitude of a star observed with the eye is called its *visual magnitude* m_v. If the magnitude is measured on an ordinary photographic plate, it is termed the *photographic magnitude* m_p. Both visual and photographic magnitudes may be apparent or absolute, depending on whether they are the observed values or the magnitudes referred to the standard distance of 10 pc.

Since the eye is more sensitive in the yellow-green region of the optical spectrum while ordinary photographic plates are most sensitive to blue light, the visual and photographic magnitudes of a star are, in general, different. The difference is called the *color index*. Thus,

Color index = $m_p - m_v$

where m_p = photographic magnitude
   m_v = visual magnitude
The redder the star, the larger the color index tends to become.

As discussed in the next chapter, the intensity of a radio source is usually measured in terms of its power flux density (watts m^{-2} cps^{-1}). The first need for defining a *radio magnitude* came through a study of external optical and radio techniques. To facilitate radio-optical comparisons of galaxies Brown and Hazard (1952) defined the *apparent radio magnitude* m_r as

$$m_r = -53.4 - 2.5 \quad \log S \tag{2-19}$$

where S = flux density of radio source at 158 Mc, watts m^{-2} cps^{-1}

To obtain the *absolute radio magnitude* M_r of a source it is necessary to know the distance. From (2-18) the absolute radio magnitude is given by

$$M_r = m_r - 5 \log r + 5 \tag{2-20}$$

where m_r = apparent radio magnitude from (2-19)

r = distance, pc

The distance r may be measured from the photographic modulus of distance $(m_p - M_p)$, as in (2-18) or, in the case of external galaxies, by observing the red shift of the optical or radio emission, as discussed in Sec. 8-11.

General References

ALLEN, C. W.: "Astrophysical Quantities," University of London Press, Ltd., London, 1963.

"American Ephemeris and Nautical Almanac," U.S. Naval Observatory, Government Printing Office, Washington, D.C. Issued yearly. Both this and the "Astronomical Ephemeris," issued in England, are unified.

"Astronomical Ephemeris," Royal Greenwich Observatory. Issued yearly.

BAKER, R. H.: "Astronomy," D. Van Nostrand Company, Inc., Princeton, N.J., 1964.

BLANCO, V. M., and S. W. McCUSKEY: "Basic Physics of the Solar System," Addison-Wesley Publishing Company, Inc., Reading, Mass., 1961. Good introductory treatment of time and coordinate systems.

DUNCAN, J. C.: "Astronomy," Harper & Row, Publishers, Incorporated, New York, 1926.

JONES, H. S.: "General Astronomy," Edward Arnold (Publishers), Ltd., London, 1961.

McCUSKEY, S. W.: "Introduction to Advanced Dynamics," Addison-Wesley Publishing Company, Inc., Reading, Mass., 1959.

McCUSKEY, S. W. "Introduction to Celestial Mechanics," Addison-Wesley Publishing Company, Inc., Reading, Mass., 1963.

NASSAU, J. J.: "Textbook of Practical Astronomy," 2d ed., McGraw-Hill Book Company, New York, 1948. Good introductory discussions of coordinate systems and basic observational measurements.

NORTON, A. P., and J. G. INGLIS: "Star Atlas," Gall & Inglis, London, 1950.

PAYNE-GAPOSCHKIN, C.: "Introduction to Astronomy," Prentice-Hall, Inc., Englewood Cliffs, N.J., 1954.

SMART, W. M.: "Textbook on Spherical Astronomy," Cambridge University Press, London, 1944.

SMART, W. M.: "Celestial Mechanics," Longmans, Green & Co., Ltd., London, 1953.

STRUVE, O., B. LYNDS, and H. PILLANS: "Elementary Astronomy," Oxford University Press, Fair Lawn, N.J., 1959.

TRUMPLER, R. J., and H. F. WEAVER: "Statistical Astronomy," University of California Press, Berkeley, Calif., 1953. Has chapters on stellar motions in the vicinity of the sun and a statistical description of the galactic system.

"The Universe," an entire issue of the *Scientific American* (vol. 195, no. 3, September, 1956) devoted to galaxies and cosmology. Eleven articles by Robertson, Fowler, Baade, Oort, Minkowski, Gamow, Hoyle, Sandage, Neyman and Scott, Ryle, and Dingle.

Special References

ABELL, G. O., and C. E. SELIGMAN: The Distribution of Clusters of Galaxies, *Astron. J.*, vol. 70, p. 317, 1965.

BAADE, W. A.: Galaxies—Present Day Problems, in "The Structure of the Galaxy," *Publ. Obs. Univ. Mich.*, vol. 10, pp. 7–17, 1951.

BLAAUW, A., C. S. GUM, J. L. PAWSEY, and G. WESTERHOUT: The New I.A.U. System of Galactic Coordinates (1958 Revision), *Monthly Notices Roy. Astron. Soc.*, vol. 121, no. 2, pp. 123–131, 1960. This paper and four others (Gum, Kerr, Westerhout; Gum and Pawsey; Blaauw; and Oort and Rougoor) constitute a series which appeared together in the *Monthly Notices*.

BLAAUW, A.: Optical Determinations of the Galactic Pole, *Monthly Notices Roy. Astron. Soc.*, vol. 121, no. 2, pp. 164–170, 1960.

BRILLOUIN, L.: "Scientific Uncertainty and Information," Academic Press, Inc., New York, 1964.

BROWN, R. H., and C. HAZARD: Extra-galactic Radio-frequency Radiation, *Phil. Mag.*, vol. 43, pp. 137–152, 1952.

BURBIDGE, E. M., and G. R. BURBIDGE: Peculiar Galaxies, *Sci. Am.*, vol. 204, pp. 50–57, February, 1961.

CAMPBELL, W. W., and J. H. MOORE: Radial Velocities of Stars, *Publ. Lick Obs.*, vol. 16, 1928.

CHANDRASEKHAR, S.: "Principles of Stellar Dynamics," Astrophysical Monographs, University of Chicago Press, Chicago, 1942. Formulates basic concept of local standard of rest.

CLEMENCE, G. M.: The Practical Use of Ephemeris Time, *Sky & Telescope*, vol. 19, pp. 148–149, January, 1960.

CURTIS, H. D.: The Nebulae, in "Handbuch der Astrophysik," vol. 7, pp. 550–563, Springer-Verlag OHG, Berlin, 1936.

DANJON, A.: chap. 8 in G. P. Kuiper and B. M. Middlehurst (eds.), "Telescopes," p. 127, University of Chicago Press, 1960.

DE VAUCOULEURS, G.: Evidence for a Local Supergalaxy, *Astron. J.*, vol. 58, no. 1205, pp. 30–32, February, 1953.

DE VAUCOULEURS, G.: The Local Supercluster of Galaxies, *Nature*, vol. 182, pp. 1478–1480, Nov. 29, 1958.

DE VAUCOULEURS, G.: Classification and Morphology of External Galaxies, in S. Flügge (ed.), "Handbuch der Physik," vol. 53, pp. 275–310, Springer-Verlag OHG, Berlin, 1959a. Discusses standard and revised systems of classification.

DE VAUCOULEURS, G.: General Physical Properties of External Galaxies, in S. Flügge (ed.), "Handbuch der Physik," vol. 53, pp. 311–372, Springer-Verlag OHG, Berlin, 1959b. Includes extensive bibliography.

EDMONDSON, F. K.: Kinematical Basis of Galactic Dynamics, in S. Flügge (ed.), "Handbuch der Physik," vol. 53, Springer-Verlag OHG, Berlin, 1959. Excellent general discussion including the local standard of rest.

GODFREDSEN, E. A.: Dynamical Stability of the Local Group, *Astrophys. J.*, vol. 134, pp. 257–261, July, 1961.

GOLDSTEIN, R. M.: Radar Investigations of the Planets, *IEEE Trans. Military Electron.*, vol. MIL-8, pp. 199–206, July–October, 1964.

GUM, C. S., F. J. KERR, and G. WESTERHOUT: A 21-cm Determination of the Principal Plane of the Galaxy, *Monthly Notices Roy. Astron. Soc.*, vol. 121, no. 2, pp. 132–149, 1960.

GUM, C. S., and J. L. PAWSEY: Radio Data Relevant to the Choice of a Galactic Coordinate System, *Monthly Notices Roy. Astron. Soc.*, vol. 121, no. 2, pp. 150–163, 1960.

HUBBLE, E. P. "The Realm of the Nebulae," Oxford University Press, London, 1936.

KRAUS, J. D.: Radio and Radar Astronomy and the Exploration of the Universe, *Trans. IEEE Military Electron.*, vol. MIL-8, pp. 232–235, July–October, 1964.

KRAUS, J. D., and H. C. KO: The Radio Position of the Galactic Nucleus, *Astrophys. J.*, vol. 122, no. 1, pp. 139–141, July, 1955.

MacRae, D. A., and G. Westerhout: "Table for the Reduction of Velocities to the Local Standard of Rest," The Observatory, Lund, Sweden, 1956. Gives components of earth's velocity vector and hydrogen-line frequency shifts for old standard of rest with apex at $\alpha = 18^h$, $\delta = +30°$ (1900.0), and $V = 20$ km sec^{-1}.

Mayall, N. U.: Comparison of Rotational Motions Observed in the Spirals M 31 and M 33 and in the Galaxy, in "The Structure of the Galaxy," *Publ. Obs. Univ. Mich.*, vol. 10, pp. 19–24, 1951.

McVittie, G. C.: Distance and Time in Cosmology: The Observational Data, in S. Flügge (ed.), "Handbuch der Physik," vol. 53, pp. 445–488, Springer-Verlag OHG, Berlin, 1959.

Newton, R. R.: Astronomy for the Non-astronomer, *IRE Trans. Space Electron. Telemetry*, vol. SET-6, pp. 1–16, March, 1960.

Neyman, J., and E. L. Scott, Large Scale Organization of the Distribution of Galaxies, in S. Flügge (ed.), "Handbuch der Physik," vol. 53, pp. 416–444, Springer-Verlag OHG, Berlin, 1959.

Ohlsson, J.: Lund Observatory Tables for Conversion of Equatorial Coordinates into Galactic Coordinates, *Lund Obs. Ann.*, 3 1932. Same as 1956 tables except gives inverse conversion, and equatorial coordinates are for epoch 1900.0.

Ohlsson, J., A. Reiz, and I. Torgard: "Lund Observatory Tables for the Conversion of Galactic into Equatorial Coordinates and for the Direction Cosines in the Equatorial System," The Observatory, Lund, Sweden, 1956. Galactic coordinates are based on old galactic pole [$\alpha = 12^h40^m$, $\delta = +28°$ (1900.0)] and equatorial coordinates are for epoch 1958.0.

Oort, J. H., and G. W. Rougoor: The Position of the Galactic Centre, *Monthly Notices Roy. Astron. Soc.*, vol. 121, no. 2, pp. 171–173, 1960.

Sandage, A. A., "The Hubble Atlas of Galaxies," Carnegie Institute of Washington, Washington, D.C., 1961.

Shapley, H.: "Galaxies," Harvard University Press, Cambridge, Mass., 1961.

Shapley, H., and A. Ames: A Survey of the External Galaxies Brighter than the 13th Magnitude, *Ann. Harvard College Obs.*, vol. 88, no. 2, 1932.

The Structure of the Galaxy, *Publ. Obs. Univ. Mich.*, vol. 10, 1951. Twelve papers presented at a symposium by W. Baade, K. G. Henize, B. Lindblad, N. U. Mayall, S. W. McCuskey, F. D. Miller, R. Minkowski, W. W. Morgan, J. J. Nassau, H. Shapley, J. Stebbins, and A. N. Vyssotsky.

Struve, O.: The Clouds of Magellan, *Sky & Telescope*, pp. 54–57, December, 1954.

van de Kamp, P.: The Nearest Stars, *Am. Scientist*, vol. 42, pp. 573–588, October, 1954.

Vyssotsky, A. N., and E. M. Janssen: An Investigation of Stellar Motions: A Determination of the Basic Solar Motion, *Astron. J.*, vol. 56, no. 1191, pp. 58–62, March, 1951.

Zwicky, F.: Multiple Galaxies, in S. Flügge (ed.), "Handbuch der Physik," vol. 53, pp. 373–389, Springer-Verlag OHG, Berlin, 1959a. Describes bridges between galaxies.

Zwicky, F.: Clusters of Galaxies, in S. Flügge (ed.), "Handbuch der Physik," vol. 53, pp. 390–415, Springer-Verlag OHG, Berlin, 1959b.

Problems

2-1. An object has coordinates of $12^h20^m24^s$ RA and $+18°45'$ dec (1900.0). Precess the coordinates to epoch 1950.0. *Ans.* $12^h22^m56^s$, $+18°28'$

2-2. An object has coordinates of $20^h00^m12^s$ RA and $-22°12'$ dec (1900.0). Precess the coordinates to epoch 1950.0. *Ans.* $20^h03^m09^s$, $-22°04'$

2-3.   What is the Julian day number at 2100 UT on January 10, 1965?

Ans. 2,438,771.375

2-4.   How far away is a star with a 10th-magnitude modulus of distance?

Ans. 1,000 pc

2-5.   What is the absolute radio magnitude of a radio source with a flux density of 10^{-25} watt m^{-2} cps^{-1} at 158 Mc if it is at a distance of 10,000 pc?   *Ans.* −5.9

2-6.   Determine the right ascension and declination of a radio source observed to transit the meridian at 0400 EST on July 15, 1964, at a longitude of $5^h30^m00^s$ (west) and latitude of +40°. The declination was observed to be +25°. Neglect refraction effects. Give the result for the epoch of date and for epoch 1950.0.

2-7.   Show that an observer moving perpendicular to a beam of light at 30 km sec^{-1} will detect a deviation of 20.6 sec of arc in the beam direction. What is this effect called?

3

Radio-Astronomy Fundamentals

3-1 Introduction The purpose of this chapter is to provide a concise introductory treatment of some of the more important basic quantities and equations used in radio astronomy. The topics include brightness, flux density, the effect of the antenna pattern on observations, blackbody radiation, emission and absorption, temperature, noise, and minimum detectable temperature. For more extensive discussions of these topics reference should be made to the excellent books and articles on these topics listed at the end of the chapter.

3-2 Basic Relations for Power, Spectral Power, and Brightness
Consider electromagnetic radiation from the sky falling on a flat horizontal area A at the surface of the earth, as in the elevation view of Fig. 3-1a and perspective view of Fig. 3-1b. The infinitesimal *power* dW from a solid angle $d\Omega$ of the sky incident on a surface of area dA is

$$dW = B \cos\theta \, d\Omega \, dA \, d\nu \tag{3-1}$$

where dW = infinitesimal power, watts
   B = brightness of sky at position of $d\Omega$, watts m^{-2} cps^{-1} rad^{-2}
   $d\Omega$ = infinitesimal solid angle of sky ($= \sin\theta \, d\theta \, d\phi$), rad^2
   θ = angle between $d\Omega$ and zenith, rad
   dA = infinitesimal area of surface, m^2
   $d\nu$ = infinitesimal element of bandwidth, cps
   The quantity B is called the sky brightness, the surface brightness, or simply the *brightness*. It is a fundamental quantity of radio and optical astronomy and is a measure of the power received per unit area per unit solid angle per unit bandwidth. The element of bandwidth lies between a particular frequency ν and $\nu + d\nu$.
   If dW is independent of the position of dA on the surface, the infinitesimal power received by the entire surface A is

$$dW = AB \cos\theta \, d\Omega \, d\nu \tag{3-2}$$

   Integrating (3-2), one can obtain the power W received over a bandwidth $\Delta\nu$ from a solid angle Ω of the sky. Thus,

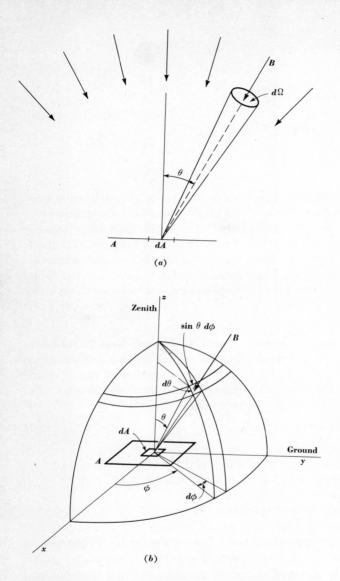

(a)

(b)

Fig. 3-1. Basic geometry for radiation of brightness B incident on a flat area shown in elevation at (a) and in perspective view at (b).

$$W = A \int_{\nu}^{\nu+\Delta\nu} \iint_{\Omega} B \cos\theta \, d\Omega \, d\nu \qquad \text{watts} \qquad (3\text{-}3)$$

In general, the brightness is a function of both the position (in the sky) and of the frequency. The variation of the brightness B with frequency is called the *brightness spectrum*. Integrating B over a bandwidth $\Delta\nu$ (extending from a frequency ν to a frequency $\nu + \Delta\nu$) gives the *total brightness B'* in this frequency band $\Delta\nu$. Thus,

$$B' = \int_{\nu}^{\nu+\Delta\nu} B \, d\nu \qquad (3\text{-}4)$$

where B' = total brightness, watts m^{-2} rad^{-2}
   B = brightness, watts m^{-2} cps^{-1} rad^{-2}
   $d\nu$ = infinitesimal bandwidth, cps
   ν = frequency, cps
   $\Delta\nu$ = finite bandwidth, cps

If the integration extends over the radio spectrum, the *total radio brightness* is obtained. Likewise, if the integration is over the optical spectrum, the *total optical brightness* is obtained.

Introducing the total brightness B', we can rewrite (3-3) as

$$W = A \iint_{\Omega} B' \cos\theta \, d\Omega \qquad \text{watts} \qquad (3\text{-}5)$$

In many situations the *power per unit bandwidth* is more pertinent than the power contained in an arbitrary bandwidth $\Delta\nu$. This power per unit bandwidth is often called the *spectral power*, since its variation with frequency constitutes the power spectrum. Its units are watts per cps. Thus, introducing the concept of spectral power, (3-1) becomes

$$dw = B \cos\theta \, d\Omega \, dA \qquad (3\text{-}6)$$

where dw = spectral power, or infinitesimal power per unit bandwidth, watts cps^{-1}

If dw is independent of the position of dA on the surface A, the spectral power received by the entire surface A is then

$$dw = AB \cos\theta \, d\Omega \qquad (3\text{-}7)$$

Integrating (3-7), one can obtain the spectral power from a solid angle Ω of sky.† Thus,

---

† The power W and the spectral power w are related by

$$W = \int_{\nu}^{\nu+\Delta\nu} w \, d\nu = w_{\text{avg}} \, \Delta\nu$$

where w_{avg} = average spectral power over bandwidth $\Delta\nu$

$$w = A \iint_{\Omega} B \cos \theta \, d\Omega \tag{3-8}$$

where w = spectral power, or power per unit bandwidth, watts cps^{-1}
Suppose that the brightness is uniform over the entire sky, then the power
per unit bandwidth from one hemisphere (solid angle = 2π steradians) is

$$w = AB \int_0^{2\pi} \int_0^{\pi/2} \cos \theta \sin \theta \, d\theta \, d\phi \qquad \text{watts cps}^{-1} \tag{3-9}$$

In (3-9) the infinitesimal solid angle $d\Omega$ has been replaced by its equivalent,
$\sin \theta \, d\theta \, d\phi$, where θ is the zenith angle and ϕ is the azimuth angle (see Fig.
3-1b).   Evaluating the integral in (3-9) yields a solid angle of π steradians
(rad^2), and (3-9) becomes

$$w = \pi AB \qquad \text{watts cps}^{-1} \tag{3-10}$$

If, in addition to the above conditions, the brightness B is uniform over a
bandwidth $\Delta\nu$, then the total power W received by the area A from a hemisphere is

$$W = AB \, \Delta\nu \iint \cos \theta \, d\Omega = \pi AB \, \Delta\nu \tag{3-11}$$

This is the power from the sky which would be received at a point on the
surface of the earth, for which case the solid angle subtended by the sky
is one hemisphere.

Example. If the brightness B of the sky at frequency ν is uniform over a bandwidth of 1 Mc and also uniform over the entire sky (solid angle 2π), find the spectral
power and the total power received by a horizontal surface of 5 m^2 area at frequency ν.
The value of B is 10^{-22} watt m^{-2} cps^{-1} rad $^{-2}$.

Solution. From (3-10) the power received per unit frequency by the 5-m^2 surface is

$$w = \pi AB = 5\pi \times 10^{-22} \text{ watt cps}^{-1}$$

and from (3-11) the total power received over the 1-Mc bandwidth is

$$W = \pi AB \, \Delta\nu = 5\pi \times 10^{-22} \times 10^6 = 5\pi \times 10^{-16} \text{ watt}$$

3-3 Brightness Distribution Since the sky brightness may vary with
direction, it is, in general, a function of angle. This may be expressed
explicitly by the symbol $B(\theta, \phi)$ for the brightness. Thus, from (3-8) the
spectral power, or power per unit bandwidth, becomes

$$w = A \iint B(\theta, \phi) \cos \theta \, d\Omega \qquad \text{watts cps}^{-1} \tag{3-12}$$

Consider the radio-astronomy situation suggested by Fig. 3-2. Here
the area A of Fig. 3-1 is replaced by the flat horizontal surface of a receiving
antenna with the power pattern of the antenna directed toward the zenith
($\theta = 0$). In this case the pertinent area is the *effective aperture* A_e of the

antenna. For antennas of large aperture it is always somewhat less than the physical aperture of the antenna (see Chap. 6). The power pattern $P_n(\theta, \phi)$ is a measure of the response of the antenna to radiation as a function of the angles θ and ϕ. It is a dimensionless quantity. It is also

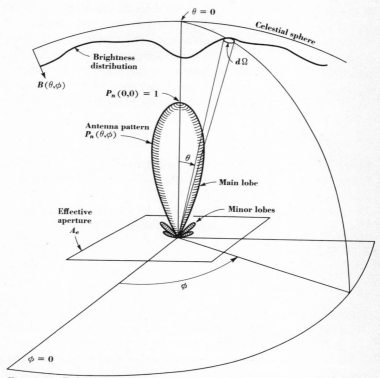

Fig. 3-2. Relation of antenna pattern to celestial sphere with associated coordinates.

normalized (maximum value unity). In the antenna case it replaces the factor $\cos \theta$ of (3-12). Thus, introducing A_e and $P_n(\theta, \phi)$ into (3-12), we have for the spectral power w or power per unit bandwidth from a solid angle Ω of the sky†

$$w = \tfrac{1}{2}\ddagger A_e \iint_\Omega B(\theta, \phi) P_n(\theta, \phi) \, d\Omega \tag{3-13}$$

† For an isotropic antenna $P_n(\theta,\phi) = 1$ at all angles. This corresponds to the case of a spherical (one-way) receiving surface of cross section A (instead of a flat one as in Fig. 3-1 or 3-2). For this case the $\cos \theta$ factor in (3-12) would then be replaced by unity, not because θ is small but because of the spherical shape of the receiving surface.

‡ The fraction $\tfrac{1}{2}$ is to be read as $\tfrac{1}{2}$ and not as $1/2$.

where w = received spectral power, watts cps^{-1}

A_e = effective aperture of antenna, m^2

$B(\theta, \phi)$ = brightness of sky, watts m^{-2} cps^{-1} rad^{-2}

$P_n(\theta, \phi)$ = normalized power pattern of antenna, dimensionless

$d\Omega$ = sin θ $d\theta$ $d\phi$ = element of solid angle, rad^2

If the radiation received is of an incoherent, unpolarized nature, only one-half of the incident power will be received, since any antenna is responsive to only one polarization component. The factor $\frac{1}{2}$ introduced in (3-13) is for this case. In the more general case where the radiation is partially polarized (and partially unpolarized), the appropriate factor may, in the limits, lie between 0 and 1. This is discussed further in Chap. 4.

If the brightness is constant, (3-13) becomes

$$w = \tfrac{1}{2}A_e B_c \iint_\Omega P_n(\theta, \phi) \, d\Omega \tag{3-14}$$

where B_c = uniform or constant brightness, watts m^{-2} cps^{-1} rad^{-2}

If the integration in (3-14) is performed over a solid angle of 4π, the integral is the *beam area* or *beam solid angle* Ω_A of the antenna (see Chap. 6). The angle might more appropriately be called the *pattern solid angle*, but beam area or beam solid angle are in wide usage.

That is

$$\Omega_A = \iint_{4\pi} P_n(\theta, \phi) \, d\Omega \tag{3-15}$$

where Ω_A = beam area, rad^2

$P_n(\theta, \phi)$ = normalized power pattern of antenna, dimensionless

$d\Omega$ = sin θ $d\theta$ $d\phi$ element of solid angle, rad^2

If the back-side response of the antenna of Fig. 3-2 is negligible and the sky has a constant brightness B_c, we have

$$w = \tfrac{1}{2}A_e B_c \Omega_A \qquad \text{watts cps}^{-1} \tag{3-16}$$

In the general case, where brightness varies with both position and frequency, the total power received is given by reexpressing (3-13) as†

$$W = \tfrac{1}{2}A_e \int_\nu^{\nu+\Delta\nu} \iint_\Omega B(\theta, \phi)P_n(\theta, \phi) \, d\Omega \, d\nu \qquad \text{watts} \tag{3-17}$$

The integration is carried out over a specified solid angle Ω of the sky and frequency band between ν and $\nu + \Delta\nu$. If the brightness is uniform over

---

† The symbol for brightness in (3-17) may be written as $B(\theta,\phi,\nu)$, indicating explicitly that the brightness is a function of both angle and frequency. Likewise the pattern symbol may be written $P_n(\theta,\phi,\nu)$. However, for simplicity the ν will be omitted, the dependence of B and P_n on frequency being implied.

the entire sky and if both brightness and antenna patterns are uniform with respect to frequency over the bandwidth $\Delta\nu$, the total power received from the sky over this bandwidth from the entire sky is

$$W = \tfrac{1}{2}A_eB_c\Omega_A \, \Delta\nu \qquad \text{watts} \tag{3-18}$$

where A_e = effective aperture of antenna, m²
$\qquad B_c$ = constant brightness, watts m⁻² cps⁻¹ rad⁻²
$\qquad \Omega_A$ = beam area of antenna, rad²
$\qquad \Delta\nu$ = bandwidth, cps

As an illustration consider the following simple example.

Example. At a frequency ν a radio-telescope antenna has a fan beam 5° by 20° wide between principal-plane half-power points. The sky at which the antenna is directed has a substantially uniform radio brightness of 10^{-21} watt m⁻² cps⁻¹ rad⁻² at the frequency ν. If the antenna has an effective aperture of 36 m², find the total power received per unit bandwidth at the frequency ν.

Solution. The beam area of the antenna is given approximately by [see (6-22)]

$$\Omega_A \simeq \tfrac{4}{3}\theta_{\rm HP}\phi_{\rm HP}$$

where Ω_A = beam area, rad² or deg²
$\qquad \theta_{\rm HP}$ = half-power beam width in one principal plane of the pattern, rad or deg
$\qquad \phi_{\rm HP}$ = half-power beam width in other principal plane of the pattern, rad or deg

Hence, for the antenna pattern in this example

$$\Omega_A = \tfrac{4}{3} \times 5° \times 20° = 133 \text{ deg}^2$$

With the aid of (3-16), the total power per unit bandwidth is

$$w = \tfrac{1}{2}A_eB_c\Omega_A = \tfrac{1}{2} \times 36 \times 10^{-21} \times 133/3,282 = 7.3 \times 10^{-22} \text{ watt cps}^{-1}$$

The fraction $133/3,283$ converts the number of square degrees (133) into steradians. Thus, 1 steradian (rad²) = $57.3^2 = 3,283$ deg², or 1 rad² = $41,253/4\pi = 3,283$ deg². If the receiver bandwidth $\Delta\nu$ is 1 Mc and if the brightness is substantially constant over this band, the total power received is

$$W = w \, \Delta\nu = 7.3 \times 10^{-22} \times 10^6 = 7.3 \times 10^{-16} \text{ watts}$$

3-4 Discrete Sources, Their Flux Density, and Brightness

A *discrete radio source* is one which is distinct or separate. Such sources may be divided into several types:

(1) Point sources
(2) Localized sources
(3) Extended sources

A *point source* is an idealization. It may be defined as one which subtends an infinitesimal solid angle. All radio sources must subtend a finite solid angle, but it is often convenient to regard sources of very small extent as point sources.† A *localized source* is a discrete source of small but finite

† A source that is small compared to the antenna beam size is effectively a point source.

extent. The term *radio star* is sometimes applied to localized or point sources, but this does not imply that they are true optical stars, i.e., hot, optically self-luminous bodies. The sun and some flare stars are the principal exceptions, being both true stars and radio sources.

An *extended source* is a discrete source of larger extent. The distinction between a localized and an extended source is arbitrary, but it has been common practice to regard sources with diameters of less than 1° as localized and sources with diameters of more than 1° as extended. A large source of many degrees angular extent may be regarded as a discrete source provided that it has a well-defined boundary.

For any discrete source, the integral of the brightness over the source yields the total source *flux density*. Thus,

$$S = \iint_{source} B(\theta,\phi) \, d\Omega \tag{3-19}†$$

where S = flux density of source, jan, or watts m^{-2} cps^{-1}

$B(\theta,\phi)$ = brightness as a function of position over source, watts m^{-2} cps^{-1} rad^{-2}

$d\Omega$ = element of solid angle ($=\sin\theta \, d\theta \, d\phi$), rad^2

The unit of flux density is watts per square meter per cycle per second (watts m^{-2} cps^{-1}). The name jansky (abbreviated jan), after the pioneer radio astronomer Karl G. Jansky, has been proposed for this unit. Thus, 1 jan = 1 watt m^{-2} cps^{-1}. The flux density of most radio sources is of the order of 10^{-26} jan and this quantity (10^{-26} watt m^{-2} cps^{-1}) is commonly called a *flux unit*.

When the source is observed with an antenna of power pattern $P_n(\theta,\phi)$, a flux density S_o is observed, as given by

$$S_o = \iint_{source} B(\theta,\phi) P_n(\theta,\phi) \, d\Omega \tag{3-21}$$

Owing to the directional pattern of the antenna the observed flux density will be less than the *actual* or *true value* as defined in (3-19). The antenna pattern $P_n(\theta,\phi)$ is like a weighting function, which makes the integrand (3-21) less than the integrand in (3-19). However, if the source is of suffi-

---

† The source flux density could also be defined as

$$S = \iint_{source} B(\theta,\phi) \cos\theta \, d\Omega \tag{3-20}$$

where θ = angle from center of source. Although this definition might appear to be more consistent with the discussion in the previous sections, the definition of (3-19) will be adopted since it makes the source flux density completely independent of the receiving device. In any event, the two definitions would differ significantly only for very extended objects, such as the Magellanic clouds.

u ... ... to ... ... that $B(\theta,\phi)$ and $I_n(\theta,\phi)$ are xpressed wrt same coordinate system.

ciently small extent, so that $P_n(\theta,\phi) \simeq 1$ over the source, the flux density by (3-21) will be nearly equal to the true value. In this case (3-21) reduces to (3-19). On the other hand, when the source is larger than the antenna-main-lobe solid angle, so that the source brightness may be regarded as constant over the main lobe, (3-21) reduces to

$$S_o = B(\theta,\phi) \iint P_n(\theta,\phi)\, d\Omega \simeq B(\theta,\phi)\Omega_M \tag{3-22}$$

where Ω_M = main-lobe solid angle

In (3-21) and (3-22) it is assumed that the antenna beam is aligned with the source. If it is not aligned further analysis is required.

For this situation, let the reference direction coincide with the x axis ($\theta = \pi/2$, $\phi = 0$), as in Fig. 3-3a, instead of with the z axis ($\theta = 0$, $\phi = 0$), as in Figs. 3-1 and 3-2. The flux density observed with a fan-beam antenna scanning a long, narrow source, as in Fig. 3-3b, will then be given by

$$S(\phi_0) = \int B(\phi)P_n(\phi-\phi_0)\, d\phi \tag{3-23}$$

where $S(\phi_0)$ = observed flux density, jan or watts m^{-2} cps^{-1}

$B(\phi)$ = source brightness distribution, watts m^{-2} cps^{-1} rad^{-1}

$P_n(\phi-\phi_0)$ = normalized antenna power pattern with respect to the angle ϕ.

ϕ_0 = displacement angle of antenna pattern

The observed flux density may also be expressed

$$S(\phi_0) = \int B(\phi)\tilde{P}_n(\phi_0-\phi)\, d\phi \tag{3-24}$$

where $\tilde{P}_n(\phi_0-\phi) = P_n(\phi-\phi_0)$

An observed *brightness distribution* is obtained by dividing (3-24) by the antenna beam angle $\phi_A \left[= \int P_n(\phi)\, d\phi \right]$. Thus,

$$B(\phi_0) = \frac{1}{\phi_A} \int B(\phi)\tilde{P}_n(\phi_0-\phi)\, d\phi \tag{3-25}$$

With a radio telescope the observed quantity is a temperature distribution $T(\phi_0)$, which is a function of the true source brightness-temperature distribution $T_s(\phi)$, so that we have (see Sec. 3-18)

$$T(\phi_0) = \frac{1}{\phi_A} \int T_s(\phi)\tilde{P}_n(\phi_0-\phi)\, d\phi \tag{3-26}$$

The one-dimensional relationship in (3-23) to (3-26), is a special case of the two-dimensional situation, where, in general, the brightness distribution and the antenna pattern are functions of both ϕ and γ ($=90°-\theta$) (see also Prob. 3-7). For the source and pattern geometry of Fig. 3-3b the problem has been simplified by considering the distributions as a function only of ϕ (units: watts m^{-2} cps^{-1} rad^{-1} instead of watts m^{-2} cps^{-1} rad^{-2}). The corresponding source brightness profile and antenna pattern as a function of ϕ are as shown in Fig. 3-4a. The observed flux density will then be as suggested in Fig. 3-4b. At any angle ϕ_0, the observed flux density $S(\phi_0)$ is proportional to the area under the product curve given by the integrand in (3-24). This area is shaded in Fig. 3-4a.

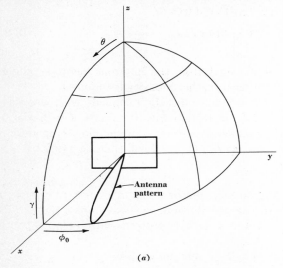

(a)

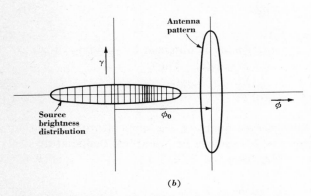

(b)

Fig. 3-3. Antenna pattern (a) and geometry for long source and fan-beam antenna pattern (b).

The flux density $S(\phi_0)$ as given in (3-23) is called the *cross-correlation function* of the brightness distribution and the antenna pattern, while as given in (3-24) it is called the *convolution*. Alternatively it may be said that $S(\phi_0)$ by (3-24) is the result of convolving B and $\tilde{P}_n$. It is to be noted that the pattern function in (3-24) is the mirror image of the actual pattern;

i.e., $\tilde{P}_n(\phi) = P_n(-\phi)$ [or $\tilde{P}_n(-\phi) = P_n(\phi)$]. For symmetrical patterns the distinction is of no consequence. Equation (3-24) may also be written

$$S(\phi_0) = \int B(\phi_0-\phi)\tilde{P}_n(\phi)\,d\phi \tag{3-27}$$

In shorthand notation (3-24) and (3-27) may be expressed as

$$S = B*\tilde{P} \tag{3-28}$$

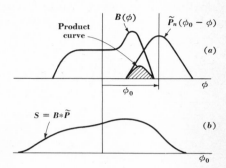

Fig. 3-4. The true brightness distribution B scanned by, or convolved with, the antenna pattern $\tilde{P}$, as in (a) yields the observed flux-density distribution S, as in (b).

Equations (3-23), (3-24), (3-27), and (3-28) are statements of the fact that, in general, an observing instrument does not provide a true picture of the observed quantity but a modified result. The effect in the case of the antenna is to smooth out the details of the distribution, as may be noted by comparing the observed profile $S(\phi_0)$ in Fig. 3-4b with the true profile $B(\phi)$ in Fig. 3-4a. The observed result will agree exactly with the original only if the instrument is perfect. For an antenna this means an infinitely sharp beam. This is a hypothetical case requiring an antenna of infinite directivity (see Chap. 6). An antenna pattern of this type may be represented with the aid of a delta function defined as follows:

$$\delta(\phi) = \begin{cases} 0 & \text{for } \phi \neq 0 \\ \infty & \text{for } \phi = 0 \end{cases} \quad \text{and} \quad \int \delta(\phi)\,d\phi = 1 \quad \text{rad} \tag{3-29}$$

For an infinitely sharp antenna pattern the condition that the brightness distribution is constant over the antenna pattern is always fulfilled (if there are no singularities in the source distribution), so that

$$S(\phi_0) = B(\phi)K_1 \int \delta(\phi_0-\phi)\,d\phi = B(\phi)\phi_A \tag{3-30}$$

In (3-30)

$$\phi_A = \int \tilde{P}_n(\phi_0-\phi)\,d\phi = K_1 \int \delta(\phi_0-\phi)\,d\phi \tag{3-31}$$

where ϕ_A = antenna beam angle
K_1 = const, dimensionless
Thus, the observed flux-density distribution is identical in form with the actual brightness distribution, as suggested in Fig. 3-5.

For the converse case of an infinitely sharp source distribution (point source) and a finite antenna beam width, the source distribution may be represented with the aid of a delta function. Thus,

$$S(\phi_0) = K_2 \int \delta(\phi)\tilde{P}_n(\phi_0-\phi) \, d\phi = S\tilde{P}_n(\phi_0) \qquad (3\text{-}32)$$

In (3-32)

$$S = \int B(\phi) \, d\phi = K_2 \int \delta(\phi) \, d\phi \qquad (3\text{-}32a)$$

where S = actual or true flux density of source

K_2 = constant, watts m^{-2} cps^{-1} rad^{-1}

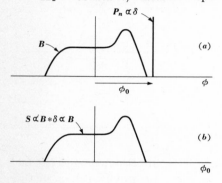

Fig. 3-5. For an infinitely sharp antenna pattern (perfect observing instrument) the observed distribution is identical with the true distribution.

Thus, the observed flux-density distribution is identical with the antenna pattern although reversed in sign, as suggested in Fig. 3-6. In fact a measurement made in this way with a point source yields a result which, by definition, is the mirror image of the true antenna pattern (far-field pattern if the distance to the source is sufficient). When the antenna pattern is aligned with the source ($\phi_0 = 0$), the observed flux density equals the actual flux density of the source.

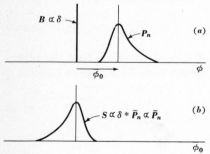

Fig. 3-6. For a point source the observed distribution is the same as the mirror image of the antenna pattern.

Convolution involves displacement, multiplication, and integration. As a further illustration, consider the uniform source distribution $B(\phi)$ and highly asymmetric triangular antenna pattern $P_n(\phi-\phi_0)$ in Fig. 3-7a. The observed flux density is

$$S(\phi_0) = \int B(\phi)\tilde{P}_n(\phi_0-\phi)\, d\phi \qquad (3\text{-}33)$$

where $\tilde{P}_n(\phi_0-\phi) = P_n(\phi-\phi_0)$

$\phi_0 = $ displacement angle

Multiplying the brightness distribution and the antenna pattern and integrating this product gives the observed flux density as a function of the displacement angle ϕ_0, as indicated in Fig. 3-7b. Graphically the result is given by moving the actual antenna pattern across the brightness distribution and for any ϕ_0 (as in Fig. 3-7a) plotting an ordinate value for flux density (in Fig. 3-7b) proportional to the area under the product curve (shaded in Fig. 3-7a).

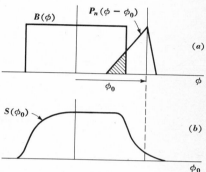

Fig. 3-7. Example of a uniform source distribution scanned by an antenna with an asymmetric pattern of triangular shape.

As suggested in Figs. 3-4, 3-6, and 3-7, an observed brightness distribution is smoother and wider than the original, or true, distribution. It may be said that the antenna produces a smoothing effect. Furthermore, it is not possible to reconstruct the brightness distribution uniquely, since an infinite number of distributions can yield the same observed distribution $S(\phi_0)$. Thus, detailed structure in the source distribution is irretrievably lost unless the antenna has an infinitely sharp pattern. The subject of antenna smoothing is discussed further in Sec. 6-9.

Returning to the two-dimensional geometry, a situation of considerable practical interest is one in which the source is smaller than the antenna main-beam area, as suggested in Fig. 3-8. Let the source be centered at $\theta = 0$ (Fig. 3-2). When the antenna is aligned with the source, as in Fig. 3-8, the flux density observed is a maximum but is less than the actual flux density. The observed or *apparent brightness* B_o is then, from (3-25),

$$B_o = \frac{\iint B(\theta,\phi)P_n(\theta,\phi)\, d\Omega}{\iint P_n(\theta,\phi)\, d\Omega} = \frac{S_o}{\Omega_A} \qquad (3\text{-}34)$$

where $B_o = $ apparent brightness, watts m^{-2} cps^{-1} rad^{-2}

$S_o = $ maximum observed flux density, watts m^{-2} cps^{-1}

$\Omega_A = $ antenna beam area, rad^2

$P_n(\theta,\phi)$ = normalized antenna power pattern, dimensionless
$B(\theta,\phi)$ = actual brightness distribution, watts m^{-2} cps^{-1} rad^{-2}
$d\Omega$ = element of solid angle, rad^2

The integration in the numerator of (3-34) is carried out over the solid angle subtended by the source and the integration in the denominator over 4π.

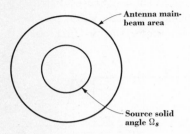

Fig. 3-8. Source of solid angle Ω_s aligned with antenna pattern.

If the source is sufficiently small compared to the beam area, so that the observed flux density S_o is substantially equal to the actual flux density S, and if the source solid angle Ω_s is known, it may be more significant to define an *actual average brightness* B_{avg}, as follows

$$B_{\mathrm{avg}} = \frac{S}{\Omega_s} = \frac{1}{\Omega_s} \iint B(\theta,\phi)\, d\Omega \qquad (3\text{-}35)$$

where S = actual source flux density, watts m^{-2} cps^{-1}
Ω_s = source solid angle, rad^2

Note that the apparent brightness involves an average over the antenna beam area while the actual average brightness involves an average over the source.

If the brightness is uniform over the source ($B = B_{\mathrm{avg}}$) and if the source is much smaller than the beam area ($\Omega_s \ll \Omega_A$), we have

$$B_o = \frac{S}{\Omega_A} = \frac{\Omega_s}{\Omega_A} B \qquad (3\text{-}36)$$

For the case where the source extent coincides with the first null of the antenna pattern

$$B_o = \frac{\Omega_M}{\Omega_A} B \qquad\qquad uoe\ \ 3\cdot 34 \qquad (3\text{-}37)$$

where Ω_M = beam area of main lobe of the antenna pattern

$$= \iint_{\substack{\text{main}\\ \text{lobe}}} P_n(\theta,\phi)\, d\Omega$$

When the source area is larger than the main-lobe area but small compared to 4π, we may write

$$B_o = \frac{\Omega_M'}{\Omega_A} B \tag{3-38}$$

where Ω_M' = beam area of main lobe and side lobes out to limits of source

$$= \iint\limits_{source} P_n(\theta,\phi) \, d\Omega$$

The antenna beam areas are related by

$$\Omega_M < \Omega_M' < \Omega_A \tag{3-39}$$

or

$$\frac{\Omega_M}{\Omega_A} < \frac{\Omega_M'}{\Omega_A} < 1 \tag{3-40}$$

The ratios in (3-40) are called beam efficiencies. The ratio Ω_M/Ω_A is the *main-beam efficiency*. This is discussed further in Chap 6.

The variation of the flux density S with frequency is called the *flux-density spectrum*. Integrating S over a bandwidth $\Delta\nu$ (extending from a frequency ν to $\nu + \Delta\nu$) yields the *total flux density* S' in this band of frequencies ($\Delta\nu$). Thus,

$$S' = \int_\nu^{\nu+\Delta\nu} S \, d\nu \tag{3-41}$$

where S' = total flux density, watts m^{-2}
  S = flux density, watts m^{-2} cps^{-1}
If the integration extends over the radio spectrum, the *total radio flux density* is obtained. Likewise, if the integration is over the optical spectrum, the *total optical flux density* is obtained.

It follows that the total (observed) power W received over a bandwidth $\Delta\nu$ from a source of extent Ω_s is given by

$$\begin{aligned}
W &= \tfrac{1}{2}A_e S_o' = \tfrac{1}{2}A_e \int_\nu^{\nu+\Delta\nu} S_o \, d\nu \\
&= \tfrac{1}{2}A_e \int_\nu^{\nu+\Delta\nu} \iint\limits_{\Omega_s} B(\theta,\phi) P_n(\theta,\phi) \, d\Omega \, d\nu
\end{aligned} \tag{3-42}$$

where W = total received power, watts
  A_e = effective aperture of antenna, m^2
  S_o' = total observed flux density, watts m^{-2}
  S_o = observed flux density, watts m^{-2} cps^{-1}
  ν = frequency, cps
  $\Delta\nu$ = bandwidth (between ν and $\nu + \Delta\nu$), cps
  $d\nu$ = infinitesimal bandwidth, cps
  $P_n(\theta,\phi)$ = antenna power pattern, dimensionless

$B(\theta,\phi)$ = brightness distribution, watts m^{-2} cps^{-1} rad^{-2}

$d\Omega$ = infinitesimal solid angle, rad^2

It is assumed in (3-42) that the antenna pattern is aligned with the source. If the flux density is constant over the bandwidth $\Delta\nu$, (3-42) reduces to

$$W = \tfrac{1}{2}A_eS_o\Delta\nu \qquad \text{watts} \tag{3-43}$$

Example 1. A discrete radio source 2° in diameter has an actual average brightness B_{avg} of 10^{-20} watt m^{-2} cps^{-1} rad^{-2} at a frequency ν. The brightness is constant over a bandwidth of 2 Mc at this frequency. Calculate (a) the actual flux density of the source at this frequency, (b) the flux density observed with an antenna having a main beam with an area of 1 deg^2, and (c) the total actual flux density over a 2-Mc bandwidth.

Solution. (a) From (3-35) the actual flux density is given by

$$S = B_{avg}\Omega_s = 10^{-20}\pi \frac{1}{3,282} = 9.6 \times 10^{-24} \text{ watt m}^{-2} \text{ cps}^{-1}$$

(b) Since the beam area is smaller in extent than the source and assuming that the brightness is uniform over the source, we have

$$S_o = B\Omega'_M$$

The beam area Ω'_M is not given. However, if it is assumed that it is very little larger than Ω_M, we have approximately that

$$S_o = 10^{-20} \frac{1}{3,282} = 3.05 \times 10^{-24} \text{ watt m}^{-2} \text{ cps}^{-1}$$

(c) The total actual flux density of the source over the 2-Mc bandwidth is, from (3-41),

$$S' = \int S \, d\nu = S \, \Delta\nu = 9.6 \times 10^{-24} \times 2 \times 10^6 = 1.9 \times 10^{-17} \text{ watt m}^{-2}$$

Example 2. If the radio-telescope antenna used to observe the source of Example 1 has an effective aperture of 150 m^2, find (a) total power received and (b) the received spectral power.

Solution. (a) The total power received from the source is given by (3-43) as

$$W = \tfrac{1}{2}A_eS_o \, \Delta\nu = \tfrac{1}{2} \times 150 \times 3.05 \times 10^{-24} \times 2 \times 10^6 = 4.6 \times 10^{-16} \text{ watt}$$

(b) The spectral power or power per unit bandwidth is

$$w = \tfrac{1}{2}A_eS_o = 2.3 \times 10^{-22} \text{ watt cps}^{-1}$$

Fig. 3-9. Radiation from a large solid angle.

Although the total flux density S' has the same dimensions as the *Poynting vector* (watts per square meter) the two are, in general, not the same. The flux density is a scalar as defined in (3-19). The Poynting vector, on the other hand, is a vector and is defined in terms of the electric and magnetic fields at a point ($= \mathbf{E} \times \mathbf{H}$). For radiation coming from a large solid angle as in Fig. 3-9, the magnitude of the Poynting vector will be less than S', but for a point source the two are equal.

3-5 Intensity or Radiance Consider that a flat surface A emits electromagnetic radiation, as suggested in Fig. 3-10. The spectral power from the surface element dA flowing out through the solid angle $d\Omega$ is then given by

$$dw = I \cos \theta d\Omega \, dA \tag{3-44}$$

where dw = spectral power or power per unit bandwidth, watts cps^{-1}
$\quad\quad I$ = intensity, watts m^{-2} cps^{-1} rad^{-2}
$\quad\quad \theta$ = angle between normal to surface and $d\Omega$, rad
$\quad\quad d\Omega$ = element of solid angle, rad^2
$\quad\quad dA$ = element of area, m^2

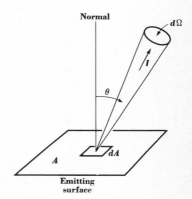

Fig. 3-10. Emitting surface of intensity, or radiance, I.

The quantity I is called the *intensity* of the radiation from the surface. In infrared terminology it is commonly called the *radiance*. The intensity, or radiance, has the dimensions of power per unit area per unit solid angle per unit bandwidth and is expressed in watts per square meter per cycle per second per steradian (watts m^{-2} cps^{-1} rad^{-2}). It is assumed in (3-44) that the radiation from the elemental area obeys Lambert's cosine law.

If the surface radiates uniformly over its entire area A, (3-44) reduces to

$$dw = AI \cos \theta \, d\Omega \tag{3-45}$$

The spectral power w radiated by the surface over a given solid angle is then expressed by the integral of (3-45), or

$$w = A \iint I \cos \theta \, d\Omega \tag{3-46}$$

where w = spectral power, watts cps^{-1}
$\quad\quad A$ = area, m^2
$\quad\quad I$ = intensity, watts m^{-2} cps^{-1} rad^{-2}
$\quad\quad \theta$ = angle between normal to A and $d\Omega$, rad
$\quad\quad d\Omega$ = element of solid angle, rad^2

The situation in Fig. 3-10, involving intensity of emission from th surface A, may be considered as the transmitting case, while the situatio in Fig. 3-1, involving the absorption of radiation from the sky by th surface A, may be considered as the receiving case. The dimensions an units for the intensity I and the brightness B are the same. Under th assumed conditions, it follows that

$$I \cos \theta \, d\Omega \, dA = -B \cos \theta \, d\Omega \, dA \qquad (3\text{-}4?$$

and

$$I = -B \qquad (3\text{-}4\!$$

3-6   Blackbody Radiation and Planck's Radiation Law

All object at temperatures above absolute zero radiate energy in the form of electr magnetic waves. Not only do objects radiate electromagnetic energy, bu they also may absorb or reflect such energy incident on them. Kirchho (1859) showed that a good absorber of electromagnetic energy is also good radiator. A perfect absorber is called a *blackbody*, and it follows tha such a body is also a perfect radiator. A blackbody absorbs all the radi tion falling upon it at all wavelengths, and the radiation from it is a fun tion of only the temperature and wavelength. Such a body is an idealiz tion, since no body having this property exists. However, some bodi made of carbon or lampblack absorb most of the incident radiation in th visible and infrared part of the spectrum and closely approximate an ide blackbody over this wavelength region.

A blackbody may be approximated by the hole in the wall of a bo or enclosure at a uniform temperature T, as suggested by Fig. 3-11. Rad ation falling on the hole from outside passes through it into the box and completely absorbed by multiple reflections from the inside walls, excep for the small fraction of radiation reflected back out the hole. Howeve this fraction will be negligible if the hole is sufficiently small compared t the other dimensions of the box. The radiation which does emerge fro the hole is characteristic of a blackbody radiator at a temperature T. T reduce edge effects the hole should be large compared to the wavelengt and the box large compared to the hole.

The brightness of the radiation from a blackbody is given by *Planck radiation law*. This law, formulated by Max Planck (1901), states that th brightness of a blackbody radiator at a temperature T and frequency ν expressed by

$$B = \frac{2h\nu^3}{c^2} \frac{1}{e^{h\nu/kT} - 1} \qquad (3\text{-}4\S$$

where B = brightness, watts m^{-2} cps^{-1} rad^{-2}
    h = Planck's constant (= 6.63 × 10^{-34} joule sec)
    ν = frequency, cps
    c = velocity of light (= 3 × 10^8 m sec^{-1})
    k = Boltzmann's constant (= 1.38 × 10^{-23} joule °K^{-1})
    T = temperature, °K

It is to be noted that $h\nu$ has the dimensions of energy. Likewise kT has the dimensions of energy, so that the ratio $h\nu/kT$ is dimensionless.

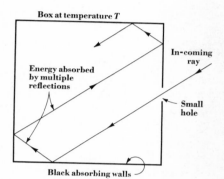

Fig. 3-11. Blackbody enclosure.

The brightness B for a blackbody radiator at four temperatures is shown graphically in Fig. 3-12. The first factor $(2h\nu^3/c^2)$ of the right-hand side of (3-49) varies as the cube of the frequency, while the second factor varies (for large ν) as $e^{-\nu}$. Hence, as indicated in Fig. 3-12, B tends to zero for large or small frequencies, with its maximum value at an intermediate frequency.

Of particular interest is the fact that the point of maximum brightness shifts to higher frequency (shorter wavelength) as the temperature is increased.

Example. Calculate the brightness of a blackbody radiator at a temperature of 6000°K and a wavelength of 0.5 micron.

Solution. Frequency and wavelength are related in vacuum by

$$\nu = \frac{c}{\lambda}$$

where ν = frequency, cps
    c = velocity of light (= 3 × 10^8 m sec^{-1})
    λ = wavelength, m

Since 1 micron equals 10^{-6} m, a wavelength of 0.5 micron corresponds to a frequency

$$\nu = \frac{c}{\lambda} = \frac{3 \times 10^8}{0.5 \times 10^{-6}} = 6 \times 10^{14} \text{ cps}$$

Introducing this value for ν in (3-49) and putting $T = 6000°$, we have

$$B = \frac{2 \times 6.62 \times 10^{-34} (6 \times 10^{14})^3}{(3 \times 10^8)^2} \frac{1}{e^{h\nu/kT} - 1}$$

where

$$\frac{h\nu}{kT} = \frac{6.62 \times 10^{-34} \times 6 \times 10^{14}}{1.38 \times 10^{-23} \times 6000} = 4.8$$

This reduces to

$$B = \frac{3.2 \times 10^{-6}}{e^{4.8} - 1} = \frac{3.2 \times 10^{-6}}{121.5 - 1} = 2.6 \times 10^{-8} \text{ watt m}^{-2} \text{ cps}^{-1} \text{ rad}^{-2}$$

If Planck's radiation-law curves for a blackbody radiator are plotted to logarithmic scales, instead of to linear scales as in Fig. 3-12, a much wider range of frequency (or wavelength) and temperature can be con-

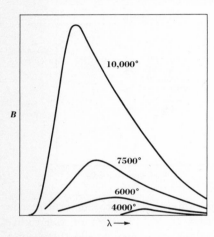

Fig. 3-12. Planck-radiation-law curves for a blackbody radiator as a function of wavelength at four temperatures.

veniently covered. Such a graph is presented in Fig. 3-13. It is apparent from this graph that when the curves for different temperatures are plotted to logarithmic scales all are identical in shape but displaced with respect to each other.

The horizontal scale, or abscissa, in Fig. 3-13 is arranged with wavelength increasing to the right (frequency decreasing). It is sometimes more convenient to reverse directions so that frequency increases to the right. A graph of this kind is shown in Fig. 3-14, which is a mirror image of the graph in Fig. 3-13.

In both Figs. 3-13 and 3-14 the brightness B in watts m^{-2} cps^{-1} rad^{-2} as calculated from (3-49) is indicated along the left-hand ordinate scale, with the frequency or wavelength along the bottom abscissa scale.

In (3-49) the brightness is expressed in power per unit area per unit bandwidth per unit solid angle (jan rad^{-2}, or watts m^{-2} cps^{-1} rad^{-2}). It is often of interest to state Planck's radiation law so that the brightness is expressed in power per unit area *per unit wavelength* per unit solid angle (watts m^{-2} wavelength^{-1} rad^{-2} or watts m^{-3} rad^{-2}). To convert (3-49) we note that frequency and wavelength are related in vacuum by

$$\nu\lambda = c \qquad\qquad\qquad (3\text{-}50)$$

where ν = frequency, cps
    λ = wavelength, m
    c = velocity of light $(3 \times 10^8 \text{ m sec}^{-1})$
It follows that

$$dv = -\frac{c}{\lambda^2}\,d\lambda \tag{3-51}$$

Also since $B\,d\nu = -B_\lambda\,d\lambda$ we have

$$B_\lambda = \frac{2hc^2}{\lambda^5}\frac{1}{e^{hc/kT\lambda} - 1} \tag{3-52}$$

where B_λ = brightness of blackbody radiator in terms of unit wavelength,
       watts m^{-2} m^{-1} rad^{-2}
    h = Planck's constant $(= 6.63 \times 10^{-34} \text{ joule sec})$
    c = velocity of light $(= 3 \times 10^8 \text{ m sec}^{-1})$
    λ = wavelength, m
    k = Boltzmann's constant $(= 1.38 \times 10^{-23} \text{ joule }°\text{K}^{-1})$
    T = temperature, °K

The symbol for the brightness in (3-52) is B_λ, the subscript λ indicating that the brightness is in terms of unit wavelength. The brightness in terms of unit bandwidth (cps), as in (3-49) is denoted by the symbol B (without subscript).

Curves of brightness as a function of frequency (or wavelength) are of the same shape. However, the maximum brightness value occurs at a different frequency (see Sec. 3-8). By providing a different frequency scale, curves of B vs. frequency can also serve as graphs of B_λ vs. frequency. This has been done in Figs. 3-13 and 3-14, where the right-hand ordinate scale is expressed in the brightness B_λ (per unit wavelength) and the corresponding frequency (or wavelength) scale is given as abscissa along the top of the chart. In Fig. 3-13 wavelength increases to the right, and in Fig. 3-14 frequency increases to the right.

3-7 Stefan-Boltzmann Law The brightness-frequency or brightness-wavelength curves of Figs. 3-13 and 3-14 are curves of the *brightness spectrum* as predicted by the Planck radiation law. Integrating the Planck radiation law over all frequencies yields the total brightness B' for a blackbody radiator. Thus,

$$B' = \frac{2h}{c^2}\int_0^\infty \frac{\nu^3}{e^{h\nu/kT} - 1}\,d\nu \tag{3-53}$$

To facilitate integration let us put (Richtmyer and Kennard, 1942)

$$x = \frac{h\nu}{kT} \tag{3-54}$$

from which it follows that

$$\nu = \frac{kT}{h} x \quad \text{and} \quad d\nu = \frac{kT}{h} dx \tag{3-55}$$

Substituting these values into (3-53) gives

$$B' = \frac{2h}{c^2} \left(\frac{kT}{h}\right)^4 \int_0^\infty \frac{x^3}{e^x - 1} dx \tag{3-56}$$

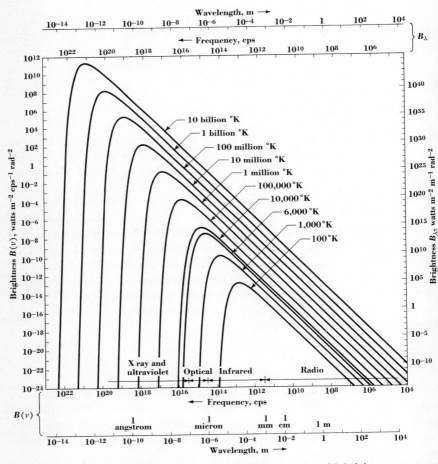

Fig. 3-13. Planck-law radiation curves to logarithmic scales with brightness expresse as a function of frequency $B(\nu)$ (left and bottom scales) and as a function of wavelengt B_λ (right and top scales). Wavelength increases to the right.

The integral is a constant. Combining this constant with the others in the equation, we obtain the Stefan-Boltzmann relation*

$$B' = \sigma T^4 \tag{3-57}$$

where B' = total brightness, watts m^{-2} rad^{-2}
    σ = constant ($= 1.80 \times 10^{-8}$ watt m^{-2} °K^{-4})
    T = temperature of blackbody, °K

The total brightness B' at a given temperature is given by the area under the Planck-radiation-law curve for that temperature. According to

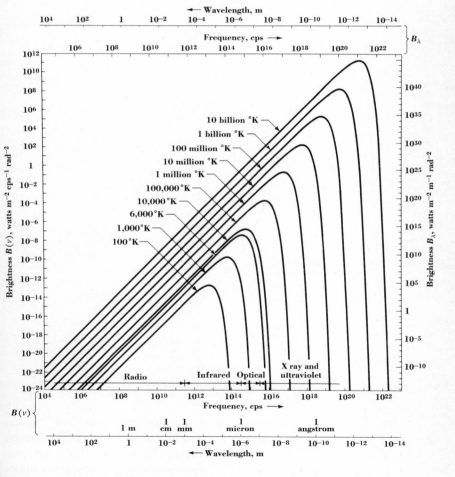

Fig. 3-14. Planck-radiation-law curves with frequency increasing to the right.

* The usual statement gives the integrated value of B' over one hemisphere as obtained by multiplying (3-57) by π. See (3-10).

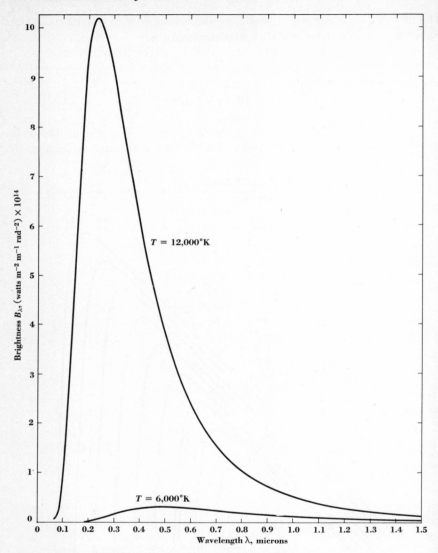

Fig. 3-15. Planck-radiation-law curves at 6000 and 12,000°K.

the Stefan-Boltzmann relation it follows that the area under the Planck-law curve for a blackbody at 12,000°K should be 16 times (= 2^4) that under the curve for 6000°K. Planck-law curves for 12,000 and 6000°K are presented (to a linear scale) in Fig. 3-15. A measurement of the areas under the curves shows that they are in the ratio of 16 to 1.

3-8 **Wien Displacement Law** An important characteristic of the Planck radiation curves (see Figs. 3-12 to 3-15) is that the peak brightness shifts to higher frequency (or shorter wavelength) with increase in temperature. To obtain a quantitative expression for this displacement let us maximize (3-49) by differentiating with respect to ν and setting the result equal to zero. Thus,

$$\frac{2h}{c^2}\left[3\nu_m{}^2(e^{h\nu_m/kT} - 1) - \nu_m{}^3\frac{h}{kT}e^{h\nu_m/kT}\right] = 0 \tag{3-58}$$

where ν_m = frequency at which B is a maximum, cps

If $e^{h\nu_m/kT} \gg 1$, (3-58) reduces approximately to

$$\frac{h\nu_m}{kT} = 3 \tag{3-59}$$

Setting $\nu_m = c/\lambda_m$, (3-59) becomes

$$\frac{hc}{kT\lambda_m} = 3 \tag{3-60}$$

or

$$\lambda_m T = \frac{hc}{3k} = 0.0048 \qquad \text{m }^\circ\text{K} \tag{3-61}$$

where λ_m = wavelength at which B is a maximum, m

This equation, in which the wavelength-temperature product $\lambda_m T$ is equal to a constant, is called the *Wien displacement law*. It indicates that the wavelength of the maximum or peak brightness varies inversely with the temperature.

Owing to the simplification of neglecting unity in comparison with $e^{h\nu_m/kT}$ in (3-58), the constant 0.0048 in (3-61) is an approximation. A more accurate value, obtained without making this simplification, is 0.0051 m °K (see Prob. 3-1).

If the brightness is expressed in terms of unit wavelength, the wavelength for the peak brightness is not the same as when brightness is expressed in terms of unit bandwidth. This may be noted by comparing the top and bottom scales in Figs. 3-13 and 3-14. The quantitative relation for the $\lambda_m T$ product when brightness is expressed in terms of unit wavelength is obtained by maximizing (3-52) and simplifying in the same way as done above in obtaining (3-61) from (3-49). The result (if the assumption is made that $e^{h\nu/kT} \gg 1$) is that

$$\lambda_m T = \frac{hc}{5k} = 0.00288 \qquad \text{m }^\circ\text{K} \tag{3-62}$$

Like (3-61), this is an approximation, and a more accurate value obtained without simplifying assumptions is 0.00290 m °K (see Prob. 3-2).

To summarize:

$\lambda_m T = 0.0051$ m °K for brightness in terms of unit frequency

$\lambda_m T = 0.0029$ m °K for brightness in terms of unit wavelength

It follows from (3-61) that if the relative brightness is plotted as a function of the wavelength-temperature product λT, a *universal Planck-radiation-law curve* is obtained which applies for any wavelength or temperature (Merritt and Hall, 1959). Such a universal curve is presented in Fig. 3-16. The ordinate scale is the relative brightness (dimensionless) as

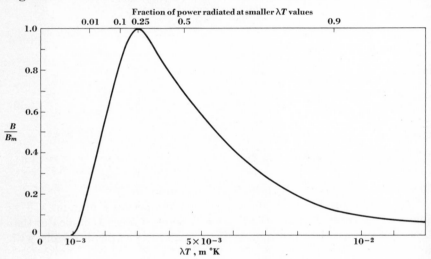

Fig. 3-16. Universal Planck-radiation-law curve giving relative brightness (B/B_m) as a function of the wavelength-temperature product expressed in meter degrees Kelvin. The scale at the top indicates the fraction of the total power radiated at λT values less than a given abscissa value. (*After Merritt and Hall, 1959.*)

given by the ratio of the brightness B to the maximum brightness B_m. The maximum relative brightness is unity. The wavelength-temperature product λT along the abscissa scale is expressed in units of meters times degrees Kelvin.

The area under the curve in Fig. 3-16 is proportional to the total power radiated over the entire spectrum. For a constant temperature the power radiated at wavelengths shorter than the wavelength for the peak of the curve is one-fourth the total power, while that above is three-fourths the total power. A scale indicating the fraction of the total power radiated below particular λT values is given along the top scale of Fig. 3-16.

9 Rayleigh-Jeans Law In the region of radio wavelengths the prod-
ct $h\nu$ may be very small compared to kT ($h\nu \ll kT$), so that the denomi-
ator of the second factor on the right side of Planck's blackbody-radiation
w (3-49) can be reexpressed as follows†

$$e^{h\nu/kT} - 1 = 1 + \frac{h\nu}{kT} - 1 = \frac{h\nu}{kT} \tag{3-63}$$

ntroducing this into (3-49) gives

$$B = \frac{2h\nu^3}{c^2}\frac{kT}{h\nu} = \frac{2\nu^2 kT}{c^2} = \frac{2kT}{\lambda^2} \tag{3-64}$$

here B = brightness, watts m^{-2} cps^{-1} rad^{-2}
 k = Boltzmann's constant ($= 1.38 \times 10^{-23}$ joule °K^{-1})
 T = temperature, °K
 λ = wavelength, m
 ν = frequency, cps
 c = velocity of light, m sec^{-1}

Equation (3-64) is the *Rayleigh-Jeans radiation law*, which is a useful
pproximation in the radio part of the spectrum.

According to the Rayleigh-Jeans radiation law, the brightness varies
versely as the square of the wavelength. On a log-log graph this relation
ppears as a straight line of negative slope. As indicated in Fig. 3-17, this
ne coincides with the Planck-law curve for the same temperature at wave-
ngths considerably longer than for the wavelength of the peak brightness.
owever, at shorter wavelengths the brightness predicted by the Rayleigh-
eans relation increases without limit, whereas the actual brightness
eaches a peak and then falls off with decreasing wavelength, as given by
ne Planck curve.

Although the Rayleigh-Jeans law is a special case of the more general
lanck law, it may be derived directly by a consideration of the classical
ehavior of a blackbody radiator. In fact, the Rayleigh-Jeans law ante-
ates the Planck law, but its failure to explain the observed variation of
diation at the shorter wavelengths was one of the things which led Planck
 postulate that a radiator possesses only a discrete set of possible energy
alues or levels. Energies intermediate to these do not occur. Further-
ore, emission or absorption of radiation of energy occurs only in discrete
mounts, or quanta, of magnitude $h\nu$ (unit: joule). A calculation of the
ean energy of the radiator having the discrete energy levels suggested by
lanck results in the Planck law as given in (3-49) instead of Rayleigh-Jeans

† The condition $h\nu \ll kT$ does not hold if the frequency is too high and the temperature
o low. Thus, $h\nu$ is approximately equal to kT for a wavelength of 1 cm and a tempera-
re of 1°K.

law of (3-64). The considerations of classical physics upon which the Rayleigh-Jeans law is based assume that the radiator may possess all energy levels and that transitions are by a continuous process. As a consequence it is to be noted that Planck's constant h does not appear in the Rayleigh-Jeans law.

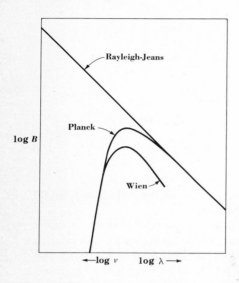

Fig. 3-17. The Rayleigh-Jeans-radiation-law curve coincides with the Planck-radiation-law curve at long wavelengths, while the Wien-radiation-law curve coincides with the Planck curve at short wavelengths.

3-10 Wien Radiation Law At shorter wavelengths, where $h\nu \gg kT$, the quantity unity in the denominator of the second factor on the right side of (3-49) can be neglected in comparison with $e^{h\nu/kT}$, so that the Planck law reduces to

$$B = \frac{2h\nu^3}{c^2}\, e^{-h\nu/kT} \tag{3-65}$$

This approximation to Planck's law is called the *Wien radiation law*. As indicated in Fig. 3-17, the Wien-law curve coincides with the Planck-law curve for the same temperature at wavelengths considerably less than the wavelength of maximum radiation.

3-11 Radiation Laws Applied to a Discrete Source If a radiating source of uniform temperature T subtends a solid angle Ω_s, the source flux densities calculated by the above radiation laws are as follows:

Planck law: $$S = \frac{2h\nu^3\Omega_s}{c^2} \frac{1}{e^{h\nu/kT} - 1} \tag{3-66}$$

Rayleigh-Jeans law: $S = \dfrac{2\nu^2 kT\Omega_s}{c^2} = \dfrac{2kT\Omega_s}{\lambda^2}$ (3-67)

$(h\nu \ll kT)$

Wien law: $S = \dfrac{2h\nu^3\Omega_s}{c^2}\,e^{-h\nu/kT}$ (3-68)

$(h\nu \gg kT)$

where S = flux density, watts m^{-2} cps^{-1}, or jan
$\quad h$ = Planck's constant ($= 6.62 \times 10^{-34}$ joule sec)
$\quad \nu$ = frequency, cps
$\quad k$ = Boltzmann's constant ($= 1.38 \times 10^{-23}$ joule °K^{-1})
$\quad T$ = temperature, °K
$\quad \Omega_s$ = solid angle subtended by source, rad^2
$\quad \lambda$ = wavelength, m
$\quad c$ = velocity of light ($= 3 \times 10^8$ m sec^{-1})

It is assumed the temperature and, hence, the brightness are constant over the source extent. If this is not the case, an integration must be performed. Thus, for a nonuniform temperature distribution the Rayleigh-Jeans law (3-67) becomes

$$S = \frac{2k}{\lambda^2} \iint T(\theta,\phi)\, d\Omega \tag{3-69}$$

where $T(\theta,\Phi)$ = temperature as a function of angle, °K
$\quad d\Omega$ = element of solid angle, rad^2

Bracewell (1962) has shown that (3-69) gives the flux density not only when $T(\theta,\phi)$ is true source temperature distribution but also when $T(\theta,\phi)$ is the observed antenna temperature (antenna lossless). Thus, the flux density is ideally an observable quantity regardless of beam width and source extent, although there is the practical difficulty of integrating the observed antenna temperature over 4π due to masking of the minor lobes by noise. However, the true temperature or brightness distribution can never be known exactly, even in principle.

3-12 Energy Density In certain discussions of the above radiation laws it may be convenient to formulate them in terms of the energy density. For example, the energy density received from a radiating source can be obtained from the brightness relation by dividing by the velocity of light and integrating over the solid angle subtended by the source. For a source of solid angle Ω_s and uniform temperature T the energy density (per unit bandwidth) is

$$\mathcal{E}_d = \frac{2h\nu^3\Omega_s}{c^3}\frac{1}{e^{h\nu/kT} - 1} \tag{3-70}$$

where $\mathcal{E}_d$ = energy density per unit of bandwidth, joules m^{-3} cps^{-1}
If the radiation is uniform from all directions (source subtends solid angle
of 4π), the energy density is given by

$$\mathcal{E}_d = \frac{8\pi h\nu^3}{c^3} \frac{1}{e^{h\nu/kT} - 1} \qquad (3\text{-}71)$$

3-13   Absorption of Electromagnetic Energy and Optical Depth

Wave propagation in an absorbing medium is accompanied by attenuation.
Let us consider a wave propagating in such an absorbing medium as suggested in Fig. 3-18a.   The flux density of the wave will decrease an amount
dS in a distance dx given by

$$dS = -S\alpha \, dx \qquad (3\text{-}72)$$

where S = flux density at point x, watts m^{-2} cps^{-1}
  dS = decrease in flux density, watts m^{-2} cps^{-1}
  α = attenuation constant, m^{-1}
  dx = element of length, m
After dividing (3-72) by S, the left side is a function only of S and the right
a function only of x.   Integrating yields

$$\int \frac{dS}{S} = -\int \alpha \, dx \qquad (3\text{-}73)$$

$$\ln S = -\alpha x + C \qquad (3\text{-}74)$$

where $\ln S$ = natural logarithm (to base e) of S, dimensionless
  α = attenuation constant, m^{-1}
  C = constant of integration, dimensionless
At the boundary of the medium ($x = 0$) the flux density is S_1.   Thus, for
this boundary condition (3-74) becomes

$$\ln S_1 = C \qquad (3\text{-}75)$$

Introducing (3-75) into (3-74) yields

$$\ln \frac{S}{S_1} = -\alpha x \qquad (3\text{-}76)$$

It follows from (3-76) that

$$\frac{S}{S_1} = e^{-\alpha x} \qquad (3\text{-}77)$$

and

$$S = S_1 e^{-\alpha x} \qquad (3\text{-}78)$$

When

$$x = \frac{1}{\alpha} \qquad (3\text{-}79)$$

(3-78) becomes

$$S = S_1 \frac{1}{e} \qquad (3\text{-}80)$$

The relation in (3-80) indicates that in traveling the distance x the wave is attenuated to $1/e$ or 36.8 percent of its initial value. Hence, this distance is sometimes called the $1/e$ *depth of penetration.* The variation of S is shown graphically in Fig. 3-18b. It is assumed that the source is so distant that the effect of inverse-distance-squared attenuation may be neglected. The

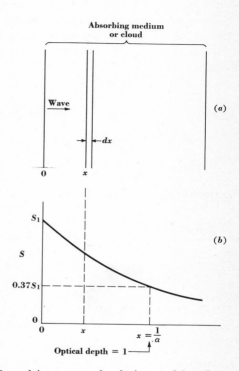

Fig. 3-18. Absorbing medium (a) and attenuation of a wave in it (b).

product αx is called the *optical depth* and is commonly designated by the symbol τ. That is,

$$\tau = \alpha x = \text{optical depth} \tag{3-81}$$

Thus, the optical depth is the product of the attenuation constant and the physical thickness (or depth) of the absorbing region. Equation (3-78) can now be written

$$S = S_1 e^{-\tau} \tag{3-82}$$

where S = observed flux density, watts m^{-2} cps^{-1}
  S_1 = actual flux density, watts m^{-2} cps^{-1}
  τ = optical depth, dimensionless

It follows that for an optical depth of unity the flux density is reduced to $1/e$ of its initial value. This is the same as saying that in an optical depth of unity the flux density has been decreased by 1 neper. Thus, from (3-82) the optical depth (or attenuation in nepers) is given by

$$\tau = \ln \frac{S_1}{S} = 2.3 \log \frac{S_1}{S} \qquad (3\text{-}83)$$

where τ = optical depth, or nepers attenuation, dimensionless
    ln = natural logarithm (base e)
    log = common logarithm (base 10)
In decibels the attenuation is given by

$$\text{Decibel attenuation} = 10 \log \frac{S_1}{S} \qquad (3\text{-}84)$$

From (3-84) and (3-83) it follows that

$$\text{Decibel attenuation} = 4.3\tau \qquad (3\text{-}85)$$

where τ = attenuation, nepers
A number of values of optical depth are compared with the corresponding decibel attenuation in Table 3-1.

Table **3-1**

Ratio of flux density to initial value S/S_1	Optical depth τ, or nepers attenuation	Decibels (dB) attenuation
1	0	0
$1/e = 0.368$	1	4.343
$1/e^2 = 0.135$	2	8.686
0.1	2.303	10
0.01	4.605	20
0.001	6.908	30
0.0001	9.210	40
0.00001	11.513	50

Thus, for example, an absorbing cloud that attenuates the flux density to 1 percent of its incident value produces 20 dB attenuation, or has an optical depth of 4.6.

If (3-82) is divided by the solid angle subtended by the source, a relation involving the brightness is obtained. Thus,

$$B = B_s e^{-\tau} \qquad (3\text{-}86$$

where B = observed brightness, watts m^{-2} cps^{-1} rad^{-2}
    B_s = actual brightness of source, watts m^{-2} cps^{-1} rad^{-2}
    τ = optical depth, dimensionless

In the case of a gaseous medium it is often convenient to express the attenuation constant α as the product of the density of the medium ρ and an absorption coefficient K which is characteristic of the medium. Thus,

$$\alpha = K\rho \tag{3-87}$$

where α = attenuation constant, m^{-1}
  K = absorption coefficient, $m^2\ kg^{-1}$
  ρ = density, $kg\ m^{-3}$

The dimensions of the absorption coefficient are a ratio (dimensionless) per unit density per unit of distance. The units are then $(kg\ m^{-3})^{-1}\ m^{-1}$ or $kg^{-1}\ m^2$. If the density of the medium is nonuniform, so that ρ is a function of x, the optical depth (or nepers attenuation) of a cloud of thickness x_1 is then

$$\tau = \int_0^{x_1} K\rho\ dx \tag{3-88}$$

It is assumed that the variation of ρ is gradual. For a constant density ρ_0, (3-88) reduces to $\tau = K\rho_0 x_1$.

3-14 Emission of Electromagnetic Energy

Consider a volume dv of matter, as shown in Fig. 3-19, which emits electromagnetic energy. Let the rate of energy emission per unit mass per unit bandwidth be given by an *emission coefficient* j (units: watts $kg^{-1}\ cps^{-1}$). The power per unit bandwidth, or spectral power, emitted by the volume is then

$$dw = j\rho\ dv \tag{3-89}$$

where dw = power emitted per unit bandwidth, watts cps^{-1}
  j = emission coefficient, watts $kg^{-1}\ cps^{-1}$
  ρ = density of matter in volume dv, $kg\ m^{-3}$
  dv = element of volume, m^3

The flux density dS observed at the point P at a distance r is

$$dS = \frac{dw}{4\pi r^2} = \frac{j\rho\ dv}{4\pi r^2} \tag{3-90}$$

while the brightness dB is

$$dB = \frac{dS}{d\Omega} = \frac{j\rho\ dv}{4\pi r^2\ d\Omega} \tag{3-91}$$

where $d\Omega$ = element of solid angle, rad^2

Since the volume $dv = r^2\ dr\ d\Omega$, (3-91) reduces to

$$dB = \frac{j\rho\ dr}{4\pi} \tag{3-92}$$

where dB = brightness, watts m^{-2} cps^{-1} rad^{-2}

  j = emission coefficient, watts kg^{-1} cps^{-1}

  ρ = density of matter in volume dv, kg m^{-3}

  dr = radial length of volume dv, m

  4π = solid angle in sphere, rad^2

For any finite depth of emitting matter between radii r_1 and r_2 the brightness B is obtained by integrating (3-92). Thus,

$$B = \frac{1}{4\pi} \int_{r_1}^{r_2} j\rho \, dr \qquad (3\text{-}93)$$

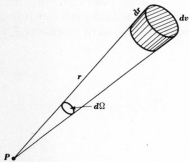

Fig. 3-19. Radiation from an emitting volume dv.

3-15 Internal Emission and Absorption Consider now the case of a cloud of matter which both emits and absorbs, and let us deduce its brightness due to its internal emission and absorption. Radiation from external sources, such as from behind the cloud, is considered to be zero.

If there were no absorption, the brightness of the emitting volume dv (Fig. 3-20) would be given as given by (3-92). However absorption between dv and the edge of the cloud (over a distance r) will result in attenuation of an amount

$$\exp\left(-\int_0^r K\rho \, dr \right) = e^{-\tau} \qquad (3\text{-}94)$$

where $\tau = \displaystyle\int_0^r K\rho \, dr$ = optical depth of cloud between edge ($r = 0$) and volume dv

Thus, the infinitesimal brightness, due to emission and absorption by the volume dv, as observed at the point P is

$$dB = \frac{1}{4\pi} j\rho \, dr \, e^{-\tau} = \frac{j}{4\pi K} e^{-\tau} K\rho \, dr \qquad (3\text{-}95)$$

The brightness due to the entire thickness r_1 of the cloud is then

$$B = \frac{j}{4\pi K} \int_0^{r_1} e^{-\tau} K\rho \, dr = \frac{j}{4\pi K} \int_0^{\tau_c} e^{-\tau} \, d\tau \qquad (3\text{-}96)$$

where B = brightness of emitting-absorbing cloud, watts m^{-2} cps^{-1} rad^{-2}

j = emission coefficient, watts kg^{-1} cps^{-1}

K = absorption coefficient, m^2 kg^{-1}

ρ = density (a function of r), kg m^{-3}

r_1 = thickness of cloud, m

$\tau_c = \displaystyle\int_0^{r_1} K\rho\,dr$ = optical depth, or nepers attenuation of entire thickness of cloud, dimensionless

$\tau = \displaystyle\int_0^{r} K\rho\,dr$ = optical depth from edge of cloud to distance r, dimensionless

$d\tau$ = infinitesimal optical depth of thickness dr of cloud, dimensionless

Integrating (3-96) yields

$$B = \frac{j}{4\pi K}(1 - e^{-\tau_c}) \tag{3-97}$$

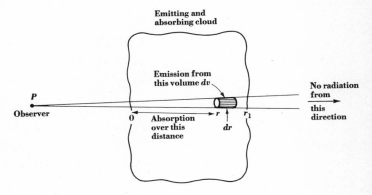

Fig. 3-20. Internal emission and absorption by a cloud.

A graph of (3-97) is presented in Fig. 3-21a, with the brightness shown as a function of the optical depth of the cloud. As the optical depth increases, the brightness approaches the value $j/4\pi K$. This quantity has the dimensions of brightness and may be regarded as the *intrinsic brightness* B_i of the cloud. Thus,

$$B = B_i(1 - e^{-\tau_c}) \tag{3-98}$$

where B = apparent or observed brightness of cloud, watts m^{-2} cps^{-1} rad^{-2}

$B_i = j/4\pi K$ = intrinsic brightness of cloud, watts m^{-2} cps^{-1} rad^{-2}

τ_c = optical thickness of cloud, dimensionless

At radio frequencies we know from the Rayleigh-Jeans law (3-64) that the brightness is proportional to the temperature. Hence, (3-98) can be written in terms of temperature as

$$T_b = T_c(1 - e^{-\tau_c}) \qquad (3\text{-}99)$$

where T_b = observed temperature of cloud, °K
T_c = actual temperature of cloud, °K
τ_c = optical depth of cloud, dimensionless

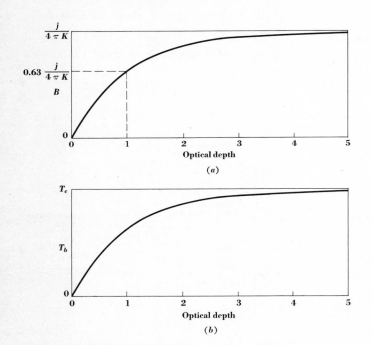

Fig. 3-21. Variation of brightness (a) and brightness temperature (b) of an emitting and absorbing cloud as a function of its optical depth τ.

The quantity T_b is sometimes called the *brightness temperature*. As the optical thickness of the cloud increases, it approaches the actual cloud temperature T_c, as shown by the graph in Fig. 3-21b.

3-16   External Irradiation with Internal Emission and Absorption and the Equation of Transfer   Next let us consider the case of a source of brightness B_s observed through a cloud which both emits and absorbs, as depicted in Fig. 3-22.   In the absence of intervening matter the bright-

ness of a source is independent of the distance at which it is observed.†
Therefore, the brightness will not be dependent on the distance a from the
observer to the cloud or the distance b from the cloud to the source. However, emission and absorption by the cloud will affect the apparent brightness of the source as observed through the cloud. The change in brightness
dB produced by a volume of length dr will then be given by (Chandrasekhar,
1950)

$$dB = -BK\rho \, dr + \frac{j}{4\pi} \rho \, dr \tag{3-100}$$

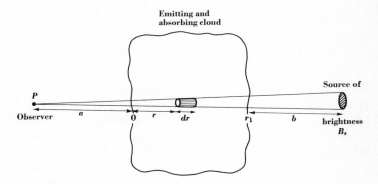

**Emitting and
absorbing cloud**

**Source of
brightness
B_s**

P

Observer a 0 r dr r_1 b

Fig. 3-22. Source of brightness B_s observed through an emitting
and absorbing cloud.

where the first term of the right-hand side of the equation represents the
loss in brightness due to absorption and the second term the gain in brightness from emission. This relation is sometimes called the *equation of
transfer*. Rearranging (3-100) gives

$$\frac{dB}{dr} + K\rho B = \frac{j\rho}{4\pi} \tag{3-101}$$

This equation is a form of Leibnitz's equation. A solution (see Prob. 3-3) is

$$B = B_s e^{-\tau_c} + \frac{j}{4\pi K} (1 - e^{-\tau_c}) \tag{3-102}$$

where B = apparent or observed brightness, watts m^{-2} cps^{-1} rad^{-2}
B_s = actual brightness of source, watts m^{-2} cps^{-1} rad^{-2}
$\tau_c = \displaystyle\int_0^{r_1} K\rho \, dr$ = optical depth, or nepers attenuation, of cloud of
physical thickness r_1, dimensionless

† This follows from the fact that the power per unit solid angle is a constant. Thus, if
the distance to the source increases, the power decreases as $1/r^2$, but the solid angle also
decreases by the same amount.

j = emission coefficient, watts kg^{-1} cps^{-1}

K = absorption coefficient, m^2 kg^{-1}

ρ = density, kg m^{-3}

dr = element of distance, m

The first term of the right-hand side of (3-102) represents the loss in brightness of the irradiating source of brightness B_s due to absorption by the cloud. The second term represents the contribution to the observed brightness due to emission and absorption by the cloud. The quantity $j/4\pi K$ is the intrinsic brightness B_i of the cloud. Thus,

$$B = B_s e^{-\tau_c} + B_i(1 - e^{-\tau_c})$$ (3-103)

where B = observed brightness, watts m^{-2} cps^{-1} rad^{-2}

B_s = brightness of source, watts m^{-2} cps^{-1} rad^{-2}

B_i = intrinsic brightness of cloud ($= j/4\pi K$)

τ_c = optical depth of cloud, dimensionless

It may be noted that this equation is equal to the sum of (3-86) for the effect of an absorbing cloud on an irradiating source and (3-98) for the effect of both emission and absorption by the cloud. In terms of temperature (3-103) becomes

$$T_b = T_s e^{-\tau_c} + T_c(1 - e^{-\tau_c})$$ (3-104)

where T_b = observed brightness temperature, °K

T_s = source temperature, °K

T_c = cloud temperature, °K

τ_c = optical depth of cloud, dimensionless

To illustrate the significance of (3-104) let us find the observed brightness temperature for a source of temperature T_s with an intervening emitting and absorbing cloud of temperature T_c, as suggested in Fig. 3-23, under four conditions:

(1) $\tau_c = 0$ cloud transparent
(2) $\tau_c \gg 1$ cloud opaque
(3) $\tau_c = 1$ intermediate case
(4) $T_c = T_s$ cloud and source temperatures equal

Under the first condition, for which the optical depth is zero, the cloud is completely transparent, and (3-104) reduces to $T_b = T_s$; i.e., the actual source temperature is observed. For the second case, where the optical depth is large, the cloud is opaque and $T_b = T_c$; i.e., only the cloud of temperature T_c is observed. For the third case, where the optical depth is unity,

$$T_b = 0.368T_s + 0.632T_c$$ (3-105)

In this case, the cloud is partially transparent, and the observer sees both the source and the cloud but at reduced temperatures. For the fourth case, where the source and cloud temperatures are assumed equal, (3-104) reduces to $T_b = T_s = T_c$. Thus, regardless of the optical thickness of the cloud, the observed temperature is always the same.

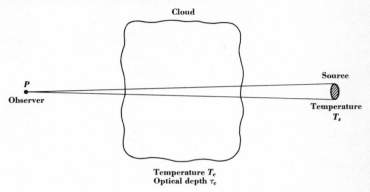

Fig. 3-23. Source of brightness temperature T_s observed through an intervening cloud of temperature T_c and optical depth τ_c.

3-17 Kirchhoff's Law Under conditions of local thermodynamic equilibrium there will be no change in brightness $(dB = 0)$ and (3-100) reduces to

$$B = \frac{j}{4\pi K} \qquad (3-106)$$

This equation, known as Kirchhoff's law, relates the brightness of an emitting-absorbing cloud to its coefficients of mass emission and absorption [see (3-102) for units]. As already mentioned in Sec. 3-15, the quantity $j/4\pi K$ is the brightness an emitting-absorbing cloud will be observed to have when its optical depth is sufficiently large.

3-18 Temperature and Noise The noise power per unit bandwidth available at the terminals of a resistor of resistance R and temperature T as in Fig. 3-24a, is given by (Nyquist, 1928)

$$w = kT \qquad (3-107)$$

where w = spectral power, or power per unit bandwidth, watts cps^{-1}

k = Boltzmann's constant $(= 1.38 \times 10^{-23}$ joule °K$^{-1})$

T = absolute temperature of resistor, °K

If the resistor is replaced by a lossless matched antenna of radiation resistance R, the impedance presented at the terminals is unchanged. However, the noise power will not be the same unless the antenna is receiving from a region at the temperature T. From (3-13) the power per unit bandwidth received by the antenna is

$$w = \tfrac{1}{2}A_e \iint B(\theta,\phi)P_n(\theta,\phi)\ d\Omega \tag{3-108}$$

If the antenna is placed inside of a blackbody enclosure at a temperature T, then the brightness will be a constant B_c in all directions. Its value according to the Rayleigh-Jeans law will be

$$B(\theta,\phi) = B_c = \frac{2kT}{\lambda^2} \tag{3-109}$$

Introducing (3-109) into (3-108) yields

$$w = \frac{kT}{\lambda^2} A_e \Omega_A \tag{3-110}$$

where w = spectral power, watts cps^{-1}
k = Boltzmann's constant ($= 1.38 \times 10^{-23}$ joule °K^{-1})
T = absolute temperature, °K
λ = wavelength, m
A_e = effective aperture of antenna, m^2
Ω_A = beam area of antenna, rad^2
But $A_e \Omega_A = \lambda^2$ (see Chap. 6), so

$$w = kT \tag{3-111}$$

which is the same noise power as for the resistor.

The actual construction of a suitable blackbody enclosure is suggested by Fig. 3-24b. For example, it might consist of a copper box lined with resistance material (such as space felt) which completely absorbs radiation of the wavelength λ. If the entire enclosure is at a uniform temperature T, the walls emit radiation of brightness B_c, as given by (3-109), which is characteristic of the temperature. The antenna has a radiation resistance R with the antenna in free space, and because of the absorbing properties of the enclosure, this resistance is not changed when the antenna is placed inside. If the enclosure contained no absorbing material, it would behave as a low-loss cavity, and the antenna radiation resistance would be greatly reduced. With the resistive material lining the enclosure, the radiation resistance of the antenna is effectively distributed throughout this absorbing-emitting material, and the temperature of the radiation resistance of the antenna is equal to the temperature T of the material.

Although the temperature of the antenna structure itself is also T in this case, it is important to note that it is *not* the temperature of the antenna structure which determines the temperature of its radiation resistance.† The temperature of the radiation resistance is determined by the temperature of the emitting region which the antenna "sees" through its directional pattern. In other words, it is the temperature of the region or regions within the antenna beam which determines the temperature of the radiation resistance. This is illustrated by the situation in Fig. 3-24c, where the enclosure has been removed and the antenna beam views an emitting region of the sky at a temperature T. Assuming that the entire antenna beam area is subtended by a sky of temperature T, the radiation resistance of the antenna will then be at the temperature T, and the received spectral power will be as given by (3-111). In such a situation the

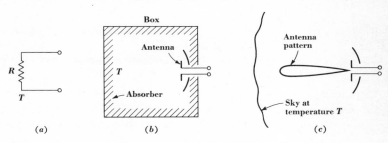

Fig. 3-24. (a) Resistor at temperature T; (b) antenna in an absorbing box at temperature T; and (c) antenna observing sky of temperature T. The same noise power is available at the terminals in all three cases.

antenna and receiver of a radio telescope may be regarded as a radiometer for determining the temperature of distant regions of space coupled to the system through the radiation resistance of the antenna. The temperature of the antenna radiation resistance is called the *antenna temperature.*

If absorbing matter is present, the temperature measured may not be the source temperature but related to it, as discussed in Sec. (3-16). Also if the source of temperature T does not extend over the entire antenna beam area ($T = 0$ elsewhere), the *measured or antenna temperature* T_A will be less. Thus, from (3-108) and (3-111) the received power per unit bandwidth is given by

$$w = \tfrac{1}{2}A_e \iint B(\theta,\phi)P_n(\theta,\phi) \, d\Omega = kT_A \tag{3-112}$$

From (3-21) and (3-112) it follows that the observed flux density of a radio source may be evaluated with the aid of the following relation

$$S_o = \frac{2kT_A}{A_e} \tag{3-113}$$

† It is assumed that the antenna is lossless.

where S_o = observed flux density of source, watts m^{-2} cps^{-1}

k = Boltzmann's constant (= 1.38×10^{-23} joule °K^{-1})

T_A = (maximum) antenna temperature due to source, °K

A_e = effective aperture of antenna, m^2

For a source of small extent compared with the beam size the observed flux density is the true flux density.

If the temperature is measured at a point remote from the antenna, the effect of transmission-line attenuation must be considered. For this case (3-113) can be reexpressed as

$$S_o = \frac{2kT_m}{\epsilon A_e} \tag{3-114}$$

where $T_m = \epsilon T_A$ = temperature due to source measured at end of transmission line of length d, °K

ϵ = efficiency factor for transmission line of length d, dimensionless, ($0 \leq \epsilon \leq 1$)

In practice T_m (and, hence, T_A) is the difference in temperature between the source plus background and the background. Expressing the brightness in terms of the temperature by the Rayleigh-Jeans law yields

$$T_A = \frac{A_e}{\lambda^2} \iint T_s(\theta,\phi)P_n(\theta,\phi)\,d\Omega \qquad \text{from } 3.112 \text{ with } B = \frac{2kT_s/\lambda}{} \tag{3-115}$$

or

$$T_A = \frac{1}{\Omega_A} \iint T_s(\theta,\phi)P_n(\theta,\phi)\,d\Omega \tag{3-116}$$

where T_A = antenna temperature due to source, °K

$T_s(\theta,\phi)$ = source temperature, °K

Ω_A = antenna beam area, rad^2

$P_n(\theta,\phi)$ = normalized antenna power pattern, dimensionless

$d\Omega$ = infinitesimal element of solid angle, rad^2

If the source is very small compared to the antenna-pattern beam area, so that $P_n(\theta,\phi) \simeq 1$, Eq. (3-116) reduces to

$$T_A = \frac{1}{\Omega_A} \iint T_s(\theta,\phi)\,d\Omega \tag{3-117}$$

$$T_A = \frac{\Omega_s}{\Omega_A} T_{\text{avg}} \tag{3-118}$$

where Ω_s = source solid angle, rad^2

Ω_A = antenna beam area or beam solid angle, rad^2 $\qquad (\Omega_A \gg \Omega_s)$

T_{avg} = average temperature over source, °K

On the other hand, if the source is of large extent compared to the antenna main beam area and at a constant temperature T_{const}, the antenna temperature is given by

$$T_A = \frac{T_{const}}{\Omega_A} \iint P_n(\theta,\phi) \, d\Omega = \frac{T_{const}}{\Omega_A} \Omega'_M \qquad (3\text{-}119)$$

For the same case the source brightness is

$$B = \frac{2kT_A}{\lambda^2} \frac{\Omega_A}{\Omega'_M} = \frac{2kT_m}{\epsilon\lambda^2} \frac{\Omega_A}{\Omega'_M} \qquad (3\text{-}120)$$

[handwritten: $\Omega_M' = $ mainbeam plus sidelobe]

The source temperature in the above equations is an *equivalent temperature*, which may be equal to the thermal temperature of the source if the radio noise power is due to thermal emission (see Chap. 8) and the source is optically thick. However, if the noise power is generated by a nonthermal mechanism, such as a plasma or a synchrotron oscillation, the equivalent temperature may be greater than the thermal temperature of the source. In this case, T_s is the temperature a blackbody radiator would need to have to give radiation equal to that observed at the wavelength λ. Hence, T_s is often called the *equivalent blackbody temperature.*

Example. Mayer, McCullough, and Sloanaker (1958a, b) at the Naval Research Laboratory measured an antenna temperature of 0.24°K at a wavelength of 3.15 cm when their radio-telescope antenna was directed at Mars. At the time of the measurements the disk of Mars subtended an angle of 18 sec of arc. Assuming that the antenna has a pencil beam of 0.116° between half-power points, find the equivalent temperature of the source (Mars).

Solution. The radius of the disk of Mars is 9 sec of arc or $9/3{,}600 = 0.0025°$. Hence, the solid angle of the disk is given by

$$\Omega_s = \pi r^2 = \pi \, (0.0025°)^2 = 2 \times 10^{-5} \text{ deg}^2$$

The beam area Ω_A of the antenna is given approximately by (see Chap. 6)

$$\Omega_A = \tfrac{4}{3}(0.116)^2 = 0.018 \text{ deg}^2$$

Hence, assuming a constant temperature over the disk, the average equivalent temperature of Mars by this measurement is, from (3-118),

$$T = T_A \frac{\Omega_A}{\Omega_s} = 0.24 \frac{0.018}{2 \times 10^{-5}} = 216°$$

3-19   Minimum Detectable Temperature and Flux Density

The minimum (antenna) temperature which a radio telescope can detect is limited by fluctuations in the receiver output caused by the statistical nature of the noise wave form. This noise is proportional to the system temperature T_{sys} of the radio telescope, which can be divided into two principal parts, that contributed by the antenna T_A and that contributed by the receiver T_R. This system noise can be reduced, in theory, to any desired extent by increasing the integration time (after detection), increasing the predetection bandwidth, or by taking the average of more than one observation. In practice, however, the integration time cannot be increased beyond the point where it begins to distort a true source profile, and the bandwidth cannot be made too wide without loss of spectral information or introduction of interfering radio signals of terrestrial origin.

The sensitivity, or minimum detectable temperature, of a radio telescope is equal to the *rms noise temperature of the system* as given by

$$\Delta T_{\min} = \frac{K_s T_{\text{sys}}}{\sqrt{\Delta \nu \, tn}} = \Delta T_{\text{rms}} \tag{3-121}$$

where $\Delta T_{\min}$ = sensitivity, or minimum detectable temperature, $^\circ$K

ΔT_{rms} = rms system noise temperature, $^\circ$K

$T_{\text{sys}} = T_A + T_{LP}[(1/\epsilon) - 1] + (1/\epsilon)T_R$ = system noise temperature (at antenna terminals), $^\circ$K

T_A = antenna noise temperature, $^\circ$K

T_R = receiver noise temperature, $^\circ$K

T_{LP} = physical temperature of transmission line (coaxial line or wave guide) between antenna and receiver, $^\circ$K

ϵ = transmission-line efficiency, dimensionless ($0 \leq \epsilon \leq 1$)

K_s = sensitivity constant, dimensionless

$\Delta \nu$ = predetection bandwidth, cps

t = postdetection integration time, sec

n = number of records averaged, dimensionless

The relation of $\Delta T_{\min}$ to the deflection of a typical radio-telescope recorder is suggested in Fig. 3-25. For a sine wave the peak-to-peak deflection is 2.83 times the rms value. For typical noise the ratio is somewhat more.

The constant K_s depends on the type of receiver and its mode of operation but is of the order of unity (see Table 7-3). The factor ϵ is unity for a lossless line and zero for an infinitely attenuating line ($\epsilon = 0.5$ for 3 dB attenuation). The temperature of the antenna-transmission-line combination may be deduced from (3-104). Thus, at the receiver the antenna acts as a source observed through an emitting and absorbing medium (transmission line) so that the temperature of the combination is $T_A \epsilon + T_{LP}(1 - \epsilon)$. The total temperature at the receiver terminals is then $T_A \epsilon + T_{LP}(1 - \epsilon) + T_R$. However, the sensitivity varies as $1/\epsilon$; so multiplying by $1/\epsilon$ yields the system temperature (at antenna terminals) as given in connection with (3-121). For further discussion see Chap. 7.

Introducing (3-121) in the Rayleigh-Jeans relation (3-64) we obtain for the *minimum detectable brightness*

$$\Delta B_{\min} = \frac{2k}{\lambda^2} \frac{K_s T_{\text{sys}}}{\sqrt{\Delta \nu \, tn}} \tag{3-122}$$

Substituting (3-121) in (3-113) yields a *minimum detectable flux density*

$$\Delta S_{min} = \frac{2k}{A_e} \frac{K_s T_{sys}}{\sqrt{\Delta \nu \, tn}} \tag{3-123}$$

where k = Boltzmann's constant ($= 1.38 \times 10^{-23}$ joule $°K^{-1}$)
A_e = effective aperture of antenna, m^2
K_s = sensitivity constant (order of unity), which depends on type of receiver.

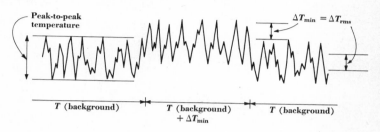

Fig. 3-25. Typical radio-telescope record showing output fluctuations due to the antenna and receiver noise temperature and due to this temperature plus the minimum detectable temperature ΔT_{min}.

The following example serves to illustrate the above relations.

Example. The Ohio State University 260-ft radio telescope had the following characteristics when operating at 1415 Mc in 1964:
System temperature = 125°K (approximately half from antenna and half from receiver)
Predetection bandwidth = 5 Mc
Postdetection integration time = 10 sec
Effective aperture of antenna = 700 m^2
The receiver was a Dicke switched type, for which $K_s = \pi / \sqrt{2}$.
Find the minimum detectable temperature and minimum detectable flux density if two records are averaged.
Solution

$$\Delta T_{min} = \frac{125\pi}{\sqrt{2}\sqrt{5 \times 10^6 \times 10 \times 2}} = 0.03°K$$

$$\Delta S_{min} = \frac{2k}{A_e} \Delta T_{min} = \frac{2 \times 1.38 \times 10^{-23}}{700} \Delta T_{min} = 10^{-27} \text{ jan}$$

The effect of averaging several records and of increasing the integration time is illustrated in Fig. 3-26. Here one record through the nucleus of the Andromeda galaxy (M 31) at 1,415 Mc with the Ohio State University 260-ft radio telescope is shown in (*a*). The telescope characteristics are substantially as indicated in the above example. By averaging four profiles the noise fluctuation is reduced by a factor of 2 ($= \sqrt{4}$), as shown in (*b*). By increasing the effective integration time from 10 to 30 sec another factor of 1.7 ($= \sqrt{3}$) reduction in noise fluctuation is achieved, as illustrated in (c). The original data for Fig. 3-26 were in digital form, which facilitates the averaging and integrating processes. Actually the increase in effective integration time in (c) was obtained by convolving the profile of (b) with a 30-sec-long square pulse. A 30-sec interval is 56 percent of the time required for the beam to drift its half-power width (10 min of arc) at a declination of 41°. This is short enough so that no information should be lost due to the smoothing (see Sec. 6-9).

When ΔT_{min} (rms system temperature) is sufficiently small, the resolution or confusion limit may be reached. This is discussed in Sec. 6-17.

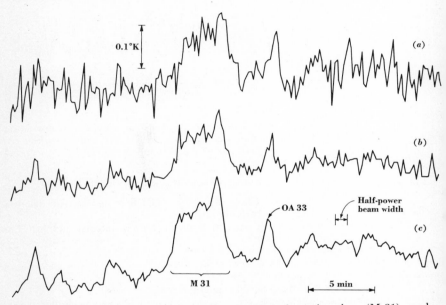

Fig. 3-26. Drift profiles through the nucleus of the Andromeda galaxy (M 31), made at 1,415 Mc with the Ohio State University 260-ft radio telescope, illustrating the reduction in noise fluctuation in going from one record (a) to the average of four records (b) and then to a threefold increase in integration time (c). In the bottom record M 31 stands out clearly with source OA 33 preceding it by several minutes.

References

ALLER, L. H.: "Astrophysics: The Atmosphere of the Sun and Stars," The Ronald Press Company, New York, 1963.

BROWN, R. H., and A. C. B. LOVELL: "The Exploration of Space by Radio," John Wiley & Sons, Inc., New York, 1958.

BELL, E. E.: Radiometric Quantities, Symbols and Units, *Proc. IRE*, vol. 47, pp. 1432–1434, September, 1959.

BRACEWELL, R. N.: Radio Astronomy Techniques, in S. Flügge (ed.), "Handbuch der Physik," vol. 54, p. 119, Springer-Verlag OHG, Berlin, 1962.

Radio-Astronomy Fundamentals   105

CHANDRASEKHAR, S.: "Radiative Transfer," Oxford University Press, Fair Lawn, N.J., 1950.
KIRCHHOFF, G. R.: Über den Zusammenhang zwischen Emission und Absorption von Licht und Wärme, *Monatsber. Akad. Wiss. Berlin*, pp. 783–787, December, 1859.
LOVELL, B., and J. A. CLEGG: "Radio Astronomy," John Wiley & Sons, Inc., New York, 1952.
MAYER, C. H., T. P. McCULLOUGH, and R. M. SLOANAKER: Measurements of Planetary Radiation at Centimeter Wavelengths, *Proc. IRE*, vol. 46, pp. 260–266, January, 1958a.
MAYER, C. H., T. P. McCULLOUGH, and R. M. SLOANAKER: Observations of Mars and Jupiter at a Wavelength of 3.15 cm, *Astrophys. J.*, vol. 127, pp. 11–16, January, 1958b.
MERRITT, T. P., and F. F. HALL, JR.: Blackbody Radiation, *Proc. IRE*, vol. 47, pp. 1435–1441, September, 1959.
NYQUIST, H.: Thermal Agitation of Electric Charge in Conductors, *Phys. Rev.*, vol. 32, pp. 110–113, 1928.
PAWSEY, J. L., and R. N. BRACEWELL: "Radio Astronomy," Oxford University Press, Fair Lawn, N.J., 1955.
PLANCK, M.: Über das Gesetz der Energieverteilung in Normalspektrum, *Ann. Physik*, vol. 4, ser. 4, pp. 553–563, January, 1901.
RICHTMYER, F. K., and E. H. KENNARD: "Introduction to Modern Physics," p. 204, McGraw-Hill Book Company, New York, 1942.
SHKLOVSKY, I. S.: "Cosmic Radio Waves," translated by R. B. Rodman and C. M. Varsavsky, Harvard University Press, Cambridge, Mass., 1960.
STEINBERG, J. L., and J. LEQUEUX: "Radio Astronomy," translated by R. N. Bracewell, McGraw-Hill Book Company, New York, 1963.
VAN DER ZIEL, A.: "Noise," Prentice-Hall, Inc., Englewood Cliffs, N.J., 1954.

Problems

3-1. According to the Wien displacement law, the wavelength-temperature product is equal to a constant. In Sec. 3-8 a value of 0.0048 m °K is obtained. This derivation involved an approximation. Show that a more accurate value, made without this approximation, is 0.0051 m °K.

3-2. If the brightness is expressed in terms of unit wavelength, the Wien displacement law gives the wavelength-temperature product equal to a constant different from that in Prob. 3-1. In Sec. 3-8 an approximation was made in obtaining the value of 0.00288 m °K. Show that a more accurate value, made without this approximation, is 0.002897 m °K.

3-3. Leibnitz's equation may be written

$$\frac{dB}{dr} + PB = Q$$

This is a differential equation of the first order and first degree. A solution is
$B = A \exp(-\int P\,dr) + \exp(-\int P\,dr)\int \exp(+\int P\,dr)Q\,dr$
where A = constant. Show that for the equation of transfer (3-101) the solution is as given in (3-102).

3-4. Calculate the brightness B_λ of a blackbody radiator at a temperature of 1000°K and a wavelength of 0.5 micron.   *Ans.* 3.18×10^{13} watt m⁻² rad⁻² λ⁻¹

3-5. Calculate the brightnesses $B(\nu)$ and B_λ of a blackbody radiator at a temperature of 10,000°K and a wavelength of 1 cm.

3-6. An idealized antenna pattern–brightness distribution is illustrated by the one-dimensional diagram in Fig. P3-6. The brightness distribution consists of a point

source of flux density S and a uniform source $2°$ wide, also of flux density S. The poi[n]
source is $2°$ from the center of the $2°$ source. The antenna pattern is triangular (sym
metrical) with a $2°$ beam width between zero points and with zero response beyon[d]
(a) Draw an accurate graph of the observed flux density as a function of angle fro[m]
the center of the $2°$ source. (b) What is the maximum ratio of the observed to the actu[al]
total flux density $(2S)$?   *Ans.* [

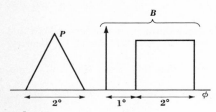

Fig. P3-6. Pattern-brightness distr[i]
bution.

3-7. For the two-dimensional geometry shown in Fig. P3-7 show that the o[b]
served flux density is given by

$$S(\gamma_0,\phi_0) = \int\int B(\gamma,\phi)\tilde{P}_n(\gamma_0 - \gamma, \phi_0 - \phi)\ d\Omega$$

where

$d\Omega = \cos\gamma\ d\gamma\ d\phi$

3-8. Show that the following forms are equivalent

$$S = \frac{2k}{\lambda^2}\int\int T_s(\theta,\phi)P_n(\theta,\phi)\ d\Omega$$

$$S = \frac{2k}{\lambda^2}T_A\Omega_A$$

$$S = \frac{2kT_A}{A_c}$$

and that when T_s is constant and $\Omega_s \ll \Omega_A$,

$$S = \frac{2k}{\lambda^2}T_s\Omega_s$$

3-9. A radio source is occulted by an intervening emitting and absorbing clou[d]
of optical depth unity and brightness temperature $100°K$. The source has a unifor[m]
brightness distribution of $200°K$ and a solid angle of 1 deg^2. The radio telescope h[as]
an effective aperture of 50 m^2. If the wavelength is 50 cm, find the antenna temper[a]
ture when the radio telescope is directed at the source. The cloud is of uniform thic[k]
ness and has an angular extent of 5 deg^2. Assume that the antenna has uniform respon[se]
over the source and cloud.   *Ans.* 23.8°

3-10. The digital output of a $1,400$-Mc radio telescope gives the following valu[e]
as a function of the sidereal time while scanning a uniform brightness region. The int[e]
gration time is 14 sec, with 1 sec idle time for print-out.

31^m30^s	234	32^m45^s	229
31 45	235	33 00	236
32 00	224	33 15	233
32 15	226	33 30	230
32 30	239	33 45	226

If the temperature calibration gives 170 units for $2.9°K$ applied, find (a) the rms noi[se]
at receiver, (b) the minimum detectable temperature, (c) the system temperatur[e]
and (d) the minimum detectable flux density. The calibration signal is introduced [at]
the receiver. The transmission line from the antenna to the receiver has 0.5 dB atte[n]

ation. The antenna effective aperture is 500 m². The receiver bandwidth is 7 Mc.
The receiver constant $K_s = 2$.

Ans. (a) 0.08°K, (b) 0.09°K, (c) 445°K, and (d) 5 × 10⁻²⁷ jan

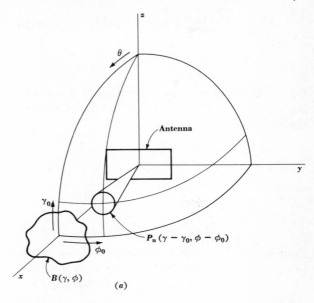

(a)

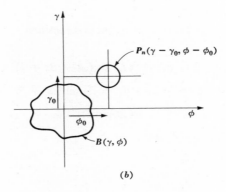

(b)

Fig. P3-7. Two-dimensional geometry.

3-11. A radio telescope has the following characteristics: antenna noise tempera-
ure 50°K, receiver noise temperature 50°K, transmission line between antenna and
receiver 1 dB loss and 270°K physical temperature, receiver bandwidth 5 Mc, receiver
integration time 5 sec, receiver constant $K_s = \pi/\sqrt{2}$, and antenna effective aperture
00 m². If two records are averaged, find (a) the minimum detectable temperature
nd (b) the minimum detectable flux density. *Ans.* (a) 0.06°K, (b) 3.2 × 10⁻²⁷ jan

4
Wave Polarization

4-1 Introduction Consider a plane wave traveling out of the page (in the positive z direction), as in Fig. 4-1, with electric-field components in the x and y directions as given by†

$$E_x = E_1 \sin(\omega t - \beta z) \tag{4-1}$$
$$E_y = E_2 \sin(\omega t - \beta z + \delta) \tag{4-2}$$

where E_1, E_2 = constants

$$\omega = 2\pi\nu$$
$$\beta = 2\pi/\lambda$$
$$\delta = \text{phase difference of } E_y \text{ and } E_x$$

Equations (4-1) and (4-2) describe two *linearly polarized waves*, one polarized in the x direction and the other in the y direction.

Combining (4-1) and (4-2) vectorially, we obtain for the total or resultant field

$$\mathbf{E} = \mathbf{x}E_x + \mathbf{y}E_y \tag{4-3}$$

where $\mathbf{x}$, $\mathbf{y}$ = unit vectors in x and y directions
It follows that

$$\mathbf{E} = \mathbf{x}E_1 \sin(\omega t - \beta z) + \mathbf{y}E_2 \sin(\omega t - \beta z + \delta) \tag{4-4}$$

At $z = 0$, $E_x = E_1 \sin \omega t$ and $E_y = E_2 \sin(\omega t + \delta)$. Expanding E_y yields

$$E_y = E_2(\sin \omega t \cos \delta + \cos \omega t \sin \delta) \tag{4-5}$$

From the relation for E_x we have

$$\sin \omega t = \frac{E_x}{E_1} \tag{4-6}$$

and

$$\cos \omega t = \sqrt{1 - \left(\frac{E_x}{E_1}\right)^2} \tag{4-7}$$

Introducing (4-6) and (4-7) in (4-5), time is eliminated, and we obtain, on rearranging, that

† For a more general discussion see Kraus (1950).

$$\frac{E_x^2}{E_1^2} - \frac{2E_xE_y\cos\delta}{E_1E_2} + \frac{E_y^2}{E_2^2} = \sin^2\delta \tag{4-8}$$

or

$$aE_x^2 - bE_xE_y + cE_y^2 = 1 \tag{4-9}$$

where $a = 1/E_1^2 \sin^2\delta$

$b = 2\cos\delta/E_1E_2 \sin^2\delta$

$c = 1/E_2^2 \sin^2\delta$

Equation (4-9) may be recognized as the equation for an ellipse in its most general form, the axes of the ellipse, in general, not coinciding with

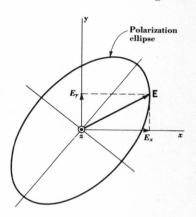

Fig. 4-1. Relation of instantaneous electric-field vector **E** to polarization ellipse.

the x or y axes. Thus, (4-4) represents the general case of elliptical polarization, the locus of the tip of the electric-field vector **E** describing an ellipse, as in Fig. 4-1.

Referring to Fig. 4-2, the line segment OA is the semimajor axis, and the line segment OB is the semiminor axis of the ellipse. The tilt angle of the ellipse is τ. The ratio of OA to OB is called the *axial ratio* (AR) *of the polarization ellipse* or simply the *axial ratio*. Thus,

$$\text{AR} = \frac{OA}{OB} \qquad 1 \le \text{AR} \le \infty \tag{4-10}$$

If $E_1 = 0$, the wave is *linearly polarized* in the y direction. If $E_2 = 0$, the wave is linearly polarized in the x direction. If $\delta = 0$ and $E_1 = E_2$, the wave is also linearly polarized but in a plane at angle of 45° with respect to the x axis. A further special case of interest occurs when $E_1 = E_2$ and $\delta = \pm 90°$. The resulting wave is *circularly polarized*. When $\delta = +90°$, the wave is said to be *left circularly polarized* and, when $\delta = -90°$, it is said to be *right circularly polarized*. Thus, from (4-4) we have for

$\delta = +90°$, at $z = 0$ and $t = 0$, that $E_x = 0$ and $\mathbf{E} = \mathbf{y}E_2$, as in Fig. 4-3a. Under the same conditions but at a later time such that $\omega t = 90°$, $E_y = 0$ and $\mathbf{E} = \mathbf{x}E_1$, as in Fig. 4-3b. The rotation of the electric-field vector is thus clockwise with the wave approaching. According to the IRE standards (1942) this sense of rotation is defined as left circular polarization. According to the older usage of classical physics this sense of rotation (clockwise with wave approaching) is defined as right circular polarization, or opposite to the IRE definition.

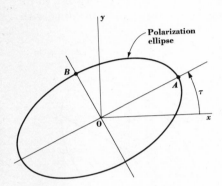

Fig. 4-2.   Polarization-ellipse geometry

If the wave is viewed receding (from negative z axis in Fig. 4-1), the electric vector appears to rotate in the opposite direction. Hence, clockwise rotation with the wave receding is the same as counterclockwise rotation with the wave approaching. In the following the IRE definition will be used, since it could also be defined (without reference to the wave direction) by means of helical-beam antennas (Kraus, 1950). Thus, a right-handed helical-beam antenna radiates or receives right circular (IRE) polarization. A right-handed helix, like a right-handed screw, is right-

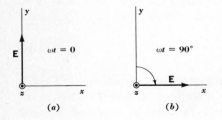

Fig. 4-3.   Change in direction of **E** for left circular polarization. Time $t = 0$ in (a) and $\omega t = 90°$ in (b).

handed regardless of the position from which the helix is viewed. There is no possibility here of ambiguity. The different definitions for the two types of circular polarization are summarized in Table 4-1.

Table 4-1†

Polarization	Classical physics usage	IRE definition (1942)	Type of helical-beam antenna for generating or receiving
Clockwise (wave approaching) or counter-clockwise (wave receding)	Right	Left	Left-handed
Counterclockwise (wave approaching) or clockwise (wave receding)	Left	Right	Right-handed

† A left circularly polarized wave (IRE) has an instantaneous electric field (as a function of space) that describes a right-handed screw.

In (4-4) the general case of an elliptically polarized wave was described in terms of two linearly polarized components. It is also possible to describe the general situation in terms of two circularly polarized waves of unequal amplitude. Thus, at $z = 0$, let

$$E_r = E_R e^{j\omega t} \tag{4-11}$$

and

$$E_l = E_L e^{-j(\omega t + \delta')} \tag{4-12}$$

where E_r = right circularly polarized wave (Fig. 4-4)
  E_l = left circularly polarized wave
  E_R, E_L = constants
  δ' = phase difference

Fig. 4-4. Right and left circularly polarized waves.

E_r E_l

Then the instantaneous linearly polarized components of the wave (E_x and E_y) are given by

$$E_x = \text{Re}\,(E_r + E_l) \tag{4-13}$$

and

$$E_y = \text{Im}\,(E_r + E_l) \tag{4-14}$$

or

$$E_x = E_R \cos \omega t + E_L \cos (\omega t + \delta') \tag{4-15}$$

and

$$E_y = E_R \sin \omega t - E_L \sin (\omega t + \delta') \tag{4-16}$$

On eliminating ωt, as done in deriving (4-8), (4-15) and (4-16) may be reduced to an equation having the form of an ellipse, demonstrating that (4-11) and (4-12) represent an elliptically polarized wave.

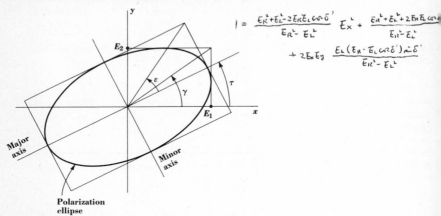

$$ | = \frac{E_R^2 + E_L^2 - 2E_R E_L \cos\delta'}{E_R^2 - E_L^2} E_x^2 + \frac{E_R^2 + E_L^2 + 2E_R E_L \cos\delta'}{E_R^2 - E_L^2} $$
$$ + 2E_x E_y \frac{E_L(E_R - E_L \cos\delta') \sin\delta'}{E_R^2 - E_L^2} $$

Fig. 4-5. Relation of amplitudes E_1 and E_2 and angles ϵ, γ, and τ to the polarization ellipse.

4-2   The Polarization Ellipse and the Poincaré Sphere   The general case of an elliptically polarized wave may be described as before (at $z = 0$) by

$$E_x = E_1 \sin \omega t \tag{4-17}$$

and

$$E_y = E_2 \sin (\omega t + \delta) \tag{4-18}$$

where δ = phase difference between E_y and E_x $(-180° \leq \delta \leq +180°)$ Referring to Fig. 4-5, let

$$\gamma = \tan^{-1} \frac{E_2}{E_1} \qquad 0° \leq \gamma \leq 90° \tag{4-19}$$

where E_2/E_1 = amplitude ratio
Also let the *tilt angle* of the polarization ellipse be designated by τ, where $0° \leq \tau \leq 180°$, and let

$$\epsilon = \cot^{-1} (\mp AR) \qquad -45° \leq \epsilon \leq +45° \tag{4-20}$$

where $AR = \dfrac{\text{major axis}}{\text{minor axis}} \qquad 1 \leq |AR| \leq \infty$

with the minus sign used for right-handed and the plus sign for left-handed (IRE) polarization.

The above quantities are interrelated by the following equations (Poincaré, 1892; Deschamps, 1951)

$$\cos 2\gamma = \cos 2\epsilon \cos 2\tau \tag{4-21}$$

$$\tan \delta = \frac{\tan 2\epsilon}{\sin 2\tau} \tag{4-22}$$

or

$$\tan 2\tau = \tan 2\gamma \cos \delta \tag{4-23}$$

$$\sin 2\epsilon = \sin 2\gamma \sin \delta \tag{4-24}$$

Knowing ϵ and τ, one can determine γ and δ using (4-21) and (4-22). Conversely, knowing γ and δ, one can find ϵ and τ by means of (4-23) and (4-24). It is convenient to describe the *polarization state* by either of the two sets of angles ϵ and τ or γ and δ. Let the polarization state as a function of ϵ and τ be designated by $M(\epsilon,\tau)$ or simply M and the polarization state as a function of γ and δ be designated by $P(\gamma,\delta)$ or simply P. Then on the Poincaré sphere, one-eighth of which is shown in Fig. 4-6, the angles ϵ, τ, γ,

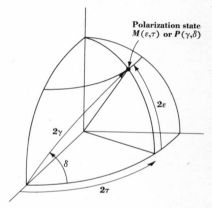

Fig. 4-6. Poincaré sphere.

and δ are related as indicated. More specifically, any point on the sphere describes a particular polarization state. In terms of $M(\epsilon,\tau)$ its coordinates are

$$2\epsilon = \text{latitude} \qquad -90° \le 2\epsilon \le +90°$$

and

$$2\tau = \text{longitude} \qquad 0° \le 2\tau \le 360°$$

while in terms of $P(\gamma,\delta)$ its coordinates are

$$2\gamma = \text{great-circle distance from origin} \qquad 0° \le 2\gamma \le 180°$$

and

$$\delta = \text{angle of great-circle line with respect to equator}$$
$$-180° \le \delta \le +180°$$

Several special cases are of interest.

Case 1. For $\delta = 0$ or $\pm 180°$, E_x and E_y are exactly in phase or out of phase. Thus, any point on the equator represents a state of linear polarization. At the origin the polarization is linear and horizontal ($\tau = 0$) as in Fig. 4-7. On the equator at 90° to the right the polarization is linear with a tilt angle of 45°, while at 180° from the origin the polarization is linear and vertical ($\tau = 90°$), etc.†

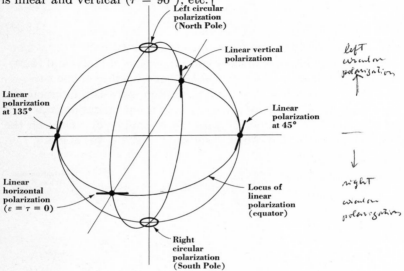

Left circular
polarization
(North Pole)

Linear vertical
polarization

Linear
polarization
at 135°

Linear
polarization
at 45°

Linear
horizontal
polarization
($\varepsilon = \tau = 0$)

Locus of
linear
polarization
(equator)

Right
circular
polarization
(South Pole)

left circular polarisation

right circular polarisation

Fig. 4-7.   Polarization at cardinal points of Poincaré sphere.

Case 2. For $\delta = \pm 90°$ and $E_2 = E_1$ ($2\gamma = 90°$), E_x and E_y have equal amplitudes but are in phase quadrature, which is the condition for circular polarization. Thus, the poles represent a state of circular polarization, the north pole representing left circular polarization and the south pole right circular polarization (IRE), as suggested in Fig. 4-7.

Cases 1 and 2 represent limiting conditions. In the general case any point in the northern hemisphere describes a left elliptically polarized wave ranging from pure left circular at the pole to linear at the equator. Likewise, any point in the southern hemisphere describes a right elliptically polarized wave ranging from pure right circular at the pole to linear at the equator.

† Strictly speaking, the term *horizontal* polarization for the $\tau = 0°$ case and *vertical* polarization for the $\tau = 90°$ case is significant only for the special case where the wave is propagating parallel to the ground so that the x- axis is horizontal and the y- axis is vertical.

4-3   The Response of an Antenna to a Wave of Arbitrary Polarization   The response of a receiving antenna, given by its terminal voltage V, when a wave of field intensity $\mathbf{E}$ is incident upon it may be expressed as

$$V = \mathbf{E} \cdot \mathbf{l} = El \cos\theta \tag{4-25}$$

where $\mathbf{l}$ = *effective length* of antenna
θ = angle between $\mathbf{E}$ and $\mathbf{l}$

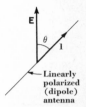

Fig. 4-8.   Dipole of effective length $\mathbf{l}$ and incident wave of field intensity $\mathbf{E}$.

In early antenna work l was commonly referred to as the *effective height*. The relation between $\mathbf{E}$ and $\mathbf{l}$ for a linearly polarized wave and a linearly polarized antenna (dipole) is suggested in Fig. 4-8. For the general case of an elliptically polarized wave the magnitude of $\mathbf{E}$ is given by

$$E = \sqrt{E_1{}^2 + E_2{}^2} \tag{4-26}$$

and the response V of a linearly polarized (dipole) antenna to such a wave may be as suggested in Fig. 4-9.

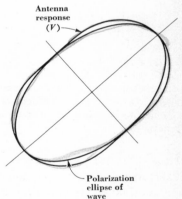

Fig. 4-9.   Response of linearly polarized antenna to an elliptically polarized wave.

Let the polarization state of the antenna be designated as $M_a(\epsilon,\tau)$ or simply M_a, and the polarization state of the wave as $M(\epsilon,\tau)$ or simply M. The polarization state of the antenna is defined as the polarization state of the wave radiated by the antenna when it is transmitting. Then we have that

$$V = El \cos\frac{MM_a}{2} \tag{4-27}$$

where MM_a is the great-circle distance between points (or polarization states) M and M_a on the Poincaré sphere.

Several special cases are of interest.

Case 1. If $MM_a = 0$, the antenna is matched to the wave, and $V = El$.

Case 2. If the wave is left circularly polarized and the antenna is right circularly polarized, $MM_a = 180°$ and $V = 0$.

Case 3. If the wave is vertically polarized and the antenna is horizontally polarized, $MM_a = 180°$ and $V = 0$.

Cases 2 and 3 are illustrations of the fact that an antenna is blind to a wave of the antipodal polarization state.

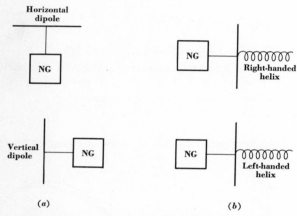

Fig. 4-10. Methods of generating a completely unpolarized wave. At (*a*) two independent noise generators are connected to vertical and horizontal dipoles (wave out of page). At (*b*) two independent noise generators are connected to helical-beam antennas of opposite hand (wave to right).

4-4 Partial Polarization and the Stokes Parameters   The foregoing sections deal with completely polarized waves, where E_1, E_2, and δ are constants (or at least slowly varying functions of time). The radiation from a monochromatic (single-frequency) transmitter is of this type. However, in general, the emission from celestial radio sources extends over a wide frequency range and within any finite bandwidth $\Delta\nu$ consists of the superposition of a large number of statistically independent waves of a variety of polarizations. The resultant wave is said to be *randomly polarized.* For such a wave we may write

$$E_x = E_1(t) \sin \omega t \tag{4-28}$$
$$E_y = E_2(t) \sin [\omega t + \delta(t)] \tag{4-29}$$

here all the time functions are independent. The time variations of
$E_1(t)$, $E_2(t)$, and $\delta(t)$ are slow compared to that of the mean frequency,
($\omega = 2\pi\nu$) being of the order of the bandwidth $\Delta\nu$.

A wave of this type could be generated by connecting one noise gen-
rator to a horizontally polarized antenna (dipole) and a second noise
enerator to a vertically polarized antenna (dipole), as in Fig. 4-10a. An
lternative scheme would be to use two noise generators and a left- and
ght-handed helical-beam antenna producing left and right circular
olarization, as in Fig. 4-10b.

The most general situation is one in which the wave is *partially
olarized*; i.e., it may be considered to be of two parts, one completely
olarized and the other completely unpolarized. A *completely unpolarized*
r completely randomly polarized) wave results if the powers radiated
om the two generators in Fig. 4-10a or b are equal. The waves emitted
y celestial radio sources are generally of the partially polarized type,
nding in many cases to completely unpolarized radiation but in other
ises to a significant amount of polarization.

To deal with partial polarization it is convenient to use the Stokes
arameters introduced by Sir George Stokes (1852) (see Chandrasekhar,

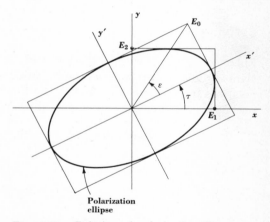

**Polarization
ellipse**

Fig. 4-11. Relation of polarization-ellipse axes
(x',y') to reference axes (x,y).

)50). As an introduction let us first consider their application to a com-
letely polarized wave.

Referring to Fig. 4-11, we can write

$$E_x = E_1 \sin(\omega t - \delta_1)$$ (4-30)
$$E_y = E_2 \sin(\omega t - \delta_2)$$ (4-31)

here $\delta_1 - \delta_2$ = phase difference of E_x and E_y

Referring to the axes coincident with the axes of the polarization ellipse (primed axes), we can also write

$$E_{x'} = E_0 \cos \epsilon \sin \omega t \qquad (4\text{-}32)$$
$$E_{y'} = E_0 \sin \epsilon \cos \omega t \qquad (4\text{-}33)$$

Now

$$E_x = E_{x'} \cos \tau - E_{y'} \sin \tau \qquad (4\text{-}34)$$
$$E_y = E_{x'} \sin \tau + E_{y'} \cos \tau \qquad (4\text{-}35)$$

from which

$$E_x = E_0 (\cos \epsilon \cos \tau \sin \omega t - \sin \epsilon \sin \tau \cos \omega t) \qquad (4\text{-}35a)$$
$$E_y = E_0 (\cos \epsilon \sin \tau \sin \omega t + \sin \epsilon \cos \tau \cos \omega t) \qquad (4\text{-}35b)$$

Expanding (4-30) and (4-31) and equating the $\sin \omega t$ and $\cos \omega t$ terms in (4-30) and (4-35a) and in (4-31) and (4-35b) yields

$$E_1 = E_0 \sqrt{\cos^2 \epsilon \cos^2 \tau + \sin^2 \epsilon \sin^2 \tau} \quad \text{and} \qquad (4\text{-}36)$$

$$E_2 = E_0 \sqrt{\cos^2 \epsilon \sin^2 \tau + \sin^2 \epsilon \cos^2 \tau} \qquad (4\text{-}37)$$

The magnitude of the total Poynting vector or flux density (watts per square meter) of the wave is[†]

$$S = S_x + S_y = \frac{E_1^2 + E_2^2}{Z} = \frac{E_0^2}{Z} \qquad (4\text{-}38)$$

where Z is the intrinsic impedance of the medium (ohms per square) (Kraus, 1953). S_x represents the Poynting vector for the wave component polarized in the x direction and S_y the Poynting vector for the wave component polarized in the y direction. Introducing (4-36) and (4-37), we have

$$S_x = \frac{E_1^2}{Z} = S(\cos^2 \epsilon \cos^2 \tau + \sin^2 \epsilon \sin^2 \tau) \qquad (4\text{-}39)$$

$$S_y = \frac{E_2^2}{Z} = S(\cos^2 \epsilon \sin^2 \tau + \sin^2 \epsilon \cos^2 \tau) \qquad (4\text{-}40)$$

The Stokes parameters I, Q, U, and V are defined as follows:

$$I = S = S_x + S_y = \frac{E_1^2}{Z} + \frac{E_2^2}{Z} \qquad (4\text{-}41)$$

$$Q = S_x - S_y = \frac{E_1^2}{Z} - \frac{E_2^2}{Z} = S \cos 2\epsilon \cos 2\tau \qquad (4\text{-}42)$$

$$U = (S_x - S_y) \tan 2\tau = S \cos 2\epsilon \sin 2\tau$$

$$= 2 \frac{E_1 E_2}{Z} \cos(\delta_1 - \delta_2) \qquad (4\text{-}43)$$

[†] In the foregoing part of this chapter E_1, E_2, E_0, E_L, and E_R are peak value. In the following, except for (4-48, 49, 71, 72), they are effective values (0.707 as much. The S values are average power.

$$V = (S_x - S_y) \tan 2\epsilon \sec 2\tau = S \sin 2\epsilon$$

$$= 2\frac{E_1 E_2}{Z} \sin(\delta_1 - \delta_2) \tag{4-44}$$

It follows that

$$I^2 = Q^2 + U^2 + V^2 \tag{4-45}$$

and that

$$\frac{U}{Q} = \tan 2\tau \tag{4-46}$$

and

$$\frac{V}{S} = \sin 2\epsilon = \frac{V}{\sqrt{Q^2 + U^2 + V^2}} \tag{4-47}$$

The above relations are for a *completely polarized wave,* and, in particular, (4-45) is a condition that is fulfilled by a completely polarized wave.

Let us consider three special cases of a completely polarized wave.

Case 1. For a left circularly polarized wave, $S_x = S_y$, AR = 1, and $\epsilon = 45°$, so that

$I = S$
$Q = 0$
$U = 0$
$V = S$

Case 2. For a right circularly polarized wave,

$I = S$
$Q = 0$
$U = 0$
$V = -S$

Case 3. For a linearly polarized wave with $\tau = 0$, $S_x = S$, $S_y = 0$, AR = ∞, and $\epsilon = 0$,

$I = S$
$Q = S$
$U = 0$
$V = 0$

Turning next to the case of a *completely unpolarized* wave, or one which is *partially polarized,* the x and y components are

$$E_x = E_1(t) \sin [\omega t - \delta_1(t)] \tag{4-48}$$
$$E_y = E_2(t) \sin [\omega t - \delta_2(t)] \tag{4-49}$$

In defining the Stokes parameters for this case it is necessary to take time averages so that

$$I = \frac{\langle E_1^2 \rangle}{Z} + \frac{\langle E_2^2 \rangle}{Z} = S_x + S_y = S \tag{4-50}$$

$$Q = \frac{\langle E_1{}^2 \rangle}{Z} - \frac{\langle E_2{}^2 \rangle}{Z} = S_x - S_y = S \langle \cos 2\epsilon \cos 2\tau \rangle \tag{4-51}$$

$$U = \frac{2}{Z} \langle E_1 E_2 \cos \delta \rangle = S \langle \cos 2\epsilon \sin 2\tau \rangle \tag{4-51a}$$

$$V = \frac{2}{Z} \langle E_1 E_2 \sin \delta \rangle = S \langle \sin 2\epsilon \rangle \tag{4-52}$$

and

$$I^2 \geq Q^2 + U^2 + V^2 \tag{4-52a}$$

where $\delta = \delta_1 - \delta_2$
and $\langle . . . \rangle$ indicates the time average. It is understood that E_1, E_2, δ, ϵ, and τ are functions of time. Thus,

$$\langle E_1{}^2 \rangle = \frac{1}{T} \int_0^T [E_1(t)]^2 \, dt \tag{4-53}$$

Referring to (4-50) through (4-52), the significance of the Stokes parameters may be expressed thus. I represents the total power (sum of x and y components). Q represents the difference of the x and y power components. U represents a power proportional to the time average of the real part of $e^{j\delta}$ and the product of the magnitudes of the x and y field components, and V is the same as U except that it involves the imaginary part of $e^{j\delta}$. One can also regard I and Q as the sum and difference of two autocorrelation functions, while U and V are like cross-correlation functions of E_1 and E_2.

For a *completely unpolarized wave* $S_x = S_y$, and E_1 and E_2 are uncorrelated, so that

$$\langle E_1 E_2 \cos \delta \rangle = \langle E_1 E_2 \sin \delta \rangle = 0$$

Thus, for this case

$$I = S$$
$$Q = 0$$
$$U = 0$$
$$V = 0$$

The condition $Q = U = V = 0$ is a requirement for a completely unpolarized wave. Thus, we note that a nonzero value for Q, U, or V indicates the presence of a polarized component in the wave. The *degree of polarization d* is defined as the ratio of the completely polarized power to the total power or

$$d = \frac{\text{polarized power}}{\text{total power}}$$

$$= \frac{\sqrt{Q^2 + U^2 + V^2}}{I} \qquad 0 \leq d \leq 1 \tag{4-54}$$

The degree of polarization d is unity for a completely polarized wave and is zero for a completely unpolarized wave. If two waves have identical Stokes parameters, the waves are identical. Further, if several *independent* waves propagating in the same direction are superimposed, the Stokes parameter of the resultant wave is the sum of the Stokes parameters of the individual waves.

The presence of the unpolarized component only in I may be brought out if the Stokes parameters are reexpressed in terms of polarized components (designated by a subscript p) and unpolarized components (designated by a subscript u). Thus,

$$I = S_u + S_p = S_u + S_{xp} + S_{yp} \tag{4-55}$$

$$Q = S_{xp} - S_{yp} \tag{4-56}$$

$$U = (S_{xp} - S_{yp}) \tan 2\tau \tag{4-57}$$

$$V = (S_{xp} - S_{yp}) \tan 2\epsilon \sec 2\tau \tag{4-58}$$

Normalizing the Stokes parameters by dividing (4-50) through (4-52) by S yields

$$s_0 = \frac{I}{S} = 1 \tag{4-59}$$

$$s_1 = \frac{Q}{S} = \frac{S_x - S_y}{S} = \langle \cos 2\epsilon \cos 2\tau \rangle \tag{4-60}$$

$$s_2 = \frac{U}{S} = \frac{2}{Z} \frac{\langle E_1 E_2 \cos \delta \rangle}{S} = \langle \cos 2\epsilon \sin 2\tau \rangle \tag{4-61}$$

$$s_3 = \frac{V}{S} = \frac{2}{Z} \frac{\langle E_1 E_2 \sin \delta \rangle}{S} = \langle \sin 2\epsilon \rangle \tag{4-62}$$

In (4-59) through (4-62) the new symbols s_0, s_1, s_2, s_3, have been introduced for the *normalized Stokes parameters*. Thus, we may express the Stokes parameters as a matrix of the column-vector type

$$S[s_i] = S \begin{bmatrix} s_0 \\ s_1 \\ s_2 \\ s_3 \end{bmatrix} \qquad i = 0, 1, 2, 3 \tag{4-63}$$

where S = total Poynting vector, watts m^{-2}

and $\quad s_0 = 1$ \hfill (4-64)

$$s_1 = \frac{\langle E_1{}^2 \rangle - \langle E_2{}^2 \rangle}{ZS} \tag{4-65}$$

$$s_2 = \frac{2}{ZS} \langle E_1 E_2 \quad \cos \delta \rangle \tag{4-66}$$

$$s_3 = \frac{2}{ZS} \langle E_1 E_2 \quad \sin \delta \rangle \tag{4-67}$$

A partially polarized wave may be regarded as the sum of a completely polarized wave and a completely unpolarized wave, so that (4-63) may be written as

$$S[s_i] = S \begin{bmatrix} s_0 \\ s_1 \\ s_2 \\ s_3 \end{bmatrix} = S \begin{bmatrix} 1-d \\ 0 \\ 0 \\ 0 \end{bmatrix} + S \begin{bmatrix} d \\ d\cos 2\epsilon \cos 2\tau \\ d\cos 2\epsilon \sin 2\tau \\ d\sin 2\epsilon \end{bmatrix} \tag{4-68}$$

where d is the degree of polarization given by

$$d = \frac{\sqrt{s_1^2 + s_2^2 + s_3^2}}{s_0} = \sqrt{s_1^2 + s_2^2 + s_3^2} \tag{4-69}$$

The first term of the last member of (4-68) gives the unpolarized power and the second term the completely polarized power. It follows that for a partially polarized wave

$$
\begin{aligned}
s_0 &= 1 \\
s_1 &= d \cos 2\epsilon \cos 2\tau \\
s_2 &= d \cos 2\epsilon \sin 2\tau \\
s_3 &= d \sin 2\epsilon
\end{aligned}
\tag{4-70}
$$

The parameters ϵ and τ in (4-68) and (4-70) refer, of course, to the completely polarized part of the wave.

The foregoing discussion of the Stokes parameters has been in terms of linearly polarized components. An alternative approach is to use circularly polarized components (Cohen, 1958). Thus, let the wave be described by

$$E_r = E_R e^{j\omega t} \tag{4-71}$$

and

$$E_l = E_L e^{-j(\omega t + \delta')} \tag{4-72}$$

where E_l and E_r are left and right circularly polarized field components. For a completely polarized wave

$$AR = \frac{E_L + E_R}{E_L - E_R} = \cot \epsilon \tag{4-73}$$

If $E_L > E_R$ the AR is positive (left-handed polarization) while if $E_R > E_L$ the AR is negative (right-handed polarization). It follows from (4-73) that

$$\cos 2\epsilon = \frac{2E_L E_R}{E_L^2 + E_R^2} = \frac{AR^2 - 1}{AR^2 + 1} \tag{4-74}$$

and

$$\sin 2\epsilon = \frac{E_L^2 - E_R^2}{E_L^2 + E_R^2} = \frac{2AR}{AR^2 + 1} \tag{4-75}$$

We also have that $\delta' = 2\tau$. Thus, the Stokes parameters may be expressed in terms of circularly polarized components as

$$I = \frac{\langle E_L{}^2 \rangle}{Z} + \frac{\langle E_R{}^2 \rangle}{Z} = S_L + S_R = S \qquad (4\text{-}76)$$

$$Q = \frac{2 \langle E_L E_R \cos \delta' \rangle}{Z} \qquad (4\text{-}77)$$

$$U = \frac{2 \langle E_L E_R \sin\delta' \rangle}{Z} \qquad (4\text{-}78)$$

$$V = \frac{\langle E_L{}^2 \rangle}{Z} - \frac{\langle E_R{}^2 \rangle}{Z} = S_L - S_R \qquad (4\text{-}79)$$

where S_L and S_R are the left and right circularly polarized Poynting vectors.

4-5 Matrix Representations for the Response of an Antenna to a Wave of Arbitrary Polarization

Following Ko (1962), let the effective aperture of a receiving antenna be represented by $A_e[a_i]$, where $[a_i]$ is a matrix of the column-vector form the same as $[s_i]$ for the wave in (4-63). Thus,

$$A_e[a_i] = A_e \begin{bmatrix} a_0 \\ a_1 \\ a_2 \\ a_3 \end{bmatrix} \quad \text{m}^2 \qquad (4\text{-}80)$$

Here a_0, a_1, a_2, and a_3 are the Stokes parameters for a wave radiated by the antenna when it is transmitting. Thus,

$$
\begin{aligned}
a_0 &= 1 \\
a_1 &= \frac{E_{1t}{}^2 - E_{2t}{}^2}{Z S_t} \\
a_2 &= \frac{2}{Z S_t} E_{1t} E_{2t} \cos \delta \\
a_3 &= \frac{2}{Z S_t} E_{1t} E_{2t} \sin \delta
\end{aligned}
\qquad (4\text{-}81)
$$

The antenna properties are described by S_t, E_{1t}, E_{2t}, and δ of the completely polarized wave which it radiates when *transmitting*. It follows that the power W available from the antenna when a wave of polarization $S[s_i]$ is incident on it is given by*

$$W = \tfrac{1}{2} S A_e [\widetilde{a_i}] \, [s_i] = \tfrac{1}{2} S A_e \sum_{i=0}^{3} a_i s_i \qquad (4\text{-}82)$$

* S is the magnitude of the total Poynting vector, i.e., the sum of both components (unit: watts m^{-2}). The relations also hold if it represents the total observed flux density (unit: watts m^{-2}, the same as for S_o' of Chap. 3) or the observed flux density (unit: jan, the same as S_o of Chap. 3, but with W in (4-82) replaced by the spectral power w).

where $[\widetilde{a_i}]$ is the transpose of $[a_i]$. Thus

$$W = \tfrac{1}{2}SA_e[a_0 \quad a_1 \quad a_2 \quad a_3] \begin{bmatrix} s_0 \\ s_1 \\ s_2 \\ s_3 \end{bmatrix} \tag{4-83}$$

or

$$W = \tfrac{1}{2}SA_e(a_0s_0 + a_1s_1 + a_2s_2 + a_3s_3) \tag{4-84}$$

This result may also be expressed as

$$W = \tfrac{1}{2}SA_e(1 + d \cos MM_a) \tag{4-85}$$

or

$$W = \tfrac{1}{2}SA_e(1 - d) + dS\,A_e\,\cos^2 \frac{MM_a}{2} \tag{4-86}$$

where MM_a is the angle between the wave and antenna polarization states on the Poincaré sphere. In (4-86) the first term on the right-hand side represents the unpolarized power and the second term the polarized power. This form brings out the fact that only one-half the incident unpolarized power is available to a receiver but all the completely polarized power is available. Let us now define four new parameters as follows:

$$\begin{aligned}
s_{11} &= \tfrac{1}{2}(s_0 + s_1) \\
s_{12} &= \tfrac{1}{2}(s_2 + js_3) \\
s_{21} &= \tfrac{1}{2}(s_2 - js_3) \\
s_{22} &= \tfrac{1}{2}(s_0 - s_1)
\end{aligned} \tag{4-87}$$

and four similar parameters a_{11}, a_{12}, etc. for the antenna. Using these as the elements of a 2×2 matrix, it may be shown that the power available to an antenna from a wave of arbitrary polarization is given by

$$W = \mathrm{Tr}\left\{ A_e \begin{bmatrix} a_{11} & a_{12} \\ a_{21} & a_{22} \end{bmatrix} \times S \begin{bmatrix} s_{11} & s_{12} \\ s_{21} & s_{22} \end{bmatrix} \right\} \tag{4-88}$$

where Tr signifies the *trace*, i.e., the sum of the diagonal elements of a square matrix, or

$$W = SA_e(a_{11}s_{11} + a_{12}s_{21} + a_{21}s_{12} + a_{22}s_{22}) \tag{4-89}$$

More concisely,

$$W = \mathrm{Tr}\{A_e[a_{ij}] \times S[s_{ij}]\} \tag{4-90}$$

According to this formulation the *response* W of the system is given by the trace of the matrix $A_e[a_{ij}]$ of the *detection system* (antenna and receiver) times the matrix $S[s_{ij}]$ of the *observable quantity* (wave). This relation is identical in form to the one used in optics, which is appropriate if we regard the antenna-receiver as a detector of radio photons. In optics the matrices in (4-88) or (4-90) are called *coherency matrices* (Born and Wolf, 1964).

The normalized Stokes parameters and coherency matrices of several waves are compared in the following examples:

Type of wave	Stokes parameters	Coherency matrix
Unpolarized wave	$\begin{bmatrix} 1 \\ 0 \\ 0 \\ 0 \end{bmatrix}$	$\frac{1}{2}\begin{bmatrix} 1 & 0 \\ 0 & 1 \end{bmatrix}$
Right circularly polarized wave	$\begin{bmatrix} 1 \\ 0 \\ 0 \\ -1 \end{bmatrix}$	$\frac{1}{2}\begin{bmatrix} 1 & -j \\ j & 1 \end{bmatrix}$
Left circularly polarized wave	$\begin{bmatrix} 1 \\ 0 \\ 0 \\ 1 \end{bmatrix}$	$\frac{1}{2}\begin{bmatrix} 1 & j \\ -j & 1 \end{bmatrix}$
Partially polarized wave, $d = \frac{1}{3}$, completely polarized part linearly polarized with $\tau = 45°$	$\begin{bmatrix} 1 \\ 0 \\ \frac{1}{3} \\ 0 \end{bmatrix}$	$\frac{1}{2}\begin{bmatrix} 1 & \frac{1}{3} \\ \frac{1}{3} & 1 \end{bmatrix}$

If several independent waves propagating in the same direction are super-imposed, the coherency matrix of the resultant wave is the sum of the coherency matrices of the individual waves. Thus, for a completely unpolarized wave we may write

$$\frac{1}{2}\begin{bmatrix} 1 & 0 \\ 0 & 1 \end{bmatrix} = \frac{1}{2}\begin{bmatrix} 1 & 0 \\ 0 & 0 \end{bmatrix} + \frac{1}{2}\begin{bmatrix} 0 & 0 \\ 0 & 1 \end{bmatrix} \tag{4-91}$$

In (4-91) the first term represents a linearly polarized wave with $\tau = 0°$ and the second term a linearly polarized wave with $\tau = 90°$. Thus, a completely unpolarized wave may be regarded as the sum of two *independent* completely polarized waves of the linearly polarized type ($\epsilon = 0$) and of equal flux density with their planes of polarization mutually perpendicular, as suggested by the generating arrangement in Fig. 4-10a. Alternatively, we may write for a completely unpolarized wave that

$$\frac{1}{2}\begin{bmatrix} 1 & 0 \\ 0 & 1 \end{bmatrix} = \frac{1}{4}\begin{bmatrix} 1 & j \\ -j & 1 \end{bmatrix} + \frac{1}{4}\begin{bmatrix} 1 & -j \\ j & 1 \end{bmatrix} \tag{4-92}$$

In this equation the first term represents a left circularly polarized wave (AR = 1) and the second term a right circularly polarized wave

(AR $= -1$). Thus, a completely unpolarized wave may be regarded as the sum of two *independent* completely polarized waves of the circularly polarized variety and of equal flux density but of opposite hand, as suggested by the generating arrangement of Fig. 4-10b.

Describing the decomposition of an unpolarized wave in terms of Stokes parameters instead of coherency matrices, we have in place of (4-91) that

$$\begin{bmatrix} 1 \\ 0 \\ 0 \\ 0 \end{bmatrix} = \frac{1}{2}\begin{bmatrix} 1 \\ 1 \\ 0 \\ 0 \end{bmatrix} + \frac{1}{2}\begin{bmatrix} 1 \\ -1 \\ 0 \\ 0 \end{bmatrix} \tag{4-93}$$

and in place of (4-92) that

$$\begin{bmatrix} 1 \\ 0 \\ 0 \\ 0 \end{bmatrix} = \frac{1}{2}\begin{bmatrix} 1 \\ 0 \\ 0 \\ 1 \end{bmatrix} + \frac{1}{2}\begin{bmatrix} 1 \\ 0 \\ 0 \\ -1 \end{bmatrix} \tag{4-94}$$

In (4-93) the first term represents a linearly polarized wave with $\tau = 0°$ and the second term a linearly polarized wave with $\tau = 90°$. In (4-94) the first term represents a left circularly polarized wave and the second term a right circularly polarized wave.

4-6　Polarization Measurements　The polarization characteristics of a wave may be measured by a wide variety of techniques. In many radio-astronomy measurements a linearly polarized antenna, such as a dipole, is rotated, giving a maximum power response $W_{\parallel}$ when the polarization direction is parallel to the major axis of the wave-polarization ellipse and a minimum response $W_{\perp}$ when the polarization direction is perpendicular to the major axis of the wave-polarization ellipse. The *degree of linear polarization* d_l of the wave is then given by

$$d_l = \frac{W_{\parallel} - W_{\perp}}{W_{\parallel} + W_{\perp}} \qquad 0 \leq d_l \leq 1 \tag{4-95}$$

This simple measurement yields only partial information regarding the wave but is of significance in determining whether a source has some linearly polarized component.

Another simple technique is to measure the left circularly polarized power response W_L and the right circularly polarized response W_R with two circularly polarized antennas of opposite hand, such as left- and right-handed helical-beam antennas. The *degree of circular polarization* d_c of the wave is then given by

$$d_c = \frac{|W_L - W_R|}{W_L + W_R} \qquad 0 \leq d_c \leq 1 \tag{4-96}$$

A measurement of $W_{\|}$ and $W_{\perp}$ is like a measurement of S_x and S_y for the case where $\tau = 0$, and hence, $U = 0$. For this case, $d = \sqrt{Q^2 + V^2}/I$, but we note from (4-51) that $Q = Id_l$ and from (4-79) that $V = Id_c$ or that

$$d = \sqrt{d_l^2 + d_c^2} \qquad 0 \le d \le 1 \tag{4-97}$$

Thus, the square of the (total) *degree of polarization* of a wave is equal to the sum of the squares of its linear and circular degrees of polarization. Noting that $d_l = \cos 2\epsilon$ and $d_c = |\sin 2\epsilon|$, the variation of the degree of linear and circular polarization with angle 2ϵ for a completely polarized wave is as shown in Fig. 4-12.

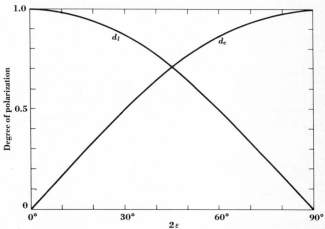

Fig. 4-12. Degree of linear polarization d_l and degree of circular polarization d_c as a function of the angle 2ϵ.

To measure the Stokes parameters a variety of techniques may be used. Suzuki and Tsuchiya (1958) used three pairs of antennas. In a typical case these might consist of two dipoles (linearly polarized) at right angles yielding power responses W_x and W_y; two more dipoles at right angles but both turned 45° with respect to the first set, as indicated in Fig 4-13, giving power responses $W_{x'}$ and $W_{y'}$; and two helical-beam (circularly polarized) antennas of left and right hand producing power responses W_L and W_R. Now the Stokes parameters for the linearly polarized antennas turned 45° with respect to the x and y axes are given by

$$
\begin{aligned}
I' &= S_{x'} + S_{y'} = I \\
Q' &= S_{x'} - S_{y'} = S \cos 2\epsilon \cos 2\tau' = U \\
U' &= (S_{x'} - S_{y'}) \tan 2\tau' = S \cos 2\epsilon \sin 2\tau' = -Q \\
V' &= S \sin 2\epsilon = V
\end{aligned}
\tag{4-98}
$$

where $\tau' = \tau - 45°$. These relations have the same form as for linearly polarized antennas parallel to the x and y axes with Q and U interchanged. For the six antennas of Fig. 4-13 it then follows that the normalized Stokes parameters are very simply obtained without phase measurements as the sum or difference of the various power responses. Thus,

$$s_0 = \frac{W_x + W_y}{W_x + W_y} = \frac{W_{x'} + W_{y'}}{W_x + W_y} = \frac{W_L + W_R}{W_x + W_y}$$

$$s_1 = \frac{W_x - W_y}{W_x + W_y}$$

$$s_2 = \frac{W_{x'} - W_{y'}}{W_x + W_y}$$ (4-99)

$$s_3 = \frac{W_L - W_R}{W_x + W_y}$$

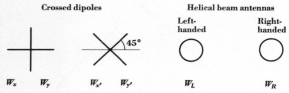

Crossed dipoles   **Helical beam antennas**

Left-handed   Right-handed

W_x W_y $W_{x'}$ $W_{y'}$ W_L W_R

Fig. 4-13. Arrangement of six antennas for measuring Stokes parameters as the sum or difference of the power received by pairs of the antennas.

The apertures of all the antennas, of course, must be equal. There is some redundancy in the above measurements. For a partially polarized wave four independent measurements are necessary, and for a completely polarized wave three are needed. Although some redundancy may be useful, the number of independent measurements can be reduced, for example, to W_x, $W_{x'}$, W_L, and W_R. In terms of these four quantities the normalized Stokes parameters are

$$s_0 = \frac{W_L + W_R}{W_L + W_R}$$

$$s_1 = \frac{2W_x - W_L - W_R}{W_L + W_R}$$

$$s_2 = \frac{2W_{x'} - W_L - W_R}{W_L + W_R}$$ (4-100)

$$s_3 = \frac{W_L - W_R}{W_L + W_R}$$

A discussion of various techniques for polarization measurements is given by Cohen (1958).

References

BORN, M., and E. WOLF: "Principles of Optics," The Macmillan Company, New York, 1964.

CHANDRASEKHAR, S.: "Radiative Transfer," Oxford University Press, Fair Lawn, N.J., 1950.

COHEN, M. H.: Radio Astronomy Polarization Measurements, *Proc. IRE*, vol. 46, pp. 172–183, January, 1958.

DESCHAMPS, G. A.: Geometrical Representation of the Polarization of a Plane Electromagnetic Wave, *Proc. IRE*, vol. 39, pp. 540–544, May, 1951.

IRE: Standards on Radio Wave Propagation, *Proc. IRE*, vol. 30, no. 7, pt. III, suppl., 1942.

KO, H. C.: On the Reception of Quasi-monochromatic Partially Polarized Radio Waves, *Proc. IRE*, vol. 50, pp. 1950–1957, September, 1962.

KRAUS, J. D.: "Antennas," chaps. 7 and 15, McGraw-Hill Book Company, New York, 1950.

KRAUS, J. D.: "Electromagnetics," p. 407, McGraw-Hill Book Company, New York, 1953.

POINCARÉ, H.: "Theorie mathématique de la lumière," G. Carré, Paris, 1892.

STOKES, G.: On the Compositional Resolution of Streams of Polarized Light from Different Sources, *Trans. Cambridge Phil. Soc.*, vol. 9, pt. 3, pp. 399–416, 1852.

SUZUKI, S., and A. TSUCHIYA: A Time-sharing Polarimeter, *Proc. IRE*, vol. 46, pp. 190–194, January, 1958.

Problems

4-1.   Three waves have the following characteristics:

(a) Unpolarized, (b) LEP, $d = \frac{1}{2}$, $|AR| = 3$, $\tau = 135°$, (c) LCP. Three antennas produce the following types of waves when transmitting: (1) LHP, (2) REP, $|AR| = 3$, $\tau = 45°$, (3) RCP. If all the antennas have unit effective aperture and all waves have unit total flux density, find the received power for the nine combinations

$$
\begin{array}{ccc}
a1 & b1 & c1 \\
a2 & b2 & c2 \\
a3 & b3 & c3 \\
\end{array}
$$

Note: LEP = left elliptically polarized, LCP = left circularly polarized, LHP = linear horizontal polarization, REP = right elliptical polarization, RCP = right circular polarization, AR = axial ratio.

Ans. Column 1: $\frac{1}{2}$, $\frac{1}{2}$, $\frac{1}{2}$; Column 2: $\frac{1}{2}$, $\frac{1}{4}$, $7/20$; Column 3: $\frac{1}{2}$, $\frac{1}{5}$, 0; unit: watt

4-2.   Ten waves have the following characteristics:

(a) $d = 0$

(b) $d = \frac{1}{8}$, AR = ∞, $\tau = 45°$

(c) $d = \frac{1}{2}$, AR = 4, $\tau = 135°$

(d) $d = \frac{1}{2}$, AR = -4, $\tau = 135°$

(e) $d = 1$, AR = 4, $\tau = 45°$

(f) $d = 1$, AR = -1

(g) $d = 1$, AR = $+1$

(h) $d = 1$, AR = ∞, $\tau = 0$

(i) $d = 1$, AR = ∞, $\tau = 90°$

(j) $d = 1$, AR = ∞, $\tau = 45°$

Find the normalized Stokes parameters and the coherency matrices for these waves.

4-3.   Starting with the basic relations (4-30) through (4-33) prove that $Q = S \cos 2\epsilon \cos 2\tau$, $U = S \cos 2\epsilon \sin 2\tau$ and $V = S \sin 2\epsilon$.

4-4. Show that
$$\cos 2\epsilon = \frac{2E_L E_R}{E_L{}^2 + E_R{}^2} = \frac{\text{AR}^2 - 1}{\text{AR}^2 + 1} \quad \text{and} \quad \sin 2\epsilon = \frac{E_L{}^2 - E_R{}^2}{E_L{}^2 + E_R{}^2} = \frac{2\text{AR}}{\text{AR}^2 + 1}$$
where AR is the axial ratio and E_L and E_R are the left and right circularly polarized field components of a wave.

4-5. Derive the Stokes parameters in terms of ϵ and τ' and the power fluxes $S_{x'}$ and $S_{y'}$ of two linearly polarized components at 45° with respect to the x and y axes. The tilt angle τ' is measured between the major axis of the polarization ellipse and the x' axis, which is at an angle of 45° with respect to the x axis.

4-6. Derive relations for the normalized Stokes parameters measured with two linearly polarized antennas at angles of 0 and 45° (responses W_x and $W_{x'}$) and two circularly polarized antennas (responses W_L and W_R).

4-7. Four linearly polarized antennas at angles 0, 45, 90, and 135° and two circularly polarized antennas left and right handed, with responses W_x, $W_{x'}$, W_y, $W_{y'}$, W_L, and W_R, have equal effective apertures. Is there a type of wave for which all six responses are equal? If so, what are the wave parameters?

4-8. A wave for which $d = \frac{1}{2}$, AR $= 4$, and $\tau = 135°$ is received by six antennas of unit effective aperture and power responses as follows:

Linear horizontal polarization: W_x
Linear vertical polarization: W_y
Linear slant polarization (45°): $W_{x'}$
Linear slant polarization (135°): $W_{y'}$
Left circular polarization: W_L
Right circular polarization: W_R

If the wave has unit flux density, find the six power responses.

4-9. Repeat Prob. 4-8 for a wave with $d = 1$, AR $= 4$, and $\tau = 135°$.

4-10. Derive (4-84) from (4-89).

4-11. A wave of 10^{-10} jan and normalized Stokes parameters 1, 0, $\frac{1}{3}$, 0 is incident on an antenna of 1,000 m² effective aperture and normalized Stokes parameters 1, 0, 1, 0. (a) Find the received (spectral) power and (b) give the wave and antenna characteristics in terms of the polarization type.
Ans.: (a) 6.7×10^{-8} watt cps^{-1}; (b) $d = \frac{1}{3}$, AR $= \infty$, $\tau = 45°$; $d = 1$, AR $= \infty$, $\tau = 45°$

4-12. Give the percentage of unpolarized wave power in each of five waves characterized by the following Stokes parameters: 1, 0, 0, 0; 1, 1, 0, 0; 1, 0, $\frac{1}{3}$, 0; 1, 0, $\frac{1}{2}$, $\frac{1}{2}$; 1, 0, $1/\sqrt{2}$, $1/\sqrt{2}$. *Ans.* 100, 0, 66.7, 29.3, 0

4-13. Derive (4-85) and (4-86).

4-14. An antenna of 100 m² effective aperture is used to receive a wave for which the degree of polarization $d = d_l = 0.5$. The tilt angle of the linear polarization is 45°. If the flux density of the wave is 10^{-12} watt m^{-2} cps^{-1} and the Stokes parameters for the antenna are 1, 0, 1, 0, what is the power received? *Ans.* 7.5×10^{-11} watt cps^{-1}

4-15. (a) A wave of unit flux density and polarization parameters 1, 0, 1, 0 is incident on a linearly polarized receiving antenna of unit effective aperture. Find the power response for antenna tilt angles of -45, -22.5, 0, 22.5, 45, 67.5, 90, 112.5 and 135°. *Ans.* 0, 0.15, 0.5, 0.85, 1.0, 0.85, 0.5, 0.15, and 0 watts cps^{-1}
(b) Repeat for the case where the wave parameters are 1, 0, $\frac{1}{2}$, 0.

4-16. A wave of unit flux density with a coherency matrix
$$\begin{bmatrix} \frac{1}{2} & \frac{1}{6} \\ \frac{1}{6} & \frac{1}{2} \end{bmatrix}$$
is incident on a right circularly polarized receiving antenna of unit effective aperture. Find the received power. *Ans.* $\frac{1}{2}$ watt cps^{-1}

5
Wave-Propagation Fundamentals

5-1 Introduction In this chapter some of the fundamental equations of electromagnetic-wave propagation are developed. Starting with Maxwell's equations, relations are developed for plane waves in lossless and conducting media. These are followed by a simple but basic treatment of waves in an ionized medium in the presence of a magnetic field. Propagation parallel and perpendicular to the magnetic field is considered, and equations for the Faraday rotation angle are developed. The chapter concludes with a brief discussion of magnetohydrodynamic waves.

5-2 Maxwell's Equations† The four equations referred to collectively as Maxwell's equations are generalizations of simpler circuit-theory relations. Thus, *Ampere's law* relating the line integral of the magnetic field **H** around a closed path to the total current I enclosed states that‡

$$\oint \mathbf{H} \cdot d\mathbf{l} = I \qquad \text{amp} \tag{5-1}$$

The total current I is equal to the surface integral of the current density **J** over the area bounded by the path of integration of **H**; so that a more general relation is

$$\oint \mathbf{H} \cdot d\mathbf{l} = \int_s \mathbf{J} \cdot d\mathbf{s} \qquad \text{amp} \tag{5-2}$$

Maxwell made this relation still more general by adding a displacement current density to account for wave motion in empty space. Thus, we have,

$$\oint \mathbf{H} \cdot d\mathbf{l} = \int_s \left(\mathbf{J} + \frac{\partial \mathbf{D}}{\partial t} \right) \cdot d\mathbf{s} \tag{5-3}$$

This relation is called Maxwell's equation as derived from Ampere's law. As stated in (5-3), it is an integral relation, the line integral of **H** being carried out over the closed path forming the periphery of the surface s over

---

† For a more complete discussion of the topics in this and the next section reference should be made to books on electromagnetic theory, e.g., Kraus (1953).
‡ See Table 5-1 for list of symbols with units.

which the current-density terms are integrated. Applying Stokes' theorem to (5-3), the corresponding differential or point relation is

$$\nabla \times \mathbf{H} = \mathbf{J} + \frac{\partial \mathbf{D}}{\partial t} \quad \text{amp} \quad \text{m}^{-2} \tag{5-4}$$

This is Maxwell's equation, as derived from Ampere's law, in its differential form.

Faraday's law relates the emf induced in a closed circuit to the time rate of decrease of the total magnetic flux linking the circuit, or

$$\upsilon = -\frac{d\Lambda}{dt} \tag{5-5}$$

The total magnetic-flux linkage Λ is equal to the surface integral of the magnetic-flux density $\mathbf{B}$ over the area bounded by the circuit, so that a more general relation is

$$\upsilon = -\frac{d}{dt}\int_s \mathbf{B} \cdot d\mathbf{s} \tag{5-6}$$

The emf induced is equal to the line integral of the electric field $\mathbf{E}$ around the circuit. A still more general relation (for stationary circuits) is then

$$\oint \mathbf{E} \cdot d\mathbf{l} = -\int_s \frac{\partial \mathbf{B}}{\partial t} \cdot d\mathbf{s} \tag{5-7}$$

Applying Stokes' theorem, the corresponding point relation is

$$\nabla \times \mathbf{E} = -\frac{\partial \mathbf{B}}{\partial t} \tag{5-8}$$

The relation (5-7) is the integral form and (5-8) the differential form of Maxwell's equation as derived from Faraday's law.

The two remaining Maxwell's equations are based on applications of *Gauss' law* to electric and magnetic fields. Thus, the integral of the electric flux density $\mathbf{D}$ over a closed surface equals the electric charge enclosed. This, in turn, is equal to the volume integral of the electric charge over the volume enclosed by the surface, or

$$\oint_s \mathbf{D} \cdot d\mathbf{s} = \int_v \rho \, dv \tag{5-9}$$

Applying Gauss' law to the magnetic field, we have

$$\oint_s \mathbf{B} \cdot d\mathbf{s} = 0 \tag{5-10}$$

The corresponding differential or point relations are

$$\nabla \cdot \mathbf{D} = \rho \tag{5-11}$$

and

$$\nabla \cdot \mathbf{B} = 0 \tag{5-12}$$

Maxwell's and other equations are summarized in Table 5-1. Equivalent forms for the curl relations are given, assuming harmonic variation with time.

Table 5-1
Maxwell's and other equations

Maxwell's equations	
Integral form	*Differential form*

$$\oint H \cdot dl = \int_s \left(J + \frac{\partial D}{\partial t} \right) \cdot ds \qquad \nabla \times H = J + \frac{\partial D}{\partial t} = J + j\omega D$$

$$\oint E \cdot dl = -\int_s \frac{\partial B}{\partial t} \cdot ds \qquad\qquad \nabla \times E = -\frac{\partial B}{\partial t} = -j\omega B$$

$$\oint D \cdot ds = \int_v \rho \, dv \qquad\qquad\qquad \nabla \cdot D = \rho$$

$$\oint B \cdot ds = 0 \qquad\qquad\qquad\qquad \nabla \cdot B = 0$$

Equations of the medium

$$D = \epsilon E \qquad B = \mu H \qquad J = \sigma E$$

Force equations

$$F = eE \qquad F = e(v \times B) \qquad F = m \frac{dv}{dt}$$

E = electric-field intensity, volts m^{-1}
D = electric flux density, coul m^{-2}
H = magnetic field, amp m^{-1}
B = magnetic flux density, webers m^{-2}
J = conduction-current density, amp m^{-2}
ρ = charge density, coul m^{-3}
dl = element of length, m
ds = element of surface, m^2
dv = element of volume, m^3
ϵ = permittivity of medium, farads m^{-1}
μ = permeability of medium, henrys m^{-1}
σ = conductivity of medium, mhos m^{-1}
F = force, newtons
v = velocity, m sec^{-1}
e = charge of particle, coul
m = mass of particle, kg
dt = element of time, sec

5-3   Plane Waves in Lossless and Conducting Media   Taking the curl of Maxwell's second curl equation (Table 5-1) and noting that $B = \mu H$, we have

$$\nabla \times \nabla \times E = -j\mu\omega(\nabla \times H) \qquad\qquad (5\text{-}13)$$

Harmonic variation with time is assumed. Introducing Maxwell's first curl equation (Table 5-1) and writing $\mathbf{J} = \sigma\mathbf{E}$ and $\mathbf{D} = \epsilon\mathbf{E}$ yields

$$\nabla \times \nabla \times \mathbf{E} = -j\mu\omega(\sigma\mathbf{E} + j\omega\epsilon\mathbf{E}) \qquad (5\text{-}14)$$

Thus, starting with Maxwell's two curl equations we have obtained in (5-14) a second-order differential equation involving only $\mathbf{E}$. It involves the space and time variation of $\mathbf{E}$ and is referred to as a *wave equation* in $\mathbf{E}$. Rearranging (5-14) yields

$$\nabla \times \nabla \times \mathbf{E} + (j\omega\mu\sigma - \omega^2\mu\epsilon)\mathbf{E} = 0 \qquad (5\text{-}15)$$

If we let

$$\gamma^2 = j\omega\mu\sigma - \omega^2\mu\epsilon \qquad (5\text{-}16)$$

(5-15) simplifies to

$$\nabla \times \nabla \times \mathbf{E} + \gamma^2\mathbf{E} = 0 \qquad (5\text{-}17)$$

For a plane wave traveling in the x direction with $\mathbf{E}$ linearly polarized in the y direction (5-17) reduces to

$$\frac{\partial^2 E_y}{\partial x^2} - \gamma^2 E_y = 0 \qquad (5\text{-}18)$$

A solution of (5-18) is given by

$$E_y = \dot{E}_0 e^{\pm\gamma x} \qquad (5\text{-}19)$$

In a lossless medium the conductivity is zero ($\sigma = 0$) so

$$\gamma = \pm j\omega\sqrt{\mu\epsilon} = \pm j\beta \qquad (5\text{-}20)$$

where $\beta = \omega\sqrt{\mu\epsilon} = \omega/v = 2\pi\nu/v\lambda = 2\pi/\lambda$ = phase constant, rad m^{-1}
$\omega = 2\pi\nu$ = radian frequency, rad sec^{-1}
$v = 1/\sqrt{\mu\epsilon}$ = velocity of wave, m sec^{-1}
ν = frequency, cps
λ = wavelength, m

Thus, for a lossless medium

$$E_y = \dot{E}_0 e^{\pm j\beta x} \qquad (5\text{-}21)$$

In this case the amplitude $\dot{E}_0$ is a phasor, since harmonic variation with time has been assumed. Thus, we can write

$$\dot{E}_0 = E_0 e^{j\omega t} \qquad (5\text{-}22)$$

from which (5-21) becomes

$$E_y = E_0 e^{j(\omega t \pm \beta x)} \qquad (5\text{-}23)$$

This solution gives the space x and time t variation of $\mathbf{E}$ for a plane wave traveling in the x direction with $\mathbf{E}$ polarized in the y direction in a lossless

medium. With a plus sign in the exponent, the solution is appropriate
for a wave traveling in the negative x direction, and with a minus sign the
solution is appropriate for a wave traveling in the positive x direction.
The wave propagates without attenuation, the exponent being a pure
imaginary quantity.

Equation (5-23) is for a lossless medium. If the medium is conducting
and the conductivity is sufficient for the condition that $\sigma \gg \omega\epsilon$ to apply,
(5-16) reduces to

$$\gamma^2 = j\omega\mu\sigma \tag{5-24}$$

and

$$\gamma = (1+j)\sqrt{\frac{\omega\mu\sigma}{2}} \tag{5-25}$$

Thus, γ has a real and imaginary part. Putting $\gamma = \alpha + j\beta$, the real part
α is associated with attenuation and the imaginary part β with phase. Thus,

$$\gamma = \alpha + j\beta = \sqrt{\frac{\omega\mu\sigma}{2}} + j\sqrt{\frac{\omega\mu\sigma}{2}} \tag{5-26}$$

where γ = propagation constant, m^{-1}
   α = attenuation constant, m^{-1}
   β = phase constant, rad m^{-1}

The solution for a plane wave traveling in the x direction with $\mathbf{E}$ polarized
in the y direction in a conducting medium is then

$$E_y = E_0 e^{j\omega t \pm \gamma x} = E_0 e^{\pm \alpha x} e^{j(\omega t \pm \beta x)} \tag{5-27}$$

In astronomical work the product αx is usually called the *optical depth*, as
discussed in Sec. 3-13. The distance required for the wave to attenuate to
$1/e$ of its original value occurs when $x = 1/\alpha$. This is called the $1/e$ *depth
of penetration* (unity optical depth).

The solution (5-27) for a conducting medium is the same as (5-23)
for a lossless medium except for the factor $e^{\pm \alpha x}$, which gives the attenuation
of the wave as a function of the distance x. In a lossless medium the
attenuation is zero ($\alpha = 0$), and (5-27) reduces to (5-23).

5-4 Plane Waves in an Ionized Medium in the Presence of a Magnetic Field†

In Sec. 5-3 wave propagation in lossless and conducting
media was discussed. Although an ionized medium may be classified as

---

† This topic is the subject of many papers and books, of which a few are Appleton (1932),
Nichols and Schelleng (1925), Hartree (1930), Mimno (1937), Mitra (1952), Ginzburg
(1961), Allis (1961), Stix (1962), and Cambel (1963). The analogous situation in a ferromagnetic medium is treated by Hogan (1952). The straightforwardness of the development of this section is due to Kouyoumjian (1964).

a conducting medium, the condition $\sigma \gg \omega\epsilon$ assumed in Sec. 5-3 may not apply. Furthermore, an ionized medium becomes anisotropic (propagation variable as a function of direction) in the presence of a steady magnetic field. This greatly complicates the situation as compared to that considered in Sec. 5-3, where the medium is assumed to be isotropic (propagation same in all directions). Consequently, a more general development will now be made for the case of a plane wave in an ionized medium in the presence of a steady magnetic field.

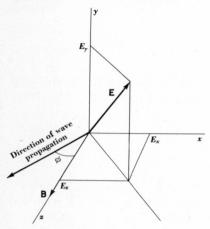

Fig. 5-1. Relation of direction of wave propagation to coordinate system. The direction of wave propagation is in the xz plane at an angle ϕ with respect to the z axis. The electric field **E** of the wave lies in a plane perpendicular to the direction of wave propagation. The steady magnetic field is in the z direction.

The medium to be considered is a plasma; i.e., it consists of equal positive and negative charges (total charge zero) but would be described as cool (thermal energies small) and tenuous (collisions unimportant). In such a plasma only the interaction of the negative charges (electrons) with the wave will be considered, the positive ions being too massive to interact appreciably.

Referring to Fig. 5-1, it will be assumed that the steady magnetic field **B** is in the z direction; i.e., $|\mathbf{B}| = B_z$. This simplifies the analysis without, however, any loss of generality. The direction of propagation of the wave is taken in the xz plane with the angle between this direction and the z axis equal to ϕ. The electric field **E** associated with the wave may have any orientation perpendicular to the direction of propagation, i.e., may be polarized in any manner.

The force equations involving a charged particle of charge e and mass m in the presence of a wave and magnetic field **B** are

$$\mathbf{F}_1 = e(\mathbf{v} \times \mathbf{B}) \tag{5-28}$$

$$\mathbf{F}_2 = e\mathbf{E} \tag{5-29}$$

$$\mathbf{F}_3 = m\mathbf{a} = m\frac{d\mathbf{v}}{dt} \tag{5-30}$$

where $\mathbf{F}$ = force, newtons

e = charge of particle, coul

m = mass of particle, kg

$\mathbf{v}$ = velocity of particle due to the wave, m sec^{-1}

$\mathbf{B}$ = steady magnetic field, webers m^{-2}

$\mathbf{E}$ = electric field of wave, volts m^{-1}

$\mathbf{a}$ = acceleration of particle, m sec^{-2}

t = time, sec

In (5-30) (Newton's second law) it is assumed that the velocity is small compared to the velocity of light; i.e., relativistic effects are neglected.†

Combining the force relations gives

$$\mathbf{F} = e(\mathbf{E} + \mathbf{v} \times \mathbf{B}) = m\frac{d\mathbf{v}}{dt} \tag{5-32}$$

Dividing by e to obtain the force per unit charge yields

$$\frac{\mathbf{F}}{e} = \mathbf{E} + \mathbf{v} \times \mathbf{B} = \frac{m}{e}\frac{d\mathbf{v}}{dt} \tag{5-33}$$

Expressing this vector equation in terms of its three (rectangular) scalar components, we have

$$\frac{F_x}{e} = E_x + v_y B_z = \frac{m}{e}\frac{dv_x}{dt} \tag{5-34}$$

$$\frac{F_y}{e} = E_y - v_x B_z = \frac{m}{e}\frac{dv_y}{dt} \tag{5-35}$$

$$\frac{F_z}{e} = E_z \qquad\quad = \frac{m}{e}\frac{dv_z}{dt} \tag{5-36}$$

Assuming harmonic motion of the wave and charged particle and solving for its component velocities, we have

$$v_x = \frac{j(m\omega/e)E_x + E_y B_z}{B_z^2 - m^2\omega^2/e^2} \tag{5-37}$$

$$v_y = \frac{j(m\omega/e)E_y - E_x B_z}{B_z^2 - m^2\omega^2/e^2} \tag{5-38}$$

$$v_z = \frac{-jeE_z}{m\omega} \tag{5-39}$$

---

† A more general statement of (5-30) is

$$\mathbf{F} = \frac{d(m\mathbf{v})}{dt} = m\frac{d\mathbf{v}}{dt} + \mathbf{v}\frac{dm}{dt} \tag{5-31}$$

where $m\mathbf{v}$ is the momentum of the particle. For $v \ll c$, $dm/dt = 0$, so that (5-31) reduces to (5-30).

where ω = radian or angular frequency ($= 2\pi\nu$ rad sec^{-1})

ν = frequency, cps

From these relations it may be inferred that the charged particle will move in a helical path with axis coincident with the z direction. For the condition

$$B_z{}^2 = \frac{m^2\omega^2}{e^2} \tag{5-40}$$

v_x and v_y become infinite, corresponding to *gyro resonance*, the effect of collisions with other particles being neglected. Calling the radian frequency at which this occurs ω_g, we have

$$\omega_g = \frac{e}{m} B_z \tag{5-41}$$

where ω_g = (radian) gyro frequency, $2\pi\nu_g$ rad sec^{-1}

This is the resonance condition in a cyclotron, and for this reason ω_g is sometimes also called the *cyclotron* (radian) *frequency*. Introducing (5-41) into (5-37) and (5-38) yields

$$v_x = \frac{e}{m} \frac{\omega_g E_y + j\omega E_x}{\omega_g{}^2 - \omega^2} \tag{5-42}$$

$$v_y = \frac{e}{m} \frac{-\omega_g E_x + j\omega E_y}{\omega_g{}^2 - \omega^2} \tag{5-43}$$

From Maxwell's equation we have that

$$\nabla \times \mathbf{H} = \mathbf{J} + j\omega\mathbf{D} = Nev + j\omega\epsilon_0\mathbf{E} \tag{5-44}$$

where N = number of charged particles per unit volume.

It is assumed that a movement of the charged particles constitutes a conduction current. Expressing (5-44) in terms of its three rectangular components,

$$(\nabla \times \mathbf{H})_x = j\omega\epsilon_0 E_x + Nev_x \tag{5-45}$$

$$(\nabla \times \mathbf{H})_y = j\omega\epsilon_0 E_y + Nev_y \tag{5-46}$$

$$(\nabla \times \mathbf{H})_z = j\omega\epsilon_0 E_z + Nev_z \tag{5-47}$$

Introducing the values of v_x, v_y, and v_z from (5-42), (5-43), and (5-39) into these relations, we obtain

$$(\nabla \times \mathbf{H})_x = j\omega\epsilon_0 E_x \left[1 + \frac{Ne^2}{\epsilon_0 m(\omega_g{}^2 - \omega^2)} \right] + \frac{Ne^2\omega_g E_y}{m(\omega_g{}^2 - \omega^2)} \tag{5-48}$$

$$(\nabla \times \mathbf{H})_y = -\frac{Ne^2\omega_g E_x}{m(\omega_g{}^2 - \omega^2)} + j\omega\epsilon_0 E_y \left[1 + \frac{Ne^2}{\epsilon_0 m(\omega_g{}^2 - \omega^2)} \right] \tag{5-49}$$

$$(\nabla \times \mathbf{H})_z = j\omega\epsilon_0 E_z \left[1 - \frac{Ne^2}{\epsilon_0 m\omega^2} \right] \tag{5-50}$$

If the frequency is much greater than the gyro frequency $(\omega \gg \omega_g)$, all three relations become identical in form. Further, if

$$\frac{Ne^2}{\epsilon_0 m \omega^2} = 1 \tag{5-51}$$

a *critical* or *plasma frequency* occurs for which the permittivity and the index of refraction of the medium become zero. Designating the critical frequency as ω_0 $(= 2\pi\nu_0)$, we have

$$\omega_0{}^2 = \frac{Ne^2}{\epsilon_0 m} \tag{5-52}$$

or

$$\nu_0 = \frac{e}{2\pi} \sqrt{\frac{N}{\epsilon_0 m}} \tag{5-53}$$

where ω_0 = critical radian frequency, $2\pi\nu_0$ rad cps^{-1}
ν_0 = critical frequency, cps
N = number of charged particles per cubic meter
e = charge of particle, coul
m = mass of particle, kg
ϵ_0 = permittivity of vacuum $(= 8.85 \times 10^{-12}$ farad m$^{-1})$

For an electron (5-53) becomes

$$\nu_0 = 9\sqrt{N} \tag{5-54}$$

where ν_0 = critical frequency cps
N = electron density, number m^{-3}

At or below the critical frequency the wave is totally reflected by the medium. For example, the value of N for the earth's ionosphere is sometimes about 10^{12} electrons per cubic meter, which from (5-54) corresponds to a critical frequency of 9 Mc.

Introducing (5-52) into (5-48) through (5-50), we may write, in general, that†

$$\nabla \times \mathbf{H} = j\omega\bar{\epsilon} \cdot \mathbf{E} \tag{5-55}$$

where $\bar{\epsilon}$ = tensor permittivity

or

$$\bar{\epsilon} = \begin{bmatrix} \epsilon_{11} & -j\epsilon_{12} & \epsilon_{13} \\ j\epsilon_{21} & \epsilon_{22} & \epsilon_{23} \\ \epsilon_{31} & \epsilon_{32} & \epsilon_{33} \end{bmatrix} \tag{5-56}$$

† From $\nabla \times \mathbf{H} = \sigma\mathbf{E} + j\omega\epsilon_0\mathbf{E}$ we could write $\nabla \times \mathbf{H} = \bar{\sigma} \cdot \mathbf{E}$, describing the medium in terms of an effective tensor conductivity $\bar{\sigma}$, *or* we could write $\nabla \times \mathbf{H} = j\omega\bar{\epsilon} \cdot \mathbf{E}$, describing the medium in terms of an effective tensor permittivity $\bar{\epsilon}$. In the present instance, the latter procedure has been chosen, as given by (5-55).

or

$$\bar{\epsilon} = \begin{bmatrix} \left(1 + \dfrac{\omega_0{}^2}{\omega_g{}^2 - \omega^2}\right)\epsilon_0 & \dfrac{-j\omega_0{}^2\omega_g\epsilon_0}{\omega(\omega_g{}^2 - \omega^2)} & 0 \\[2ex] \dfrac{j\omega_0{}^2\omega_g\epsilon_0}{\omega(\omega_g{}^2 - \omega^2)} & \left(1 + \dfrac{\omega_0{}^2}{\omega_g{}^2 - \omega^2}\right)\epsilon_0 & 0 \\[2ex] 0 & 0 & \left(1 - \dfrac{\omega_0{}^2}{\omega^2}\right)\epsilon_0 \end{bmatrix} \qquad (5\text{-}57)$$

In the absence of a magnetic field ($\mathbf{B} = 0$) or where $\omega \gg \omega_g$, (5-57) reduces to

$$\bar{\epsilon} = \begin{bmatrix} \left(1 - \dfrac{\omega_0{}^2}{\omega^2}\right)\epsilon_0 & 0 & 0 \\[2ex] 0 & \left(1 - \dfrac{\omega_0{}^2}{\omega^2}\right)\epsilon_0 & 0 \\[2ex] 0 & 0 & \left(1 - \dfrac{\omega_0{}^2}{\omega^2}\right)\epsilon_0 \end{bmatrix} \qquad (5\text{-}58)$$

and

$$\nabla \times \mathbf{H} = j\omega\left(1 - \dfrac{\omega_0{}^2}{\omega^2}\right)\epsilon_0\mathbf{E} \qquad (5\text{-}59)$$

The quantity in parentheses is equivalent to the relative permittivity ϵ_r of the medium. The index of refraction is then

$$\eta = \sqrt{\epsilon_r} = \sqrt{1 - (\omega_0/\omega)^2} = \sqrt{1 - (\nu_0/\nu)^2} \qquad (5\text{-}60)$$

When $\nu \geq \nu_0$, the index of refraction is real and the wave is propagated. However, η is less than unity, so that refraction is opposite to that which occurs when waves enter a denser medium. When $\nu \leq \nu_0$, the index of refraction is zero or imaginary and the wave is not propagated but reflected.

When a cloud of charged particles is perturbed, it may tend to oscillate at the plasma frequency. Certain types of radiation from the solar atmosphere or corona are believed to be due to such plasma oscillations. The electron density in the solar corona decreases with height, and from (5-54) the plasma frequency will also decrease. Hence, at a given frequency ν, radiation is received only from the corona at heights above the height for which $\nu = \nu_0$. Actually the radiation appears to come mostly from a relatively thin layer just above the height for which $\nu = \nu_0$.

These comments apply to the quiet sun (see Sec. 8-7a). In the case of disturbances on the sun, such as flares, more dynamic phenomena may be involved, in which high-velocity jets or shock waves rise into the solar corona. Plasma oscillations of electron clouds excited by such mechanisms appear to be responsible for relatively narrow band radiation which drifts downward in frequency as the disturbance rises through the corona into regions of lower electron density.   These topics are discussed in more detail in Sec. 8-7c.

Returning to the general situation (with **B** present), we have on taking the curl of Maxwell's equation

$$\nabla \times \mathbf{E} = -j\omega\mu_0\mathbf{H} \tag{5-61}$$

for the case $\mu = \mu_0$ (magnetic effects assumed negligible) and introducing $\nabla \times \mathbf{H}$ from (5-55), that

$$\nabla \times \nabla \times \mathbf{E} = \omega^2\mu_0\bar{\epsilon} \cdot \mathbf{E} \tag{5-62}$$

This is a wave equation in **E**.   Neglecting the effect of collisions (attenuation assumed zero), let a solution be assumed of the form

$$\mathbf{E} = \mathbf{E}_0 e^{j(\omega t - \beta r)} \tag{5-63}$$

where β = phase constant (in r direction)

   r = distance measured in direction of wave propagation (see Fig. 5-1)

It may then be shown that

$$\beta^2 = \omega^2\mu_0\{(\epsilon_{11}{}^2 - \epsilon_{12}{}^2 - \epsilon_{11}\epsilon_{33})\sin^2\phi + 2\epsilon_{11}\epsilon_{33} \pm [(\epsilon_{11}{}^2 - \epsilon_{12}{}^2 - \epsilon_{11}\epsilon_{33})^2\sin^4\phi$$
$$+ 4\epsilon_{12}{}^2\epsilon_{33}{}^2\cos^2\phi]^{1/2}\}/2[(\epsilon_{11} - \epsilon_{33})\sin^2\phi + \epsilon_{33}] \tag{5-64}$$

where ϕ is the angle between **B** (or z) and wave direction (see Fig. 5-1), and where ϵ_{11}, ϵ_{12}, etc., are as given by comparing (5-56) and (5-57).   For the case where ω is much larger than ω_g, so that $\epsilon_{11} \simeq \epsilon_{33}$, but not so large that ϵ_{12} (or ϵ_{21}) can be neglected, (5-64) reduces to

$$\beta^2 = \omega^2\mu_0\epsilon_{11}\left\{1 - \frac{\epsilon_{12}{}^2}{2\epsilon_{11}{}^2}\left[\sin^2\phi \pm \sqrt{\sin^4\phi + \left(\frac{2\epsilon_{11}\cos\phi}{\epsilon_{12}}\right)^2}\right]\right\} \tag{5-65}$$

When the wave propagates parallel to the magnetic field ($\phi = 0$), the condition is called *longitudinal propagation*, and (5-65) reduces for this case to

$$\beta = \omega\sqrt{\mu_0(\epsilon_{11} \pm \epsilon_{12})} \tag{5-66}$$

From (5-66) it can be shown that the wave may consist of two circularly polarized components of opposite hand.

When the wave propagates perpendicular to **B** ($\phi = 90°$), the condition is called *transverse propagation*, and (5-65) reduces for this case to

$$\beta = \omega\sqrt{\mu_0\left(\epsilon_{11} - \frac{\epsilon_{12}{}^2}{\epsilon_{11}}\right)} \tag{5-67}$$

when the electric field **E** of the wave is linearly polarized *perpendicular* to **B** and to

$$\beta = \omega\sqrt{\mu_0\epsilon_{33}} \tag{5-68}$$

when **E** is *parallel to* **B**. In this latter case propagation is the same as though no magnetic field were present. This wave is sometimes referred to as the *ordinary ray*, while in the case described by (5-67), where **E** is perpendicular to **B**, the wave is referred to as the *extraordinary ray*.

More general relations than the above may be deduced for *quasi-longitudinal* and *quasi-transverse* conditions of propagation. For the former case, ϕ is considered sufficiently small that $\sin^2 \phi$ and $\sin^4 \phi$ terms can be neglected but cos ϕ not equated to unity. For the latter case ϕ is considered to be sufficiently near to 90° that cos ϕ terms can be neglected. We then have for the *quasi-longitudinal* case that

$$\beta = \omega \sqrt{\mu_0(\epsilon_{11} \pm \epsilon_{12} \cos \phi)} \tag{5-69}$$

and for the *quasi-transverse* case that

$$\beta = \omega \sqrt{\mu_0 \left[\epsilon_{11} - (1 \pm 1) \frac{\epsilon_{12}^2}{2\epsilon_{11}} \sin \phi \right]} \tag{5-70}$$

The above propagation relations are summarized in Table 5-2.

Table 5-2
Propagation relations†

Quasi-longitudinal propagation (ϕ small)	$\beta = \omega \sqrt{\mu_0(\epsilon_{11} \pm \epsilon_{12} \cos \phi)}$
Quasi-transverse propagation (ϕ near 90°)	$\beta = \omega \sqrt{\mu_0 \left[\epsilon_{11} - (1 \pm 1) \frac{\epsilon_{12}^2}{2\epsilon_{11}} \sin \phi \right]}$
Longitudinal propagation ($\parallel$ to **B**; $\phi = 0$)	$\beta = \omega \sqrt{\mu_0(\epsilon_{11} \pm \epsilon_{12})}$
Transverse propagation ($\perp$ to **B**; $\phi = 90°$):	
Ordinary ray; **E**$\parallel$**B**	$\beta = \omega \sqrt{\mu_0 \epsilon_{11}}$
Extraordinary ray; **E**$\perp$**B**	$\beta = \omega \sqrt{\mu_0 \left(\epsilon_{11} - \frac{\epsilon_{12}^2}{\epsilon_{11}} \right)}$

$$\epsilon_{11} = \epsilon_{33} = \left(1 - \frac{\omega_0^2}{\omega^2} \right)\epsilon_0 \qquad \epsilon_{12} = \frac{-\omega_0^2 \, \omega_g \, \epsilon_0}{\omega^3}$$

ω = radian frequency of wave = $2\pi\nu$ rad sec^{-1}

ω_g = radian gyro frequency = $\dfrac{e}{m} B$

ω_0 = radian plasma frequency ($= e\sqrt{N/\epsilon_0 m}$)

B = magnetic flux density, webers m^{-2}

e = charge on particle, coul

m = mass of particle, kg

ϵ_0 = permittivity of vacuum ($= 8.85 \times 10^{-12}$ farad m^{-1})

β = phase constant ($= 2\pi/\lambda$)

λ = wavelength, m

μ_0 = permeability of vacuum ($= 4\pi \times 10^{-7}$ henry m^{-1})

ϕ = angle between wave direction and magnetic field **B**

† See text for assumptions involved.

Kouyoumjian (1964) has shown that the axial ratio of the polarization ellipse of the wave can be expressed by the relation

$$AR = \frac{\beta^2 \epsilon_{11} - \omega^2 \mu_0 (\epsilon_{11}{}^2 - \epsilon_{12}{}^2)}{\beta^2 \epsilon_{12} \cos^2 \phi} \tag{5-71}$$

where ϕ is the angle between the direction of the wave and the magnetic field, as before, but in this case it is not restricted.

-5   Faraday Rotation†

A linearly polarized wave may be regarded as the resultant of two circularly polarized waves of equal amplitude and opposite hand (see Chap. 4). If the two circularly polarized waves have different phase constants, the plane of polarization of the resultant linearly polarized wave rotates as the wave propagates. Thus, consider two circu-

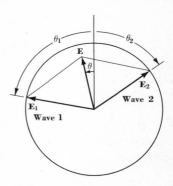

Fig. 5-2. Resolution of linearly polarized wave into two circularly polarized waves (1 and 2) of opposite rotation direction.

larly polarized waves traveling out of the page, as in Fig. 5-2, for the quasi-longitudinal case (ϕ small). The elemental angular rotation of one wave is given by

$$d\theta_1 = \beta^- \, dr \tag{5-72}$$

and of the other wave by

$$d\theta_2 = \beta^+ \, dr \tag{5-73}$$

where β^- and β^+ are as given by (5-69) for the cases where the sign in the parentheses is minus and plus, respectively. The net rotation angle is then

$$d\theta = \frac{\beta^- - \beta^+}{2} \, dr \tag{5-74}$$

The name derives from Faraday's early investigations of the phenomenon, in which he observed the rotation of the plane of polarization of light passing through a crystal having an applied magnetic field.

Introducing (5-69) and the values of ϵ_{11} and ϵ_{12} from (5-56) and (5-57) for $\omega \gg \omega_g$, $\omega \gg \omega_0$, and small ϕ, we obtain approximately that

$$d\theta = \frac{Ne^3B\lambda^2 \cos \phi \, dr}{8\pi^2c^3\epsilon_0 m^2} \qquad (5\text{-}75)$$

The total *Faraday rotation* for the *quasi-longitudinal case* (ϕ small) is then

$$\theta = \frac{e^3\lambda^2}{8\pi^2c^3\epsilon_0 m^2} \int_0^r NB \cos \phi \, dr \qquad \text{rad} \qquad (5\text{-}76)$$

where e = charge of particle, coul
     m = mass of particle, kg
     ω = radian frequency = $2\pi\nu$ rad sec^{-1}
     ϵ_0 = permittivity of vacuum ($= 8.85 \times 10^{-12}$ farad m^{-1})
     c = velocity of light ($= 3 \times 10^8$ m sec^{-1})
     N = number of particles, number m^{-3}
     B = magnetic flux density, webers m^{-2}
     ϕ = angle between **B** and direction of wave propagation
     r = distance in direction of propagation, m

If B and ϕ are constant, (5-76) becomes

$$\theta = \frac{e^3B\lambda^2 \cos \phi}{8\pi^2c^3\epsilon_0 m^2} \int_0^r N \, dr \qquad (5\text{-}77)$$

If B and ϕ are known, a measurement of θ at a wavelength λ permits determination of the total number of charged particles in a column of 1 m cross section between the source and observer given by

$$N_t = \int_0^r N \, dr \qquad (5\text{-}78)$$

where N_t = total number of particles in column of length r and cross section 1 m^2

For the *quasi-transverse case* (ϕ near 90°) we have

$$d\theta = \frac{Ne^4\lambda^3B^2 \sin^2 \phi \, dr}{32\pi^3c^4m^3\epsilon_0} \qquad (5\text{-}79)$$

and for the total Faraday rotation

$$\theta = \frac{e^4\lambda^3}{32\pi^3c^4m^3\epsilon_0} \int_0^r NB^2 \sin^2 \phi \, dr \qquad (5\text{-}80)$$

The ratio of the longitudinal and transverse rotations is given by

$$\frac{d\theta(\text{long})}{d\theta(\text{trans})} \simeq \frac{4\pi cm}{eB\lambda} = \frac{2m\omega}{Be} \qquad (5\text{-}81)$$

In the case of electrons in the earth's ionosphere we have at a frequency of 100 Mc, and taking the earth's field at 5×10^{-5} webers m^{-2}, that the ratio of longitudinal to transverse rotation is of the order of 100. At Columbus, Ohio, the earth's magnetic-field geometry is as suggested in Fig. 5-3. The propagation path to Columbus from an artificial earth satellite, traveling from north to south as indicated, would change from a quasi-transverse to a quasi-longitudinal condition during the pass. From (5-81) one would expect the Faraday-rotation fading rate of signals from a satellite-borne transmitter to increase significantly as the satellite moves from north to south. Such a transition is, in fact, observed.

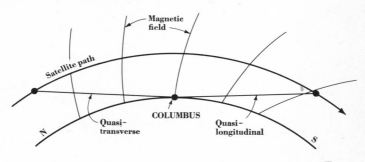

Fig. 5-3. Transition of transmission condition from quasi-transverse to quasi-longitudinal during pass of artificial earth satellite.

In the interstellar medium B is even less (order of 10^{-5} gauss or 10^{-9} weber m^{-2}), so that the rotation when propagation is parallel to **B** becomes dominant, particularly at the higher frequencies. Thus, the rotation effects may be described entirely in terms of the quasi-longitudinal case.

In radio-astronomy observations of the polarization of radio sources, the *position angle* θ of the electric-field vector of the linearly polarized component of the radiation is usually measured relative to north with θ increasing counterclockwise, to the east. Since the position angle of the electric-field vector varies as the square of the wavelength [see (5-76)], a θ versus λ^2 graph should show a straight-line relationship over a range of wavelengths. By extrapolating the straight line to zero wavelength the *polarization angle at the source*, or *intrinsic polarization*, can be determined. The slope of the line $(= \theta/\lambda^2)$ has been called the *rotation measure*. The rotation measure is positive when the magnetic-field direction is toward the observer. There is an ambiguity of $n\pi$ (n = integer) in the position angle, which is resolved if measurements are made sufficiently close in frequency so that θ changes less than $\pi/2$ between measurements.

Results of polarization measurements on Taurus A by Gardner and Whiteoak (1963), Mayer, McCullough, and Sloanaker (1964), and Hollinger, Mayer, and Mennella (1964) are presented in Fig. 5-4, with the position-angle variation above and the degree of linear polarization in

percent below.   Extrapolating the position-angle to zero wavelength gives
an intrinsic polarization angle of approximately 150°.   The rotation measure is about −25 rad m⁻².   Assuming a constant electron density, magnetic
field, and angle ϕ, (5-76) reduces for electrons to †

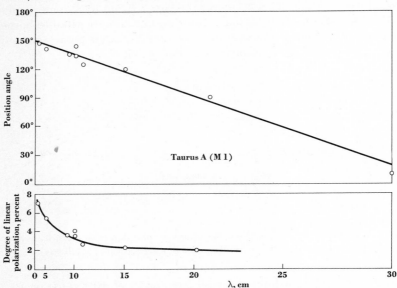

Fig. 5-4.   Position angle and degree of linear polarization as a function of wavelength for the radio source Taurus A (Crab nebula). Distance along the horizontal scale is proportional to the square of the wavelength.

$$\Delta\theta = 2.6 \times 10^{-13} \, NB\lambda^2 \cos\phi \, \Delta r \qquad \text{rad} \qquad (5\text{-}82)$$

where N = number of electrons, m⁻³

B = magnetic flux density, webers m⁻²

λ = wavelength, m

ϕ = angle between wave direction and **B**

Δr = path length, m

Converting to units frequently used in astronomical calculations, (5-82)
becomes

$$\Delta\theta = 8.1 \times 10^5 \, NB \, \lambda^2 \cos\phi \, \Delta r \qquad \text{rad} \qquad (5\text{-}83)$$

† For the case of a transmitter in an artificial earth satellite traveling toward or away
from the observer along a line nearly parallel to **B**, it is convenient to replace $\Delta\theta$ by
and Δr by $v \, \Delta T$, where v is the satellite velocity (in meters per second) and ΔT the time
(in seconds) between nulls in the signal as measured with a linearly polarized receiving
antenna. Assuming that B is known, the average N can then be determined. It is assumed that the fading (nulls) is entirely due to Faraday-rotation effects and not to rotation of the satellite (with a linearly polarized antenna).

where N = number of electrons, cm^{-3}

B = magnetic flux density, gauss

λ = wavelength, m

Δr = path length, pc

Assuming $\phi = 0$, $B = 10^{-5}$ gauss, and $\Delta r = 1,100$ pc, (5-83) yields in the case of Taurus A ($|\theta/\lambda^2| = 25$) an average density of about 3×10^{-3} electrons per cubic centimeter for the interstellar medium. If there are reversals in the direction of B over the path length, a larger electron density and/or value of B would be indicated.

Differences in $\int NB \cos \phi \, dr$ over paths to various parts of a radio source will cause differences in Faraday rotation which will increase as the square of the wavelength, tending to depolarize the radiation, i.e., decrease the degree of linear polarization. Most polarized radio sources exhibit this tendency, presumably for this reason. The depolarization in some sources, such as Cygnus A, is especially rapid. As discussed in Chap. 8, the degree of linear polarization for this source decreases from about 8 to 1.5 percent between 3 and 5 cm wavelength.

To observe even a small degree of linear polarization from sources that are small compared to the beam size indicates a relatively high degree of regularity of the magnetic field in the source. With antenna directivity sufficient to resolve the source into components it might be expected that at least some of the components would exhibit a higher degree of linear polarization than the source as a whole.

5-6 Magnetohydrodynamic Waves In Sec. 5-4 no motion of the medium other than a harmonic motion of the charged particles was considered. Suppose now that a layer or column in a conducting medium with applied magnetic field is mechanically moved or displaced. No incident waves are considered but rather the effect of the motion of the medium in generating a disturbance or wave. This situation has been investigated by Alfvén (1950). The waves produced by the motion are referred to as *magnetohydrodynamic waves*, *mhd waves*, or *Alfvén waves*.

It should be noted that the medium being considered here differs significantly from that considered in Secs. 5-4 and 5-5. The medium considered here is much denser, so that collisions are important. It could consist of a conducting fluid such as mercury, or it might consist of a very dense, hot plasma.

Let a slab in a conducting fluid, as in Fig. 5-5, be moved with a velocity v in the y direction. The slab is infinite in extent in the $\pm y$ directions. A steady magnetic field B_0 is applied in the z direction. Since v is in the y direction and B_0 in the z direction, we have from

$$\mathbf{E} = \mathbf{v} \times \mathbf{B} \tag{5-84}$$

that an electric field **E** will be induced in the x direction by the motion, as suggested in Fig. 5-5. This field produces a current in the slab in the x direction, which has its return paths through the medium above and below the slab. This current, in turn, produces forces which tend to impede the motion of the slab but which impart an acceleration to the fluid above and below the slab in the y direction. Thus, the original motion initiates

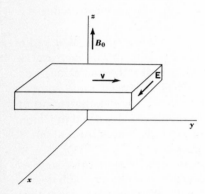

Fig. 5-5.   Relation of slab with velocity **v** to coordinate system.

motion of the fluid, which propagates as a wave parallel to the magnetic field.   The pertinent equations are

$$\nabla \times \mathbf{H} = \mathbf{J} + \frac{\partial \mathbf{D}}{\partial t} \tag{5-85}$$

$$\nabla \times \mathbf{E} = -\frac{\partial \mathbf{B}}{\partial t} \tag{5-86}$$

$$\mathbf{J} = \sigma(\mathbf{E} + \mathbf{v} \times \mathbf{B}) \tag{5-87}$$

$$\mathbf{B} = \mu \mathbf{H} \tag{5-88}$$

$$\frac{d\mathbf{v}}{dt} = \mathbf{G} + \frac{1}{\rho}(\mathbf{J} \times \mathbf{B} - \nabla p) \tag{5-89}$$

where **v** = velocity of displacement, m sec^{-1}

J = current density, amp m^{-2}

ρ = mass density, kg m^{-3}

p = pressure, kg m^{-1} sec^{-2}

G = parameter involving nonelectromagnetic forces, m sec^{-2}

Equations (5-85) and (5-86) are Maxwell curl equations, (5-87) is a current equation, (5-88) is an equation characterizing the medium, and (5-89) is a hydrodynamic equation with the dimensions of acceleration.

Consider the case of plane waves in an incompressible fluid (density constant).   From (5-87) we have that

$$\frac{\mathbf{J}}{\sigma} = \mathbf{E} + \mathbf{v} \times \mathbf{B} \tag{5-90}$$

From the geometry $J_y = J_z = 0$, so that

$$E_x = \frac{J_x}{\sigma} - v_y B_0 \tag{5-91}$$

and $E_y = E_z = 0$. From (5-86) and (5-88)

$$\frac{\partial E_x}{\partial z} = -\mu \frac{\partial H_y}{\partial t} \tag{5-92}$$

Introducing (5-91) for E_x in (5-92) yields

$$\mu \frac{\partial H_y}{\partial t} = B_0 \frac{\partial v_y}{\partial z} - \frac{1}{\sigma} \frac{\partial J_x}{\partial z} \tag{5-93}$$

Neglecting **G**, we have from (5-89)

$$\frac{\partial v_y}{\partial t} = -\frac{1}{\rho} J_x B_0 \tag{5-94}$$

It is assumed that ∇p can have no component perpendicular to z. Introducing (5-85) in (5-94) we have (neglecting $\partial \mathbf{D}/\partial t$ in comparison with $\mathbf{J}$)

$$\frac{\partial v_y}{\partial t} = \frac{B_0}{\rho} \frac{\partial H_y}{\partial z} \tag{5-95}$$

Eliminating v_y between (5-93) and (5-95), we obtain a wave equation in H_y for a wave propagating in the z direction or

$$\frac{\partial^2 H_y}{\partial t^2} = \frac{B_0^2}{\mu\rho} \frac{\partial^2 H_y}{\partial z^2} + \frac{1}{\mu\sigma} \frac{\partial^3 H_y}{\partial t \, \partial z^2} \tag{5-96}$$

For very large conductivity the last term may be neglected. The velocity of the wave, or *Alfvén velocity*, is then given by

$$v = \frac{B_0}{\sqrt{\mu\rho}} \tag{5-97}$$

where v = velocity of wave, m sec^{-1}
  B_0 = magnetic flux density, webers m^{-2}
  μ = permeability of medium, henrys m^{-1}
  ρ = density of medium, kg m^{-3}

A physical insight into this result may be obtained by applying Faraday's elastic-string representation of a magnetic field. The analogy is oversimplified but instructive. Thus, as done by Alfvén, let us consider that the conductivity of the medium is infinite, so that the lines of magnetic field are "frozen in." The field and medium must then move together, and the fluid can be regarded as attached or glued to the lines of force of the magnetic field. Hence, the lines of the field can be thought of as stretched elastic strings with mass equal to the mass of the fluid per line of force. D'Alembert's wave equation for the wave motion on a stretched string extending in the z direction is

$$\frac{\partial^2 y}{\partial t^2} = \frac{S}{m}\frac{\partial^2 y}{\partial z^2} \qquad (5\text{-}98)$$

where y = transverse displacement, m

S = tension or force, kg m sec^{-2}

m = mass per unit length, kg m^{-1}

By dimensional analysis of (5-98) the factor S/m is seen to be equal to the square of the velocity v of the wave. Thus,

$$v = \sqrt{\frac{S}{m}} \qquad (5\text{-}99)$$

To calculate S for the case of a magnetic-field line, let S be the force per weber of flux or H (newton weber^{-1}). By dimensional analysis of (5-99) the quantity m in the magnetic-field case is found to be mass per unit length per weber of flux density. Thus,

$$m = \frac{\text{kg}}{\text{meter weber}} = \frac{\text{kg}}{\text{meter}^3\, \dfrac{\text{weber}}{\text{meter}^2}} = \frac{\rho}{B}$$

and

$$v = \sqrt{\frac{HB}{\rho}} = \sqrt{\frac{B^2}{\mu\rho}} = \frac{B}{\sqrt{\mu\rho}} \qquad (\text{m sec}^{-1}) \qquad (5\text{-}100)$$

which is the same result obtained in (5-97).

It is of interest to evaluate (5-97) for the case of the solar photosphere where appropriate values are

$$\rho = 2 \times 10^{-4} \text{ kg m}^{-3} \qquad \text{and} \qquad B_0 = 0.1 \text{ weber m}^{-2}$$

Taking $\mu = \mu_0 = 4\pi \times 10^{-7}$ henry m^{-1} yields a velocity of 6.3 km sec^{-} for a magnetohydrodynamic wave in the photosphere. In the solar corona at about 1 solar radius above the photosphere the value of ρ is much less and a velocity close to that of the velocity of light is obtained.

References

ALFVÉN, H.: "Cosmical Electrodynamics," Oxford University Press, Fair Lawn, N.J. 1950.

ALFVÉN, H., and C. G. FALTHAMMAR: "Cosmical Electrodynamics," 2d ed., Oxford University Press, Fair Lawn, N.J., 1963.

ALLIS, W. P.: Propagation of Waves in a Plasma in a Magnetic Field, *IRE Trans Microwave Theory Tech.*, vol. MTT-9, p. 79, January, 1961.

APPLETON, E. V.: Wireless Studies of the Ionosphere, *J. Inst. Elec. Engrs., London,* vol. 71, pp. 642–650, 1932.

CAMBEL, A. B.: "Plasma Physics and Magnetofluidmechanics," McGraw-Hill Book Company, New York, 1963.

GARDNER, F. F., and J. B. WHITEOAK: Polarization of Radio Sources and Faraday Rotation Effects in the Galaxy, *Nature*, vol. 197, pp. 1162–1164, Mar. 23, 1963.

GINZBURG, V. L.: "Propagation of Electromagnetic Waves in Plasma," Gordon and Breach, Science Publishers, Inc., New York, 1961.

HARTREE, D. R.: The Propagation of Electromagnetic Waves in a Refracting Medium in a Magnetic Field, *Proc. Cambridge Phil. Soc.*, vol. 27, pp. 143–162, 1930–1931.

HOGAN, C. L.: The Ferromagnetic Faraday Effect at Microwave Frequencies and Its Applications: The Microwave Gyrator, *Bell System Tech. J.*, vol. 31, pp. 1–31, January, 1952.

HOLLINGER, J. P., C. H. MAYER, and R. A. MENNELLA: Polarization of Cygnus A and Other Sources at 5 cm, *Astrophys. J.*, vol. 140, pp. 656–665, Aug. 15, 1964.

JORDAN, E. C.: "Electromagnetic Waves and Radiating Systems," Prentice-Hall, Inc., Englewood Cliffs, N.J., 1950.

KOUYOUMJIAN, R. G.: Lecture notes for electromagnetic theory course, Ohio State University, and private communication, 1964.

KRAUS, J. D.: "Electromagnetics," McGraw-Hill Book Company, New York, 1953.

MAYER, C. H., T. P. MCCULLOUGH, and R. M. SLOANAKER: Linear Polarization of the Centimeter Radiation of Discrete Sources, *Astrophys. J.*, vol. 139, pp. 248–268, Jan. 1, 1964.

MIMNO, H. R.: The Physics of the Ionosphere, *Rev. Mod. Phys.*, vol. 9, pp. 1–43, January, 1937.

MITRA, S. K.: The Upper Atmosphere, *Asiatic Soc. Calcutta*, 1952.

NICHOLS, H. W., and J. C. SCHELLENG: Propagation of Electric Waves over the Earth, *Bell System Tech. J.*, vol. 4, pp. 215–234, 1925.

SPITZER, L.: "Physics of Fully Ionized Gases," Interscience Publishers, Inc., New York, 1956.

STIX, T. H.: "The Theory of Plasma Waves," McGraw-Hill Book Company, New York, 1962.

Problems

5-1. A particle with a negative charge of 10^{-18} coul and a mass of 10^{-24} kg is at rest in a field-free space. If a uniform electric field $\mathbf{E} = 100$ volt m^{-1} is applied, find the velocity of the particle 1 μsec later. *Ans.* 100 m sec^{-1}

5-2. What is the energy (in Mev) for protons moving in a circular path with a radius of 50 cm in a magnetic field of 1 weber m^{-2}? *Ans.* 12 Mev

5-3. A linear conductor carries a current of 100 amp in the positive x direction. If the flux density everywhere is equal to 2 webers m^{-2} in a direction parallel to the xy plane and at an angle of 45° with respect to the x axis, find the vector force on a 2-m length of the conductor. *Ans.* 282.8 newtons in the $+z$ direction

5-4. Prove (5-37) to (5-39).

5-5. Prove (5-57).

5-6. Prove (5-64).

5-7. Prove (5-75). *Hint:* Note that $\sqrt{1 + \delta} = 1 + (\delta/2)$, where δ is a small quantity, i.e., $\delta \ll 1$.

5-8. What is the electron density if a Faraday-rotation fading rate of three per minute is observed at 108 Mc from an artificial earth satellite traveling 7 km sec^{-1} toward the observer substantially parallel to the earth's magnetic field? Take the earth's field as 1 gauss. *Ans.* 1.1×10^{11} m^{-3}.

5-9. (a) Find $NB \cos \phi \, \Delta r$ if $\Delta\theta = 60°$ from polarization measurements at 10 and 20 cm wavelength. (b) If $B = 10^{-4}$ gauss and $\Delta r = 10$ pc, find N, assuming $\phi = 0$.

5-10. Calculate the velocity of an Alfvén wave in the solar corona assuming a magnetic field of 50 gauss and an electron density of 10^{12} m^{-3}. Assume that the medium is hydrogen and fully ionized.

6
Radio-Telescope Antennas

6-1 Introduction An antenna may be defined as the region of transition between a free-space wave and a guided wave (receiving case) or vice versa (transmitting case.)† The antenna of a radio telescope acts as a collector of radio waves. The antenna is analogous to the lens or mirror of an optical telescope.

The response of an antenna as a function of direction is given by the antenna *pattern*. By reciprocity this pattern is the same for both receiving and transmitting conditions.

The pattern commonly consists of a number of lobes, as suggested in Fig. 6-1a. The lobe with the largest maximum is called the *main lobe*, while the smaller lobes are referred to as the minor lobes or side and back lobes.

If the pattern is measured at a sufficient distance from the antenna so that an increase in the distance causes no change in the pattern, the pattern is the *far-field pattern*. Measurements at lesser distances yield *near-field patterns*, which are a function of both angle and distance. The pattern may be expressed in terms of the field intensity (*field pattern*) or in terms of the Poynting vector or radiation intensity (*power patterns*). Figure 6-1a is a power pattern in polar coordinates. To show the minor-lobe structure in more detail the pattern can be plotted on a logarithmic or decibel scale (decibels below main-lobe maximum). Figure 6-1b is an example of a pattern on a decibel scale in rectangular coordinates. The pattern in Fig. 6-1b is the same as the one in Fig. 6-1a.

A single pattern, as in Fig. 6-1, would be sufficient to completely specify the variation of radiation with angle provided the pattern is symmetrical. This would mean, in the case of Fig. 6-1a, that the three-dimensional pattern is a figure of revolution of the one shown around the pattern axis. If the pattern is not symmetrical, a three-dimensional diagram or a contour map is required to show the pattern in its entirety. However, in practice two patterns, one like that in Fig. 6-1a through the narrowest

† For a more general discussion of antennas and their basic properties see, for example, Kraus (1950).

part of the lobe and another perpendicular to it through the widest part of the lobe, may suffice. These mutually perpendicular patterns through the main-lobe axis are called the *principal-plane patterns*. The above statement assumes that the antenna is linearly polarized in one of the principal planes. If this is not the case, more patterns may be required.

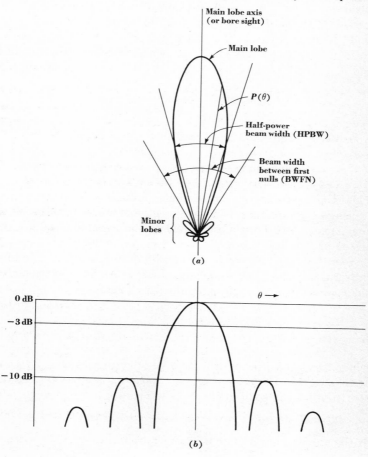

Fig. 6-1. (*a*) Antenna pattern in polar coordinates and linear power scale; (*b*) antenna pattern in rectangular coordinates and decibel power scale.

As an example, the dominant radiation from an antenna might be linearly polarized in one principal plane, but the radiation from some minor lobes might be cross-polarized, i.e., linearly polarized in the principal plane at right angles. Or the antenna might be elliptically polarized. A discussion of such patterns and their measurement is given by Kraus (1950, Chap. 15).

6-2 Beam Width, Beam Solid Angle, Directivity, and Effective Aperture

A useful numerical specification of the pattern can be made in terms of the angular width of the main lobe at a particular level. The angle at the half-power level or *half-power beam width* (HPBW) is the one most commonly used. The *beam width between first nulls* (BWFN) or the beam widths at -10 or -20 dB below the pattern maximum are also useful.

Another significant way of describing the pattern is in terms of *solid angle*. Let the relative antenna power pattern as a function of angle be given by $P(\theta,\phi)$ [$= E(\theta,\phi)E^*(\theta,\phi)$, where $E(\theta,\phi)$ is the far-field pattern] and its maximum value by $P(\theta,\phi)_{max}$†. Then (Kraus, 1950),

$$\Omega_A = \iint_{4\pi} P_n(\theta,\phi)\, d\Omega \tag{6-1}$$

where Ω_A = beam solid angle, rad²
$P_n(\theta,\phi) = P(\theta,\phi)/P(\theta,\phi)_{max}$ = normalized antenna power pattern, dimensionless
$d\Omega$ = elemental solid angle ($= \sin\theta\, d\theta\, d\phi$), rad²

The *beam solid angle* Ω_A is the angle through which all the power from a transmitting antenna would stream if the power (per unit solid angle) were constant over this angle and equal to the maximum value. This is suggested in Fig. 6-2.

In (6-1) the integration is carried out over a solid angle of 4π. If the integration is restricted to the main lobe, as bounded by the first minimum, the *main-beam solid angle* is obtained.‡ Thus,

$$\Omega_M = \iint_{\substack{main \\ lobe}} P_n(\theta,\phi)\, d\Omega \tag{6-2}$$

where Ω_M = main-beam or main-lobe solid angle, rad²
It follows that the *minor-lobe solid angle* Ω_m is given by the difference of the (total) beam solid angle and the main-beam solid angle. That is,

$$\Omega_m = \Omega_A - \Omega_M \tag{6-3}$$

If the antenna has no minor lobes ($\Omega_m = 0$), we have $\Omega_A = \Omega_M$.

---

† $E^*(\theta,\phi)$ is the complex conjugate of $E(\theta,\phi)$. $P(\theta,\phi)$ is proportional to the Poynting vector $S(\theta,\phi) = E(\theta,\phi)E^*(\theta,\phi)/Z$, where Z = intrinsic impedance of the medium.

‡ Ω_A, the beam solid angle, should not be confused with Ω_M, the solid angle of the main lobe or main beam. Ω_A is the solid angle of the entire antenna pattern, so that *pattern solid angle* might be a more appropriate name for Ω_A. However, beam solid angle has been in wide use for many years for Ω_A.

In patterns for which no clearly defined minimum exists the extent of the main lobe may be somewhat indefinite, and an arbitrary level such as -20 dB can be used to delineate it.

Another important antenna parameter is the *directivity*, which may be defined as the ratio of the maximum radiation intensity (antenna transmitting) to the average radiation intensity, or

$$D = \frac{U(\theta,\phi)_{max}}{U_{avg}} \tag{6-4}$$

where $U(\theta,\phi)_{max}$ = maximum radiation intensity, watts rad^{-2}
U_{avg} = average radiation intensity, watts rad^{-2}

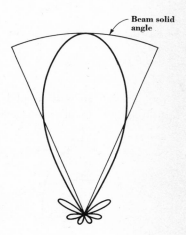

Fig. 6-2. Relation of beam solid angle to antenna pattern.

The average radiation intensity is given by the total power W radiated divided by 4π, and the total power is equal to the radiation intensity $U(\theta,\phi)$ integrated over 4π. Hence,

$$D = \frac{U(\theta,\phi)_{max}}{W/4\pi} = \frac{4\pi U(\theta,\phi)_{max}}{\displaystyle\iint_{4\pi} U(\theta,\phi)\, d\Omega} \tag{6-5}$$

or

$$D = \frac{4\pi}{\displaystyle\iint_{4\pi} \frac{U(\theta,\phi)}{U(\theta,\phi)_{max}}\, d\Omega} \tag{6-6}$$

Since the radiation intensity is proportional to the Poynting vector, we note from (6-1) that (6-6) can be expressed as

$$D = \frac{4\pi}{\displaystyle\iint_{4\pi} P_n(\theta,\phi)\, d\Omega} = \frac{4\pi}{\Omega_A} \tag{6-7}$$

Thus, the directivity of an antenna is equal to the solid angle of a sphere (4π) divided by the antenna beam solid angle.

The directivity of an antenna is a fixed numerical (dimensionless) quantity. Multiplying the directivity by the normalized power pattern yields the *directive gain*, a quantity which is a function of angle. Thus,

$$DP_n(\theta,\phi) = D(\theta,\phi) \tag{6-8}$$

where $D(\theta,\phi)$ = directive gain, dimensionless
Since $P_n(\theta,\phi)_{max} = 1$, it follows that

$$D = D(\theta,\phi)_{max} \tag{6-9}$$

From (6-7) and (6-8) it is also clear that

$$\iint\limits_{4\pi} D(\theta,\phi)\, d\Omega = 4\pi \tag{6-10}$$

Antenna patterns may be plotted in terms of directive gain, as in Fig. 6-3. For a nondirectional antenna the pattern would be everywhere equal to the

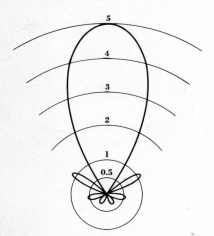

Fig. 6-3. Directive-gain pattern.

level $D(\theta,\phi) = 1$. This is called the *isotropic level*. In specifying the minor-lobe structure of an antenna the isotropic level is often a convenient reference.

In the foregoing discussion the directivity has been expressed entirely as a function of the antenna pattern with no reference to the size or geometry of the antenna. To show that the directivity is a function of the antenna size consider the far electric-field intensity E_r at a distance r in a direction broadside to a radiating aperture, as in Fig. 6-4. If the field intensity in the aperture is constant and equal to E_a (volts per meter), the power W radiated is given by

$$W = \frac{|E_a|^2}{Z} A \qquad (6\text{-}11)$$

where A = antenna aperture, m²

$\quad\;\; Z$ = intrinsic impedance of the medium, ohms square⁻¹

The power radiated may also be expressed in terms of the field intensity E_r (volts m⁻¹) at a distance r by

$$W = \frac{|E_r|^2}{Z} r^2 \Omega_A \qquad (6\text{-}12)$$

where Ω_A = beam solid angle of antenna, rad²

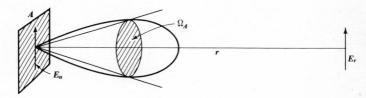

Fig. 6-4. Radiation from aperture A with uniform field E_a.

It is shown in Sec. 6-7 that the field intensities E_r and E_a are related by

$$|E_r| = \frac{|E_a|A}{r\lambda} \qquad (6\text{-}13)$$

where λ = wavelength, m

Substituting (6-13) in (6-12) and equating (6-11) and (6-12) yields

$$\lambda^2 = A\Omega_A \qquad (6\text{-}14)$$

where λ = wavelength, m

$\quad\;\; A$ = antenna aperture, m²

$\quad\;\; \Omega_A$ = beam solid angle, rad²

In (6-14) the aperture A is the physical aperture A_p if the field is uniform over the aperture, as assumed, but in general A is the *effective aperture A_e*. Thus, more generally,

$$\lambda^2 = A_e\Omega_A \qquad (6\text{-}15)$$

where A_e = effective aperture, m²

According to this important relation, the product of the effective aperture of the antenna and the antenna beam solid angle is equal to the wavelength squared. From (6-15) and (6-7) we have that

$$D = \frac{4\pi}{\lambda^2} A_e \qquad (6\text{-}16)$$

Three expressions have been given in this section for the directivity D of the antenna. They are

$$D = \frac{U(\theta,\phi)_{\max}}{U_{\text{avg}}} \tag{6-17}$$

$$D = \frac{4\pi}{\Omega_A} \tag{6-18}$$

$$D = \frac{4\pi}{\lambda^2} A_e \tag{6-19}$$

The resolution of an antenna in one plane is about equal to the half-power beam width in that plane. It then follows that the number of sources distributed uniformly over the sky which an antenna can resolve is given approximately by

$$N_r = \frac{4\pi}{\Omega_M} \tag{6-20}$$

Since $\Omega_M \leq \Omega_A$, a more conservative value would be

$$N_r = \frac{4\pi}{\Omega_A} = D \tag{6-21}$$

This is, however, an idealization. In practice the number which can be resolved unequivocally is probably an order of magnitude less than given by either (6-20) or (6-21), so that the directivity should be regarded (ideally) only as an upper limit value for the number of sources which an antenna can resolve.

A simple, useful relation for the directivity may be derived from (6-18). Thus,

$$D = \frac{4\pi}{\Omega_A} = \frac{4\pi\epsilon_M}{\Omega_M} = \frac{4\pi\epsilon_M}{k_p\,\theta_{\text{HP}}\,\phi_{\text{HP}}} = \frac{41{,}253\epsilon_M}{k_p\,\theta_{\text{HP}}{}^\circ\,\phi_{\text{HP}}{}^\circ} \tag{6-22}$$

where ϵ_M = beam efficiency (see Sec. 6-3 and 6-16); in most large antennas
$\epsilon_M = 0.75 \pm 0.15$
θ_{HP} = half-power beam width in θ plane, rad
ϕ_{HP} = half-power beam width in ϕ plane, rad
$\theta_{\text{HP}}{}^\circ$ = half-power beam width in θ plane, deg
$\phi_{\text{HP}}{}^\circ$ = half-power beam width in ϕ plane, deg
k_p = factor depending on pattern shape (see Sec. 6-16); typically
$k_p = 1.05 \pm 0.05$

6-3 Beam and Aperture Efficiencies The ratio of the main-beam solid angle to the (total) beam solid angle is called the main-beam efficiency or simply the *beam efficiency* ϵ_M. Thus,

$$\epsilon_M = \frac{\Omega_M}{\Omega_A} \tag{6-23}$$

The ratio of the minor-lobe solid angle to the (total) beam solid angle has been referred to as the *stray factor* ϵ_m, or

$$\epsilon_m = \frac{\Omega_m}{\Omega_A} \tag{6-24}$$

It follows that

$$\epsilon_M + \epsilon_m = 1 \tag{6-25}$$

(It is sometimes useful to define a beam efficiency that involves the beam solid angle of the main beam plus the near side lobes. This would result in a beam efficiency $\epsilon_M{}'$, somewhat greater than ϵ_M).

The ratio of the effective aperture to the physical aperture is the *aperture efficiency*, as given by

$$\epsilon_{ap} = \frac{A_e}{A_p} \tag{6-26}$$

The aperture efficiency is, in general, different from the beam efficiency. Their ratio is given by

$$\frac{\epsilon_{ap}}{\epsilon_M} = \frac{A_e \Omega_A}{A_p \Omega_M} = \frac{\lambda^2}{A_p \Omega_M} \tag{6-27}$$

The subject of efficiencies is discussed further in Sec. 6-16.

6-4 Array Theory: Two Point Sources Consider two in-phase point sources separated by a distance L, as in Fig. 6-5a. A point source is an idealization representing here an isotropic radiator occupying zero volume. By reciprocity the pattern of arrays of such sources (transmitting case) will be identical with the pattern when the array is used as a receiving antenna. Taking the reference point for phase halfway between the sources, the far field in the direction ϕ is given by

$$E = E_2 e^{i\psi/2} + E_1 e^{-i\psi/2} \tag{6-28}$$

where $\psi = \beta L \sin\phi = \dfrac{2\pi L}{\lambda} \sin\phi$ $\beta = \dfrac{2\pi}{\lambda}$

If $E_1 = E_2 = E_0$,

$$E = 2E_0 \frac{e^{i\psi/2} + e^{-i\psi/2}}{2} = 2E_0 \cos\frac{\psi}{2} \tag{6-29}$$

For a spacing $L = \lambda/2$ the pattern is as shown in Fig. 6-5.

It was assumed that each point source was isotropic (completely nondirectional). If the individual point sources have directional patterns which are identical, the resultant pattern is given by (6-29), where E_0 is now also a function of angle [$E_0 = E(\phi)$]. The pattern $E(\phi)$ may be called the *primary pattern* and $\cos \psi/2$ in (6-29) the *secondary pattern* or *array factor*. This is an example of the principle of *pattern multiplication*, which may be stated in more general terms as follows (Kraus, 1950):

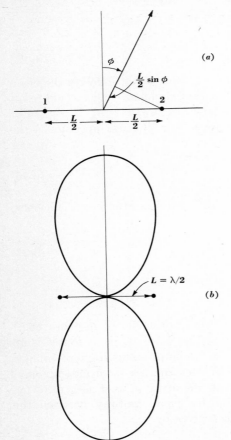

Fig. 6-5. (*a*) Geometry for array of two isotropic point sources, (*b*) field pattern of two in-phase isotropic point sources with one-half wavelength spacing.

"The total field pattern of an array of nonisotropic but similar sources is the product of the individual source pattern and the pattern of an array of isotropic point sources each located at the phase center of the individual source and having the same relative amplitude and phase, while the total phase pattern is the sum of the phase patterns of the individual source and the array of isotropic point sources."

If the reference for phase for the two sources in Fig. 6-5 had been taken at source 1, the resultant far field would be

$$E = E_1 + E_2 e^{j\psi} \tag{6-30}$$

and if $E_1 = E_2 = E_0$, as before,

$$E = 2E_0 \left(\cos \frac{\psi}{2} \right) e^{j\psi/2} = 2E_0 \cos \frac{\psi}{2} \, \underline{/\psi/2} \qquad (6\text{-}31)$$

The field (amplitude) pattern is the same as before, but the phase pattern is not. This is because the reference was taken at the *phase* center (midpoint of array) in developing (6-29) but at one end of the array in developing (6-31).

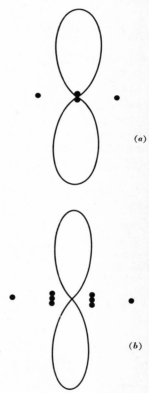

(a)

(b)

Fig. 6-6. (*a*) Binomial array with source amplitudes 1:2:1; (*b*) binomial array with source amplitudes 1:3:3:1. The spacing between sources is one-half wavelength.

6-5 Array Theory: The Binomial Array The relative far-field pattern of two equal in-phase isotropic point sources spaced one-half wavelength apart is given by:

$$E = \cos \left(\frac{\pi}{2} \sin \phi \right) \qquad (6\text{-}32)$$

as shown in Fig. 6-5*b*. This pattern has no minor lobes. If a second identical array of two sources is placed one-half wavelength from the first, the arrangement shown in Fig. 6-6*a* is obtained. The two sources at the center should be superimposed but are shown separated for clarity. By the principle of pattern multiplication the resultant pattern is given by

$$E = \cos^2 \left(\frac{\pi}{2} \sin \phi \right) \tag{6-33}$$

as shown in Fig. 6-6a. If this three-source array with amplitudes 1:2:1 is arrayed with an identical one at a spacing of one-half wavelength, the arrangement of Fig. 6b is obtained with the pattern

$$E = \cos^3 \left(\frac{\pi}{2} \sin \phi \right) \tag{6-34}$$

as shown in Fig. 6-6b. This array has effectively four sources with amplitudes 1:3:3:1, and it also has no minor lobes.

Continuing this process it is possible to obtain a pattern with arbitrarily high directivity and no minor lobes if the amplitudes of the sources in the array correspond to the coefficients of a binomial series (Stone). These coefficients are conveniently displayed by Pascal's triangle (Table 6-1). Each internal integer is the sum of the adjacent ones above. The pattern of the array is then

$$E = \cos^{n-1} \left(\frac{\pi}{2} \sin \phi \right) \tag{6-35}$$

where n is the total number of sources.

Table 6-1
Pascal's triangle

					1							
				1		1						
			1		2		1					
		1		3		3		1				
	1		4		6		4		1			
1		5		10		10		5		1		
1	6		15		20		15		6		1	

Although the above array has no minor lobes, its directivity is less than that of an array of the same size with equal amplitude sources. In practice most arrays are designed as a compromise between these extreme cases (Kraus, 1950).

6-6   Array Theory: n Sources of Equal Amplitude and Spacing
The binomial array of Sec. 6-5 is a nonuniform array. Turning now to a uniform array, as in Fig. 6-7, with n isotropic sources of equal amplitude and spacing, the far field is

$$E = E_0[1 + e^{j\psi} + e^{j2\psi} + \cdot \cdot \cdot + e^{j(n-1)\psi}] \tag{6-36}$$

or

$$E = E_0 \sum_{n=1}^{n=N} e^{j(n-1)\psi} \tag{6-37}$$

where $\psi = \beta d \sin \phi + \delta$
    d = spacing between sources
    δ = progressive phase difference between sources

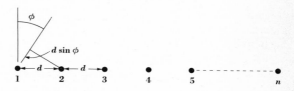

Fig. 6-7. Array of n isotropic sources of equal amplitude and spacing.

Multiplying (6-36) by $e^{j\psi}$ yields

$$Ee^{j\psi} = E_0(e^{j\psi} + e^{j2\psi} + e^{j3\psi} + \cdot \cdot \cdot + e^{jn\psi}) \tag{6-38}$$

Subtracting (6-38) from (6-36), we have

$$E = E_0 \frac{1 - e^{jn\psi}}{1 - e^{j\psi}} = E_0 \frac{\sin (n\psi/2)}{\sin (\psi/2)} \bigg/ (n - 1) \frac{\psi}{2} \tag{6-39}$$

If the center of the array is chosen as the reference for phase, instead of source 1, the phase angle, $(n - 1)\psi/2$, is eliminated. If the sources are nonisotropic but similar, E_0 will represent the primary or individual source pattern, while $\sin (n\psi/2)/\sin (\psi/2)$ is the array factor.

For isotropic sources and the center of the array as reference the pattern is

$$E = E_0 \frac{\sin (n\psi/2)}{\sin (\psi/2)} \tag{6-40}$$

As $\psi \to 0$, (6-40) reduces to

$$E = nE_0 \tag{6-40a}$$

This is the maximum value of the field, which is n times the field from a single source. In the direction of the maximum the condition $\beta d \sin \phi = -\delta$ is satisfied. For a broadside array (maximum at $\phi = 0$) the sources must be in phase ($\delta = 0$). Dividing (6-40) by (6-40a) yields the normalized field pattern

$$E_n = \frac{E}{nE_0} = \frac{1}{n}\frac{\sin{(n\psi/2)}}{\sin{(\psi/2)}} \tag{6-41}$$

Referring to (6-39), the null directions of the pattern occur for $e^{jn\psi} = 1$ provided $e^{j\psi} \neq 1$. This requires that $n\psi = \pm 2k\pi$ or

$$\pm\frac{2k\pi}{n} = \beta d \sin\phi_0 + \delta \tag{6-42}$$

or

$$\phi_0 = \sin^{-1}\left[\left(\pm\frac{2k\pi}{n} - \delta\right)\frac{1}{d_r}\right] \tag{6-43}$$

where ϕ_0 = null angle
$d_r = \beta d = 2\pi d/\lambda$
$k = 1, 2, 3, \ldots$ (but $k \neq mn$, where $m = 1, 2, 3, \ldots$)

In a broadside array ($\delta = 0$) the null angles are given by

$$\phi_0 = \sin^{-1}\left(\pm\frac{2k\pi}{nd_r}\right) = \sin^{-1}\left(\pm\frac{k\lambda}{nd}\right) \tag{6-44}$$

If the array is large so that $nd \gg k\lambda$,

$$\phi_0 \simeq \pm\frac{k\lambda}{nd} = \pm\frac{k}{nd_\lambda} \simeq \pm\frac{k}{L_\lambda} \tag{6-45}$$

where $L_\lambda = L/\lambda$ = length of array, wavelengths
$L_\lambda = (n-1)d_\lambda \simeq nd_\lambda$ if n is large

The first nulls (ϕ_{01}) occur when $k = 1$. Hence, the beam width between first nulls (BWFN) is

$$\text{BWFN} = 2\phi_{01} = \frac{2}{L_\lambda}\text{ rad} = \frac{114\overset{\circ}{.}6}{L_\lambda} \tag{6-46}$$

The more commonly used parameter is the half-power beam width (HPBW), which is about one-half (more nearly 0.44) the beam width between first nulls (BWFN) of a long uniform array.† Thus

$$\text{HPBW} \simeq \frac{\text{BWFN}}{2} = \frac{1}{L_\lambda}\text{ rad} = \frac{57\overset{\circ}{.}3}{L_\lambda} \tag{6-47}$$

Consider a two-dimensional array of area L_1L_2. If $\Omega_A = \theta_{\text{FN}}\phi_{\text{FN}}/4$, we have that

$$D = \frac{4\pi}{\Omega_A} = \frac{4\pi}{\lambda^2}L_1L_2 = \frac{4\pi}{\lambda^2}A_p \tag{6-48}$$

† See Appendix 6 for table of beam-width relations.

where L_1 = length of array in θ direction

L_2 = length of array in ϕ direction

θ_{FN} = beam width between first nulls in θ direction

ϕ_{FN} = beam width between first nulls in ϕ direction

A_p = physical aperture of the broadside array ($= L_1 L_2$)

From (6-48) we also have for the uniform array that

$$\Omega_A A_p = \lambda^2 \qquad (6\text{-}49)$$

In this calculation for the uniform (equal-amplitude) broadside array, it is assumed that the array is unidirectional, radiating only in one broadside direction.†

The effect of removing sources in a random manner from a large array has been treated by Maher and Cheng (1963) and by Lo (1963). A reduction in the number of sources can reduce the cost of the array. The effect of the reduction is expressed in terms of the probabilities that the beam width or side-lobe level will not increase by more than given amounts.

6-7 Continuous Aperture Distribution Consider now a continuous-current sheet or field distribution over an aperture as in Fig. 6-8. Assuming that the current or field is perpendicular to the page (y direction) and is uniform with respect to y, the electric field at a distance r from an elemental aperture $dx\,dy$ is‡

$$dE = -j\omega\,d\mathcal{A}_y = -\frac{j\omega\mu}{4\pi r}\frac{E(x)}{Z}\,e^{-j\beta r}\,dx\,dy \qquad (6\text{-}50)$$

where $\mathcal{A}_y$ = vector potential $\left(= \dfrac{\mu}{4\pi}\iiint\dfrac{J_y}{r}\,dv, \text{ in general}\right)$, volts sec m^{-1}

J_y = current density, amp m^{-2}

$E(x)$ = aperture electric-field distribution, volts m^{-1}

Z = intrinsic impedance of medium, ohms square^{-1}

ω = $2\pi\nu$ (ν = frequency), rad sec^{-1}

μ = permeability of medium, henrys m^{-1}

For an aperture with a uniform dimension y_1 perpendicular to the page and with the field distribution over the aperture a function only of x, the

---

† It is to be noted that the condition $\Omega_A = \theta_{\text{FN}}\phi_{\text{FN}}/4$ in (6-48) yields $\Omega_A = 1.3\theta_{\text{HP}}\phi_{\text{HP}}$. From (6-22) we have in a typical case about the same result for Ω_A.

‡ Note that $E(x)/Z = E_y(x)/Z = H_x = J_y z$, where z is the thickness of the current sheet. Also $dE = dE_y$.

electric field as a function of ϕ at a large distance from the aperture $(r \gg a)$, is, from (6-50),

$$E(\phi) = \frac{-j\omega\mu y_1 e^{-j\beta r_0}}{4\pi r_0 Z} \int_{-a/2}^{+a/2} E(x)e^{j\beta x \sin \phi} \, dx \tag{6-51}$$

The magnitude of $E(\phi)$ is then

$$|E(\phi)| = \frac{y_1}{2r_0\lambda} \int_{-a/2}^{+a/2} E(x)e^{j\beta x \sin \phi} \, dx \tag{6-52}$$

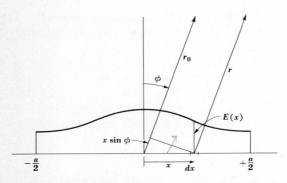

Fig. 6-8. Aperture of width a and amplitude distribution $E(x)$.

where $\beta = 2\pi/\lambda$. For a uniform aperture distribution $[E(x) = E_a]$ (6-52) reduces to

$$|E(\phi)| = \frac{y_1 E_a}{2r_0\lambda} \int_{-a/2}^{+a/2} e^{j\beta x \sin \phi} \, dx \tag{6-53}$$

on axis $(\phi = 0)$ we have

$$|E(\phi)| = \frac{E_a a y_1}{2r_0\lambda} = \frac{E_a A}{2r_0\lambda} \tag{6-54}$$

where A = aperture area $(= ay_1)$

   E_a = electric field in aperture plane

For unidirectional radiation from the aperture (in direction $\phi = 0$ but not in direction $\phi = 180°$) $|E(\phi)|$ is twice the value given in (6-54), or the same as used in (6-13).

   Integration of (6-53) gives

$$|E(\phi)| = k_0 \frac{\sin\left[(\beta a/2)\sin\phi\right]}{(\beta a/2)\sin\phi} \tag{6-55}$$

where

$$k_0 = \frac{AE_a}{2r_0\lambda} \tag{6-56}$$

Introducing $\beta d \sin \phi$ for ψ in (6-40), the field of a large array of n discrete sources of spacing d is

$$E = nE_0 \frac{\sin [(\beta a'/2) \sin \phi]}{(\beta a'/2) \sin \phi} \tag{6-57}$$

where the length of the large array is $a' = (n - 1)d \simeq nd$. It is also assumed in (6-57) that ϕ is restricted to small angles. This is not an undue restriction if the array is large and only the main lobe and first side lobes are of interest. Under these conditions it is clear that the field pattern (6-57) of the large array of discrete sources is the same as the pattern (6-55) for the continuous array of the same length ($a = a'$).

6-8 Fourier-transform Relations between the Far-field Pattern and the Aperture Distribution A one-dimensional† aperture distribution $E(x_\lambda)$ and its far-field distribution $E(\sin \phi)$ are reciprocal Fourier transforms (Booker and Clemmow, 1950) as given by

$$E(\sin\phi) = \int_{-\infty}^{\infty} E(x_\lambda)e^{j2\pi x_\lambda \sin \phi} \, dx_\lambda \tag{6-58}$$

and

$$E(x_\lambda) = \int_{-\infty}^{\infty} E(\sin \phi)e^{-j2\pi x_\lambda \sin \phi} \, d(\sin \phi) \tag{6-59}$$

where $x_\lambda = x/\lambda$. For real values of ϕ, $|\sin \phi| \leq 1$, the field distribution represents radiated power, while for $|\sin \phi| > 1$ it represents reactive or stored power (Rhodes, 1963). The field distribution $E(\sin \phi)$, or angular spectrum, refers to an angular distribution of plane waves. Except for $|\sin \phi| > 1$ the angular spectrum for a finite aperture is the same as the far-field pattern $E(\phi)$ (the far-field condition $r \gg a$ does not hold for an infinite aperture, i.e., where $a = \infty$). Thus, for a finite aperture the Fourier-integral representation of (6-58) may be written

$$E(\phi) = \int_{-a_\lambda/2}^{+a_\lambda/2} E(x_\lambda)e^{j2\pi x_\lambda \sin \phi} \, dx_\lambda \tag{6-60}$$

This is identical with (6-51) except for constant factors. Equation (6-51) is an absolute relation, whereas (6-60) is relative. Examples of the far-field patterns $E(\phi)$ for several aperture distributions $E(x_\lambda)$ of the same extent are presented in Fig. 6-9. (See also Appendix 6).

Taking the uniform distribution as reference, the more tapered distributions (triangular and cosine) have larger beam widths but smaller minor lobes, while the most gradually tapered distributions (cosine squared and

---

† For the more general two-dimensional case see Bracewell (1962a)

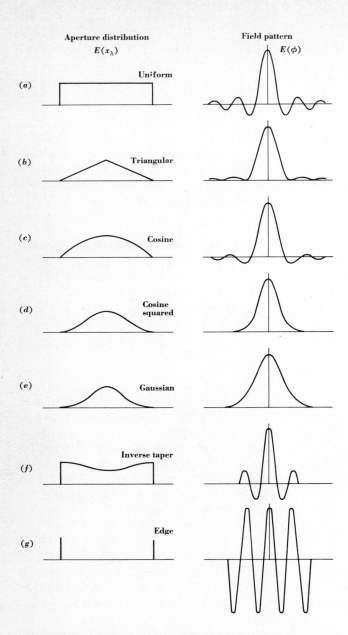

Fig. 6-9. Different aperture distributions with associated antenna patterns.

Gaussian) have still larger beam widths but no minor lobes. On the other hand, an inverse taper (less amplitude at the center than at the edge), such as shown in Fig. 6-9f, yields a smaller beam width but larger minor lobes than for the uniform distribution. Such an inverse taper might inadvertently result from aperture blocking due to a feed structure in front of the aperture. Carrying the inverse taper to its extreme limit results in the edge distribution of Fig. 6-9g. This distribution is equivalent to that of a two-element interferometer and has a beam width one-half that of the uniform distribution but side lobes equal in amplitude to the main lobe.

A useful property of (6-60) is that the distribution may be taken as the sum of two or more component distributions, $E_1(x_\lambda)$, $E_2(x_\lambda)$, etc., with the resulting pattern the sum of the transforms of these distributions. Thus,

$$E_1(\phi) + E_2(\phi) + \cdots$$
$$= \int_{-a_\lambda/2}^{+a_\lambda/2} E_1(x_\lambda)e^{j2\pi x_\lambda \sin \phi} \, dx_\lambda + \int_{-a_\lambda/2}^{+a_\lambda/2} E_2(x_\lambda)e^{j2\pi x_\lambda \sin\phi} \, dx_\lambda + \cdots$$

$$(6\text{-}61)$$

6-9 Spatial Frequency Response and Pattern Smoothing It has been shown (Booker and Clemmow, 1950) that the Fourier transform of the antenna power pattern is proportional to the complex autocorrelation function of the aperture distribution. Thus,

$$\bar{P}(x_{\lambda_0}) \propto \int_{-\infty}^{\infty} E(x_\lambda - x_{\lambda_0})E^*(x_\lambda) \, dx_\lambda \qquad (6\text{-}62)$$

where $\bar{P}(x_{\lambda_0})$ = Fourier transform of antenna power pattern $P_n(\phi) \propto$ autocorrelation function of aperture distribution

$E(\phi)$ = field pattern

$E(x_\lambda)$ = aperture distribution

$x_\lambda = x/\lambda$ = distance, wavelengths

$x_{\lambda_0} = x/\lambda_0$ = displacement, wavelengths

The autocorrelation function involves displacement x_{λ_0}, multiplication, and integration. The situation for a uniform aperture distribution is illustrated by Fig. 6-10. The aperture distribution is shown at (b) and as displaced by x_{λ_0} at (a). The autocorrelation function, as shown at (c), is proportional to the area under the product curve of the upper two distributions or, in this case, to the area of overlap. It is apparent that the autocorrelation function is zero for values of x_{λ_0} greater than the aperture width $a_\lambda[\bar{P}(x_{\lambda_0}) = 0$ for $|x_{\lambda_0}| > a_\lambda]$.

As discussed in Sec. 3-4, the observed response of a radio-telescope antenna to a sky brightness distribution is proportional to the convolution of the antenna power pattern and the brightness distribution. Thus,

$$S(\phi_0) = \int_{-\infty}^{\infty} B(\phi)\bar{P}_n(\phi_0 - \phi) \, d\phi \qquad (6\text{-}63)$$

where $S(\phi_0)$ = observed flux-density distribution
  $B(\phi)$ = true source brightness distribution
  $\tilde{P}_n(\phi)$ = mirror image of normalized antenna power pattern
  ϕ_0 = displacement, hour angle
It follows that

$$\bar{S}(x_\lambda) = \bar{B}(x_\lambda)\bar{P}(x_{\lambda_0}) \tag{6-64}$$

where the bars mean the Fourier transform. Since $\bar{P}(x_{\lambda_0})$ varies as the auto-correlation function of the aperture distribution, it follows that $\bar{S}(x_\lambda)$ and $S(\phi_0)$ are zero where $\bar{P}(x_{\lambda_0}) = 0$. This means that there is a cutoff for all

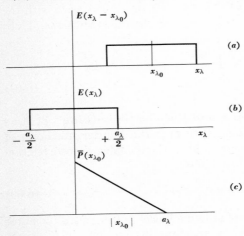

Fig. 6-10. The autocorrelation function of the aperture distribution yields the Fourier transform of the antenna pattern.

values of x_{λ_0} greater than a_λ (Bracewell and Roberts, 1954). The quantity x_{λ_0} is called the *spatial frequency*† and a_λ its *cutoff value*. Thus,

$$x_{\lambda_c} = a_\lambda \tag{6-65}$$

where x_{λ_c} = spatial-frequency cutoff
The reciprocal of x_{λ_c} gives an angle

$$\phi_c = \frac{1}{a_\lambda} \text{ rad} = \frac{57.3}{a_\lambda} \text{ deg} \tag{6-66}$$

Comparing (6-66) with (6-46), it follows that this (cutoff) angle ϕ_c is equal to one-half the beam width between first nulls for a uniform aperture

† Wavelengths per aperture.

distribution ($\phi_c = $ BWFN/2), and is 12 percent greater than the beam width at half power ($\phi_c = 1.12$ HPBW). The significance of ϕ_c is that structure in the sky brightness distribution having a period of less than BWFN/2 will not appear in the observed response. Thus, the antenna tends to smooth the true brightness distribution (Bracewell and Roberts, 1954). This is illustrated in Fig. 6-11a. Half of the beam width between

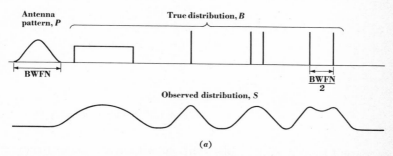

Fig. 6-11a. Smoothed distribution S observed with antenna pattern P.

Fig. 6-11b. Observed half-power width as a function of source width in half-power beam widths for a large uniform linear-aperture antenna and a uniform one-dimensional source.

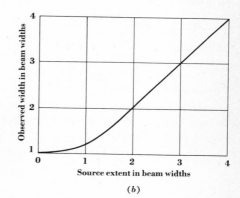

first nulls (BWFN/2) is equal to the *Rayleigh resolution*. Thus, two point sources separated by this distance will be just resolved, as indicated at the right in Fig. 6-11a.

The observed half-power width as a function of the source width in half-power beam widths for a large uniform linear-aperture antenna and a uniform one-dimensional source is shown in Fig. 6-11b. A source of half-power width equal to the antenna half-power beam width produces about 20 percent beam broadening, or an observed width of 1.2 beam widths. For larger source widths the observed width approaches the actual source width. Thus, from the amount of broadening an estimate may be made of the equivalent source extent.

6-10   The Simple (Adding) Interferometer   The resolution of a radio telescope can be improved, for example, by increasing the aperture a. However, this may not be economically feasible.   A less expensive approach to the problem is to use two antennas spaced a distance s apart, as in Fig. 6-12.   If each antenna has a uniform aperture distribution of width a, the resulting autocorrelation function is as shown in Fig. 6-13.   It is

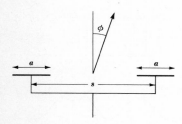

Fig. 6-12.   Simple interferometer.

apparent that by making observations with spacings out to s_λ it is possible to obtain higher spatial-frequency components in the observed pattern to a cutoff

$$x_{\lambda_c} = s_\lambda + a_\lambda \qquad (6\text{-}67)$$

and a smaller resolution angle

$$\phi_c = \frac{1}{s_\lambda + a_\lambda} \text{ rad} = \frac{57.3}{s_\lambda + a_\lambda} \text{ deg} \qquad (6\text{-}68)$$

In the following analysis it will be shown that if observations are made to sufficiently large spacings, it is possible, in principle, to deduce the true brightness distribution.

The normalized far-field pattern of the two-element array from (6-29) is

$$E(\phi) = E_n(\phi) \cos \frac{\psi}{2} \qquad (6\text{-}69)$$

where $E_n(\phi)$ = normalized field pattern of individual array element
$\psi = 2\pi s_\lambda \sin \phi$
The relative power pattern is equal to the square of $|E(\phi)|$, or

$$P(\phi) = |E(\phi)|^2 = |E_n(\phi)|^2 \cos^2 \frac{\psi}{2}$$

$$= |E_n(\phi)|^2 (1 + \cos \psi) \qquad (6\text{-}70)$$

For large spacings the pattern has many lobes, which, in optics, are referred to as fringes.   The first null occurs when $\psi = \pi$, from which the beam width between first nulls, or *fringe spacing*, is

$$\text{BWFN} = \frac{1}{s_\lambda}\text{ rad} = \frac{57.3}{s_\lambda}\text{ deg} \tag{6-71}$$

This is one-half the BWFN value for a continuous array of aperture width $\lambda = s_\lambda$ or a large array of discrete sources of the same length ($L_\lambda = s_\lambda$) as given by (6-46).

The pattern maxima occur when $\psi = 2\pi n$, where n ($=0, 1, 2, 3, \dots$) is the *fringe order*. Thus,

$$\phi_{\max} = \frac{n}{s_\lambda}\text{ rad} = \frac{57.3n}{s_\lambda}\text{ deg} \tag{6-72}$$

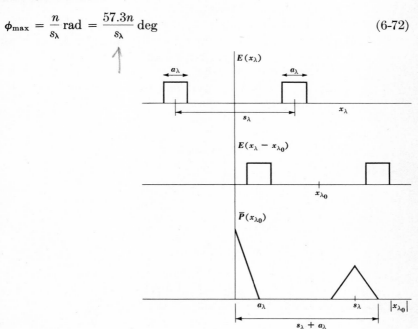

Fig. 6-13. Autocorrelation function of aperture distribution of simple interferometer.

Referring to Fig. 6-14, the first factor in (6-70) represents the individual-element pattern, as shown in (a), and the second factor the pattern of the array of two elements, as in (b). The product of the two factors gives the interferometer pattern, as indicated in (c). In these patterns a point source is implied. In the general case, for a source of angular extent α the observed flux density is the convolution of the true source distribution and the antenna power pattern, as discussed in Sec. 3-4. Assuming that the source extent is small compared to the individual-element pattern, so that $|E_n(\phi)|$ is essentially constant across the source, we have in the one-dimensional case that

$$S(\phi_0, s_\lambda) = |E_n(\phi)|^2 \int_{-\alpha/2}^{+\alpha/2} B(\phi)\{1 + \cos\left[2\pi s_\lambda \sin\left(\phi_0 - \phi\right)\right]\}d\phi$$

$$= |E_n(\phi)|^2 \Big\{ \int_{-\alpha/2}^{+\alpha/2} B(\phi)\, d\phi$$

$$+ \int_{-\alpha/2}^{+\alpha/2} B(\phi)\cos\left[2\pi s_\lambda \sin\left(\phi_0 - \phi\right)\right] d\phi \Big\}$$

$$= |E_n(\phi)|^2 \Big\{ S_0 + \int_{-\alpha/2}^{+\alpha/2} B(\phi)\cos\left[2\pi s_\lambda \sin\left(\phi_0 - \phi\right)\right] d\phi \Big\} \quad (6\text{-}73$$

where $S(\phi_0, s_\lambda)$ = observed flux-density distribution, watts m^{-2} cps^{-1}
$B(\phi)$ = true source brightness distribution, watts m^{-2} cps^{-1} rad^{-1}
ϕ_0 = displacement angle ($=$ hour angle), rad
α = source extent, rad
s_λ = s/λ (where s = interferometer element spacing)
S_0 = flux density of source

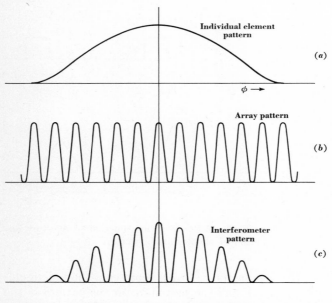

Fig. 6-14. (*a*) Individual-element pattern; (*b*) array pattern; and (*c*) the resultant interferometer pattern for the case of a point source.

The observed distribution as a function of hour angle is shown i
Fig. 6-15 for three cases: Fig. 6-15*a*, source extent very small compare
with the lobe spacing ($\alpha \ll 1/s_\lambda$), the same as in Fig. 6-14; Fig. 6-15*b*
source extent comparable to, but smaller than, the lobe spacing ($\alpha < 1/s_\lambda$)

and Fig. 6-15c, source distribution uniform and equal in extent to the lobe spacing ($\alpha = 1/s_\lambda$).

Assuming that the observations are made at the meridian or that

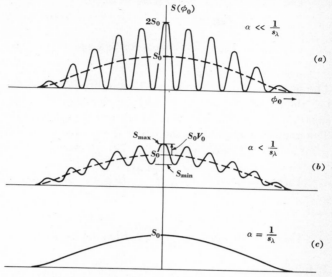

Fig. 6-15. Interferometer pattern (a) for point source; (b) for a uniform extended source of angle $\alpha < 1/s_\lambda$; and (c) for a uniform extended source of angle $\alpha = 1/s_\lambda$.

the source is tracked by the individual array elements, so that $|E_n(\phi)|^2 = 1$, (6-73) becomes

$$S(\phi_0,s_\lambda) = S_0 + \int_{-\alpha/2}^{+\alpha/2} B(\phi) \cos\left[2\pi s_\lambda \sin(\phi_0 - \phi)\right] d\phi \qquad (6\text{-}74)$$

If the source is small, so that $\phi_0 - \phi \ll \pi$, we may write

$$S(\phi_0,s_\lambda) = S_0 + \cos 2\pi s_\lambda \phi_0 \int_{-\alpha/2}^{+\alpha/2} B(\phi) \cos 2\pi s_\lambda \phi \, d\phi$$
$$+ \sin 2\pi s_\lambda \phi_0 \int_{-\alpha/2}^{+\alpha/2} B(\phi) \sin 2\pi s_\lambda \phi \, d\phi \qquad (6\text{-}75)$$

Or $S(\phi_0,s_\lambda)$ may be expressed as the sum of a constant term and a variable term (sum of two terms). Thus,

$$S(\phi_0,s_\lambda) = S_0[1 + V(\phi_0,s_\lambda)] \qquad (6\text{-}76)$$

where

$$V(\phi_0,s_\lambda) = \frac{1}{S_0} \cos 2\pi s_\lambda \phi_0 \int_{-\alpha/2}^{+\alpha/2} B(\phi) \cos 2\pi s_\lambda \phi \, d\phi$$
$$+ \frac{1}{S_0} \sin 2\pi s_\lambda \phi_0 \int_{-\alpha/2}^{+\alpha/2} B(\phi) \sin 2\pi s_\lambda \phi \, d\phi \qquad (6\text{-}77)$$

The variable term may also be expressed as a cosine function with a displacement $\Delta\phi_0$. Thus,

$$V(\phi_0,s_\lambda) = V_0(s_\lambda) \cos [2\pi s_\lambda(\phi_0 - \Delta\phi_0)] \tag{6-78}$$

or

$$V(\phi_0,s_\lambda) = V_0(s_\lambda)(\cos 2\pi s_\lambda\phi_0 \cos 2\pi s_\lambda\Delta\phi_0 + \sin 2\pi s_\lambda\phi_0 \sin 2\pi s_\lambda\Delta\phi_0) \tag{6-79}$$

The quantity $V_0(s_\lambda)$ represents the amplitude of the observed lobe pattern, i.e., the *fringe amplitude*. It is also called the *fringe visibility* or simply the *visibility*. As a function of s_λ, it may be referred to as the *visibility function*. The angle $\Delta\phi_0$ represents the fringe displacement from the position with a point source. From (6-77) and (6-79) we have

$$V_0(s_\lambda) \cos 2\pi s_\lambda\Delta\phi_0 = \frac{1}{S_0} \int_{-\alpha/2}^{+\alpha/2} B(\phi) \cos 2\pi s_\lambda\phi \, d\phi \tag{6-80}$$

and

$$V_0(s_\lambda) \sin 2\pi s_\lambda\Delta\phi_0 = \frac{1}{S_0} \int_{-\alpha/2}^{+\alpha/2} B(\phi) \sin 2\pi s_\lambda\phi \, d\phi \tag{6-81}$$

It follows that

$$V_0(s_\lambda)e^{j2\pi s_\lambda\Delta\phi_0} = \frac{1}{S_0} \int_{-\alpha/2}^{+\alpha/2} B(\phi)e^{j2\pi s_\lambda\phi} \, d\phi \tag{6-82}$$

The quantity $V_0(s_\lambda)e^{j2\pi s_\lambda\Delta\phi_0}$ is called the *complex visibility function*. If the source is contained within a small angle, the limits can be extended to infinity without appreciable error, giving

$$V_0(s_\lambda)e^{j2\pi s_\lambda\Delta\phi_0} = \frac{1}{S_0} \int_{-\infty}^{+\infty} B(\phi)e^{j2\pi s_\lambda\phi} \, d\phi \tag{6-83}$$

According to (6-83), the complex visibility function is equal to the Fourier transform of the source brightness distribution (times $1/S_0$). By the inverse Fourier transform we obtain

$$B(\phi_0) = S_0 \int_{-\infty}^{+\infty} V_0(s_\lambda)e^{j2\pi s_\lambda\Delta\phi_0} e^{-j2\pi s_\lambda\phi_0} \, ds_\lambda \tag{6-84}$$

or

$$B(\phi_0) = S_0 \int_{-\infty}^{+\infty} V_0(s_\lambda)e^{-j2\pi s_\lambda(\phi_0-\Delta\phi_0)} \, ds_\lambda \tag{6-85}$$

According to (6-84) and (6-85), the true brightness distribution of a source may be obtained, in principle, as the Fourier transform of the complex visibility function (an observable quantity).

To do this in practice requires observations at suitable intervals out
to sufficiently large spacings, a high source signal-to-noise ratio, and no
other (confusing) sources of significant flux density in the individual-
element response pattern. Thus, there are practical limits to the detail
with which the source distribution can be determined. According to
Bracewell (1958), the spacing interval need be no smaller than $1/\alpha$, where
α is the full source extent.

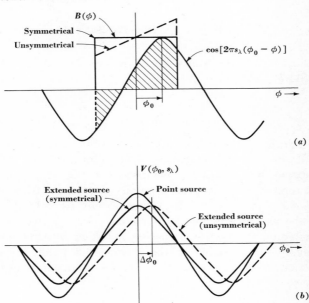

Fig. 6-16. Interferometer patterns for symmetrical and
unsymmetrical source distributions.

Referring to Fig. 6-15b, the visibility may be read from the observed
record as

$$V_0(s_\lambda) = \frac{S_{\max} - S_{\min}}{S_{\max} + S_{\min}} \qquad (6\text{-}86)$$

where

$$V_0(s_\lambda) = \text{visibility } [0 \leq V_0(s_\lambda) \leq 1]$$

Referring to Fig. 6-16a, the value of the integral in (6-74) is proportional
to the net shaded area, the areas above the ϕ axis being positive and the
areas below negative. This integral (times $1/S_0$) is the variable quantity
$V(\phi_0, s_\lambda)$, and its variation with respect to ϕ_0 for a fixed s_λ is a cosine function,

as suggested by the solid curves in Fig. 6-16b, one for a point source ($\alpha \to 0$) and the other for an extended source. For symmetrical source distributions (even functions) the fringe displacement is zero or one-half fringe ($\Delta\phi_0 = \frac{1}{2} s_\lambda$). For unsymmetrical sources, such as the one shown by the dashed lines in Fig. 6-16a, the fringes will have a displacement $\Delta\phi$ as suggested in Fig. 6-16b.

For symmetrical sources the visibility is from (6-80)

$$V_0(s_\lambda) = \pm \frac{1}{S_0} \int_{-\alpha/2}^{+\alpha/2} B(\phi) \cos 2\pi s_\lambda \phi \, d\phi \qquad (6\text{-}87)$$

For a uniform source [$B(\phi)$ = constant] and noting that $\alpha B(\phi) = S_0$ (6-87) reduces to

$$V_0(s_\lambda) = \pm \frac{\sin 2\pi s_\lambda(\alpha/2)}{2\pi s_\lambda(\alpha/2)} \qquad (6\text{-}88)$$

A graph of the visibility $V_0(s_\lambda)$ as a function of s_λ is presented in Fig. 6-1 for the case where the source is uniform and 1° in width. As the source extent becomes small compared to the fringe spacing ($\alpha \ll 1/s_\lambda$), the visibility $V_0(s_\lambda)$ approaches unity, but as the fringe spacing becomes very small compared to the source extent ($\alpha \gg 1/s_\lambda$), $V_0(s_\lambda)$ tends to zero. The visibility is also zero (for a uniform source) when the source extent is equal to the fringe spacing ($1/s_\lambda$) or integral multiples thereof. For symmetrical sources we have from (6-85) that the source brightness distribution is given by the Fourier cosine transform of the visibility function or

$$B(\phi_0) = S_0 \int_{-\infty}^{+\infty} V_0(s_\lambda) \cos 2\pi s_\lambda \phi_0 \, ds_\lambda \qquad (6\text{-}89)$$

Also from (6-79)

$$V(\phi_0, s_\lambda) = V_0(s_\lambda) \cos 2\pi s_\lambda \phi_0 \qquad (6\text{-}90)$$

so that another form for (6-89) is

$$B(\phi_0) = S_0 \int_{-\infty}^{+\infty} V(\phi_0, s_\lambda) \, ds_\lambda = 2S_0 \int_0^{+\infty} V(\phi_0, s_\lambda) \, ds_\lambda \qquad (6\text{-}91)$$

Curves of $V(\phi_0, s_\lambda)$ as a function of s_λ for several values of ϕ_0 are also shown in Fig. 6-17 for the uniform 1° source.

As an example suppose that the visibility of a uniform, 1°-wide source is observed with a simple interferometer. Then from (6-89) it should be possible to reobtain the original distribution. Thus, substituting (6-88) in (6-89)

$$B(\phi_0) = 2S_0 \int_0^\infty \cos 2\pi s_\lambda \phi_0 \frac{\sin 2\pi s_\lambda(\alpha/2)}{2\pi s_\lambda(\alpha/2)} \, ds_\lambda \qquad (6\text{-}92)$$

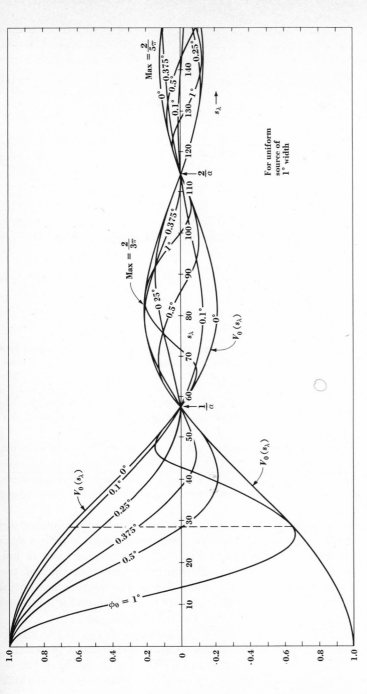

Fig. 6-17. Visibility functions vs. interferometer spacing s_λ for a uniform source of 1° width. At a fixed spacing s_λ, $V(\phi_0, s_\lambda)$ varies between the maximum and minimum curves $V_0(s_\lambda)$ as a function of the fringe drift or displacement ϕ_0. Thus, for $s_\lambda = 28.6$ (dashed line) $V(\phi_0, s_\lambda)$ is about 0.65 for $\phi_0 = 0°$, about 0.25 for $\phi_0 = 0°.375$ for $\phi_0 = 0°.5$, and about -0.65 for $\phi_0 = 1°$.

179

This equation has the form

$$B(\phi_0) = \frac{2S_0}{\alpha\pi} \int_0^\infty \cos mx \, \frac{\sin x}{x} \, dx \qquad (6\text{-}93$$

where $x = 2\pi s_\lambda \alpha/2$
$\quad\quad\; m = 2\phi_0/\alpha$

The definite integral (6-93) is well known and yields

$$B(\phi_0) = \frac{2S_0}{\alpha\pi} \frac{\pi}{2} = \frac{S_0}{\alpha} \qquad \text{watts m}^{-2} \text{ cps}^{-1} \text{ rad}^{-1}$$

for $-1 < m < +1$ and $B(\phi_0) = 0$ for $m < -1$ and $m > +1$. For a uniform source of width $\alpha = 1°$ the resulting source brightness $B_0 =$

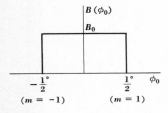

$(m = -1)$ $(m = 1)$

Fig. 6-18. Source distribution reconstructed from visibility functions.

$S_0/\alpha = 57.3 \, S_0$ watts m^{-2} cps^{-1} rad^{-1}, where S_0 = source flux density. The source has this brightness in the range $-\frac{1}{2}° < \phi_0 < +\frac{1}{2}° \, (-1 < m < +1$ and is zero outside, as indicated in Fig. 6-18.

In this hypothetical example we have gone full circle. Thus, it was shown that by scanning a source with an interferometer it was decomposed into a quantity called the visibility. Then by taking the Fourier transform of the visibility the original source brightness distribution was reconstructed (Fig. 6-18).

Examples of visibility functions for other source distributions are given in Fig. 6-19. The case of a uniform source of extent α with a symmetrical hole of extent β is shown in Fig. 6-19a for several cases of hole width. When β approaches α, the distribution approaches that of two point sources with a separation α. When $\beta = 0$, the source distribution is uniform. In Fig. 6-19b the visibility function is presented for a uniform source with a bright center of 4 times the side brightness.

It is to be noted that the visibility functions of Fig. 6-19a can be obtained as the visibility for a uniform source distribution of width α minus the visibility of a uniform source of width β, while in Fig. 6-19b the result can be obtained as the sum of two uniform distributions of width α and β.

Comparing (6-83) with (6-58) and (6-60), it is apparent that the complex visibility function is related to the source brightness distribution in the same manner that the far-field pattern of an antenna is related to the antenna aperture distribution. Accordingly, the graphs of Fig. 6-9 may also be interpreted as giving the visibility functions for various source distributions and field patterns for various aperture distributions in Fig. 6-19. The restriction holds that the source extent is small and the aperture extent large.

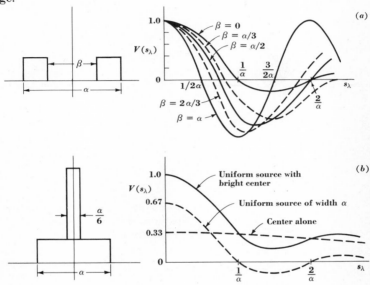

Fig. 6-19. (*a*) Visibility functions for a source of uniform brightness with holes of various widths β; (*b*) visibility function for a uniform source with a bright center.

For simplicity only one-dimensional distributions have been considered. This case is of considerable practical importance. The principles may be extended also to the more general two-dimensional case. A thorough treatment of the two-dimensional problem is given by Bracewell (1958).

In the above discussion monochromatic radiation at a single frequency was assumed. If the antennas and receiver respond uniformly over a bandwidth $\nu_0 \pm \Delta\nu/2$ with the radiation considered to be made up of mutually incoherent monochromatic components, the result for a point source is similar in form to the one above for a uniform source of width α, but with $\alpha/2$ replaced by $\Delta\nu/2$ (see Prob. 6-19). A result of too wide a bandwidth is that the higher-order fringes may be obliterated.

6-11   The Phase-switched (Multiplying) Interferometer   The in
terferometer of Fig. 6-12 discussed in Sec. 6-10 is of the simplest type, in
which the voltages of the two antenna elements are continuously added
If the phase of one of the elements is periodically reversed and the out
put of the receiver reversed in synchronism, as suggested in Fig. 6-20, a
phase-switched interferometer results, which gives the visibility without an

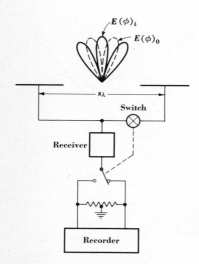

Fig. 6-20.   Phase-switched interferometer with
patterns shown for in-phase (solid) and out-of
phase (dashed) conditions.

additive constant in the output (Ryle, 1952).   One method of reversing
the phase of an array element is to insert (or remove) one-half wavelength
of transmission line.

When the antenna elements are in phase, the far-field pattern is

$$E(\phi)_i = E_0(\phi)[e^{j(\psi/2)} + e^{-j(\psi/2)}] \tag{6-94}$$

where $E_0(\phi)$ = normalized field pattern of individual array element

$$\psi = 2\pi s_\lambda \sin \phi$$

In the phase-reversed, or out-of-phase, condition the pattern is

$$E(\phi)_o = E_0(\phi)[e^{j(\psi/2)} - e^{-j(\psi/2)}] \tag{6-95}$$

The relative power patterns in the two cases are

$$P(\phi)_i = |E_0(\phi)|^2 [e^{j(\psi/2)} + e^{-j(\psi/2)}] [e^{j(\psi/2)} + e^{-j(\psi/2)}]^* \tag{6-96}$$

and

$$P(\phi)_o = |E_0(\phi)|^2 [e^{j(\psi/2)} - e^{-j(\psi/2)}] [e^{j(\psi/2)} - e^{-j(\psi/2)}]^* \tag{6-96a}$$

The system output is then proportional to the difference of the two patterns, so that the recorded pattern is given by

$$P(\phi) = P(\phi)_i - P(\phi)_o = 2|E_0(\phi)|^2(e^{j\psi} + e^{-j\psi}) \qquad (6\text{-}97)$$

or (normalized) by

$$\begin{aligned} P_n(\phi) &= |E_n(\phi)|^2 \cos \psi \\ &= |E_n(\phi)|^2 \cos (2\pi s_\lambda \sin \phi) \end{aligned} \qquad (6\text{-}98)$$

For $|E_n(\phi)|^2 \simeq 1$ the recorded flux density for a source of brightness distribution $B(\phi)$ and extent α is then

$$S(\phi_0, s_\lambda) = \int_{-\alpha/2}^{+\alpha/2} B(\phi) \cos [2\pi s_\lambda \sin (\phi_0 - \phi)] \, d\phi \qquad (6\text{-}99)$$

Thus, the phase-switched interferometer produces a fluctuating output of average value zero. In other words, there is no constant additive term ($= S_0$) as in (6-74). The outputs of a simple (unswitched) and a phase-switched interferometer are compared in Fig. 6-21.

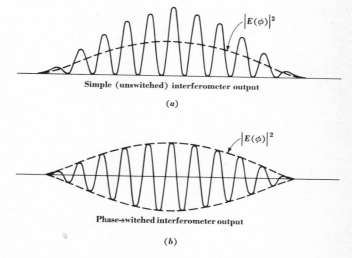

Simple (unswitched) interferometer output

(a)

Phase-switched interferometer output

(b)

Fig. 6-21. (a) Output of simple, unswitched interferometer; (b) output of phase-switched interferometer.

The autocorrelation function of the aperture distribution for a phase-switched interferometer is indicated in Fig. 6-22. This type of interferometer has the spatial-frequency characteristics of a bandpass filter. By contrast, the simple unswitched interferometer has the properties of a low-pass plus bandpass filter (Fig. 6-13), while a continuous-aperture antenna is like a low-pass filter (Fig. 6-10).

The elements in the interferometers discussed thus far have been assumed to be identical. Let us now consider the more general case of two dissimilar elements, 1 and 2, as in Fig. 6-23. For the case of phase switching, it can be readily shown that the relative power pattern is given by

$$P(\phi) = E_1(\phi)E_2^*(\phi)e^{j\psi} + E_1^*(\phi)E_2(\phi)e^{-j\psi} \tag{6-100}$$

where $E_1(\phi)$ = far-field pattern of element 1
$E_2(\phi)$ = far-field pattern of element 2
$\psi = 2\pi s_\lambda \sin\phi$
If the elements are identical $[E_1(\phi) = E_2(\phi)]$, (6-100) reduces to the same pattern as in (6-98). If the elements are symmetrical, so that $E_1(\phi) = E_1^*(\phi)$ and $E_2(\phi) = E_2^*(\phi)$ (all lobes of phase 0 or π), the relative power pattern (6-100) becomes

$$P(\phi) = E_1(\phi)E_2(\phi)\cos\psi \tag{6-101}$$

This is the general relation for a two-element phase-switched interferometer with symmetrical but dissimilar elements. The power pattern is seen to be proportional to the product of the field patterns of the individual elements. Hence, the phase-switched interferometer is sometimes referred to as a

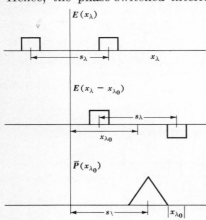

Fig. 6-22. The autocorrelation function of the aperture distribution of a phase-switched interferometer has the spatial frequency response $\bar{P}(x_{\lambda_0})$ of a bandpass filter.

multiplying type. The voltage delivered by one element is proportional to the square root of its aperture area, so that the received power and the overall system aperture are proportional to the geometric mean of the two element aperture areas. Thus,

$$P \propto \sqrt{A_1 A_2} \tag{6-102}$$

where P = received power
A_1 = aperture area of element 1
A_2 = aperture area of element 2

If the spacing between the elements is reduced to zero ($s_\lambda = 0$ in Fig. 6-23), so that the elements form a cross (Mills and Little, 1953; Mills, 1963), we have, neglecting constants, that (6-100) reduces to

$$P(\phi) = E_1(\phi)E_2^*(\phi) + E_1^*(\phi)E_2(\phi)$$
$$= \text{Re } E_1(\phi) \text{ Re } E_2(\phi) + \text{Im } E_1(\phi) \text{ Im } E_2(\phi) \tag{6-103}$$

If the elements are symmetrical, the patterns are entirely real, so that (6-103) reduces to

$$P(\phi) = E_1(\phi)E_2(\phi) \tag{6-104}$$

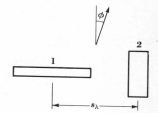

Fig. 6-23. Interferometer with two dissimilar elements.

6-12 The Multielement, or Grating, Interferometer

In Sec. 6-6, an array of n sources of equal amplitude and spacing was discussed, and the far-field pattern was shown to be

$$E(\phi) = E_0(\phi) \frac{\sin (n\psi/2)}{\sin (\psi/2)} \tag{6-105}$$

where $\psi = 2\pi\, d_\lambda \sin \phi$

d_λ = spacing between elements, wavelengths

ϕ = angle from perpendicular to array

n = number of sources

For a long array (nd_λ large) the beam width between first nulls (BWFN) of the main lobe (at $\phi = 0$) is given by

$$\text{BWFN} = \frac{2}{nd_\lambda} \qquad \text{rad} \tag{6-106}$$

If d_λ exceeds unity, side lobes appear which are equal in amplitude to the main lobe. These so-called *grating lobes* have a spacing from the main lobe of

$$\phi_G = \sin^{-1} \frac{1}{d_\lambda} \qquad \text{rad} \tag{6-107}$$

If $d_\lambda \gg 1$, this reduces approximately to

$$\phi_G = \frac{1}{d_\lambda} \qquad \text{rad} \tag{6-108}$$

Now if ϕ_G is greater than the source extent, only one large lobe of the pattern will be on the source at a time, as suggested in Fig. 6-24. To avoid confusion effects there must be no other sources of significant strength in the grating lobes. This multielement, or grating type of interferometer arrangement has been used by Christiansen and Warburton (1953) for the sun, the effect of a succession of single-lobe scans being obtained as the sun drifts through the pattern. The sun is a strong source, so that the grating lobes off the sun contribute relatively little to the received power.

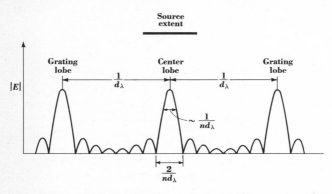

Fig. 6-24. Grating interferometer with grating-lobe spacing ϕ_G larger than source extent.

With a two-element interferometer of spacing s_λ equal to nd_λ, lobes or fringes would be obtained with a BWFN $= 1/nd_\lambda$, or half the width in (6-106). However, the next large lobe would be at the same angular distance $(= 1/nd_\lambda)$. The grating interferometer has a BWFN $= 2/nd_\lambda$ and a HPBW about half this, the same as a continuous array of length nd_λ. The grating-lobe spacing is $1/d_\lambda$, or n times as large as the HPBW. The source extent α, in the case of the sun, is about $\frac{1}{2}°$. Hence, for a grating interferometer d should be no more than 114 wavelengths. Taking $d = 100\lambda$ and $n = 32$ yields BWFN $= 2/nd_\lambda = 0°.036$ and HPBW $\simeq 1/nd_\lambda = 0°.018$. Thus, in this case the beam width is sufficiently small compared to the source extent to reveal considerable detail.

It is assumed that the directivity of the individual elements is not enough to affect the pattern near $\phi = 0$. To suppress all grating lobes including the first would require an individual-element aperture of the order of d_λ. This would put the first null on the first grating lobe, but the array is now equivalent to a continuous aperture. The advantage of the grating interferometer, also sometimes called a *suppressed-lobe interferometer*, is the economy of a partially filled aperture to give the narrow beam width (high resolution) of a long continuous aperture for the case of strong sources of extent less than $1/d_\lambda$. However, the sensitivity is less than for the continuous aperture, the effective aperture of the system being equal only to the sum of the effective apertures of the individual elements.

6-13   The Compound, Cross, and Other Interferometers Several interferometers are discussed in this section which are basically of the phase-switched type with dissimilar elements. In the first of these, shown in Fig. 6-25, element 1 consists of a continuous array of length $s_\lambda/2$, while

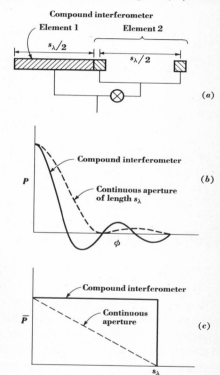

Fig. 6-25. (a) Compound interferometer consisting of continuous aperture of length $s_\lambda/2$ and simple interferometer of spacing $s_\lambda/2$; (b) power pattern of this interferometer compared with power pattern of uniform continuous aperture of length s_λ; and (c) spatial-frequency response of the compound interferometer compared with the uniform continuous aperture.

element 2 is itself a simple (unswitched) interferometer with element spacing $s_\lambda/2$. Such *compound interferometers* have been developed by Covington (1960; Covington and Harvey, 1959). Assuming that the elements of the simple interferometer have zero width, so that the entire length of the compound interferometer is s_λ, the relative power pattern is

$$P(\phi) = \frac{\sin(\psi/4)}{\psi/4} \cos \frac{\psi}{4} \cos \frac{\psi}{2} \qquad (6\text{-}109)$$

or

$$P(\phi) = \frac{\sin(\psi/2)}{\psi/2} \cos\frac{\psi}{2} = \frac{\sin\psi}{\psi} \tag{6-110}$$

where $\psi = 2\pi s_\lambda \sin\phi$

In (6-109) the first factor is the field pattern of the continuous aperture, the second factor the field pattern of the simple interferometer, and the third factor [$\cos(\psi/2)$] the separation pattern.

If this compound interferometer is replaced by a continuous uniform aperture of length s_λ, the relative power pattern (6-57) is

$$P(\phi) = \left|\frac{\sin(\psi/2)}{\psi/2}\right|^2 \tag{6-111}$$

where $\psi = 2\pi s_\lambda \sin\phi$

The power patterns of the compound interferometer (6-110) and the continuous array (6-111) are compared in Fig. 6-25*b* with their respective spatial-frequency characteristics in Fig. 6-25*c*. The compound interferometer (like phase-switched types in general) has positive and negative power-pattern side lobes. On actual records what appears to be the first null corresponds to the maximum of the first side lobe (negative). Regarding this point as the first null, the

$$\text{BWFN} = \frac{3}{2s_\lambda} \quad \text{rad} \tag{6-112}$$

For the continuous uniform aperture of the same overall length (s_λ)

$$\text{BWFN} = \frac{2}{s_\lambda} \quad \text{rad} \tag{6-113}$$

Thus, the compound interferometer's effective beam width is three-fourths that of the continuous aperture. But its side lobes are larger than for the continuous aperture. From the spatial-frequency characteristics in Fig. 6-25*c* it is apparent, however, that the pass band of the compound interferometer is uniform, whereas that for the continuous aperture is triangular. With a uniform pass band the Fourier components out to the cutoff value s_λ are unattenuated, and the observed profile gives the Fourier approximation (or principal solution) for a finite range of spatial frequencies. From (6-64) we have

$$\bar{S}(s_\lambda) = \bar{B}(s_\lambda)\bar{P}(s_\lambda) \tag{6-114}$$

so that if all the Fourier transform of $B(\phi)$ [$= \bar{B}(s_\lambda)$] is contained within the cutoff s_λ, the true source brightness distribution may be recovered. However, if $\bar{B}(s_\lambda)$ extends beyond the cutoff, as suggested in Fig. 6-26, the higher-frequency components will be lost in the result; i.e., the observed pattern will be smoothed (Bracewell and Roberts, 1954). Even more smoothing occurs with the uniform aperture, since its spatial-frequency response is triangular (Fig. 6-26). Another method of analysis of the interferometer of Fig. 6-25 is discussed in Prob. 6-18.

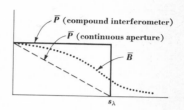

Fig. 6-26. Fourier transform $\bar{P}$ of antenna power pattern for compound interferometer (solid) and uniform-aperture antenna (dashed) compared with Fourier transform $\bar{B}$ of brightness distribution (dotted).

Another compound interferometer built by Covington (1960) is shown in Fig. 6-27. In this type a continuous array as element 1 is phase-switched with respect to a four-unit grating interferometer operating as element 2. Assuming that the elements of the grating interferometer have zero width, the relative power pattern is given by

$$P(\phi) = \frac{\sin (\psi/8)}{\psi/8} \cos \frac{\psi}{8} \cos \frac{\psi}{4} \cos \frac{\psi}{2}$$

$$= \frac{\sin \psi}{\psi} \qquad (6\text{-}115)$$

where $\psi = 2\pi s_\lambda \sin \phi$

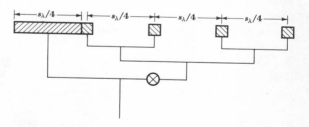

Fig. 6-27. Compound interferometer consisting of continuous aperture and four-unit grating interferometer.

In (6-115) (top line) the first factor is the field pattern of the continuous aperture, the second and third factors the field pattern of the grating interferometer, and the fourth factor the separation pattern. This pattern (6-115) is the same as given by (6-110) for the interferometer of Fig. 6-25, even though the total aperture area is half what it was for the system of Fig. 6-25.

Defining the ratio of actual aperture area to that occupied by a continuous aperture of the same overall length as the *filling factor,* we may say that for the system of Fig. 6-25 the filling factor is 50 percent, while for that of Fig. 6-27 it is 25 percent. It is assumed that the areas of the elements of the grating interferometer are small enough to be negligible. Although such an arrangement provides desirable pattern characteristics, the sensitivity may be unacceptably low if the total aperture of the grating interferometer is too small. Thus, from (6-102), the effective aperture of the system is proportional to the square root of the *product* of the aperture areas of the grating interferometer and the continuous aperture. Hence, in practice, apertures of appreciable size would be desirable for the grating-interferometer elements, resulting in filling factors in excess of 50 percent for the system of Fig. 6-25 and in excess of 25 percent for the system of Fig. 6-27. By using a grating interferometer of more elements the filling factor could be further reduced without affecting the pattern. It is necessary only for the continuous-aperture length to be equal to the element spacing of the grating interferometer.

Reducing the spacing between the dissimilar elements of the interferometer of Fig. 6-23 to zero results in a cross arrangement. Such interferometers, commonly called Mills crosses (Mills, 1963), are used extensively. The power pattern (6-103) has no separation factor, and for symmetrical elements the pattern is simply the product of the field patterns of the individual elements (6-104). If the elements are long and narrow, as suggested in Fig. 6-28a, so that the east-west element has a narrow beam and the north-south element a very broad beam in the east-west plane, the resulting *power pattern* for the system in the east-west plane is effectively the same as the *field pattern* of the east-west element. Hence, the beam width will be larger and the minor lobes higher than for a single aperture of the same size as the east-west element.

Omitting half of one element of a cross (north-south element in Fig. 6-28b) forms a T arrangement. Since the east-west element is symmetrical, the imaginary term in (6-103) is zero, and the pattern is given by

$$P(\phi) = \text{Re } E_1(\phi) \text{ Re } E_2(\phi)$$
$$= E_1(\phi) \text{ Re } E_2(\phi) \tag{6-116}$$

where $E_1(\phi)$ = far-field pattern of east-west element
$E_2(\phi)$ = far-field pattern of north-south element
Unless the aperture width of the north-south element is increased, the effective aperture of the T is less than that of the full cross, but the beam width is about the same.

The antenna structure may be further reduced, for example, if the north-south element of the T is replaced by a small movable element, as in Fig. 6-28c. A scan of the sky at each position of the movable element yields one Fourier (or visibility) component of the sky brightness distribution. By Fourier inversion a sky brightness map may then be obtained. This system is one of several variants developed and used by Ryle and his associates at Cambridge, England, and is referred to as an *aperture-synthesis method* (Ryle and Hewish, 1960). In its most basic form aperture synthesis may be considered as a method whereby the information-gathering capability of a large aperture is synthesized by measurements with an interferometer of two elemental apertures, one fixed and the other

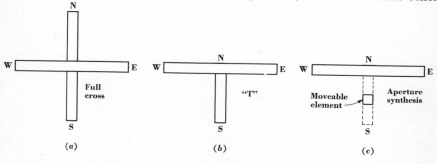

Fig. 6-28. (a) Full-cross interferometer; (b) partial cross or T; and (c) arrangement used in aperture synthesis.

movable to all positions within the large aperture. With the large complete aperture the incoming waves falling on each elemental area are added in phase at a focal point. In the aperture-synthesis method a large number of measurements are made with the movable elemental aperture at different positions and the results summed by a computer, which may also introduce suitable weighting and phasing for controlling the synthesized antenna pattern.

A two-element interferometer observing a point source has a power response which varies as $1 + \cos \psi$ for an adding type and as $\cos \psi$ for a multiplying type, where $\psi = 2\pi s_\lambda \sin \phi$. For a constant spacing s_λ between elements the output is a function of the direction angle ϕ of the source. If ϕ is constant (but not zero), a fluctuating output may still be obtained by varying s_λ. This may be done for a fixed physical separation s by sweeping the frequency, it being assumed that the source has a broad spectrum. By using unequal, instead of equal, lengths of cable to each element

$$\psi = 2\pi \frac{\nu}{c} \left(s \sin \phi + \frac{\Delta l}{p} \right) \tag{6-117}$$

where Δl = difference in cable lengths
  p = relative phase velocity in cables ($= v/c$)

A fluctuating output may be obtained even when $\phi = 0$, provided Δl is sufficiently large.   The fringe separation is given by

$$\Delta \nu = \frac{c}{s \sin \phi + (\Delta l/p)} \qquad (6\text{-}118)$$

where $\Delta \nu$ = frequency difference between fringes (on output-vs.-frequency
        presentation)

From (6-118) the position angle ϕ of the source is given, for small angles, by

$$\phi = \frac{1}{s}\left(\frac{c}{\Delta \nu} - \frac{\Delta l}{p}\right) \qquad (6\text{-}119)$$

Interferometers using this *swept-frequency principle* have been developed by Wild and Sheridan (Sheridan, 1963) for locating transient sources on the sun.

Fig. 6-29.   Three-helix swept-lobe array for planetary observations between 25 and 35 Mc at the Ohio State–Ohio Wesleyan Radio Observatory.

Another method of sweeping the interferometer lobe pattern is by continuously varying the length of the cable to one interferometer element. This may be accomplished, in effect, by inserting a continuous phase shifter in one cable so that

$$\psi = 2\pi s_\lambda \sin \phi + \phi_s \qquad (6\text{-}120)$$

where ϕ_s = angle introduced by phase shifter

If $\phi_s = -2\pi s_\lambda \sin \phi$, the interferometer pattern can be held fixed with respect to a moving source at the angle ϕ. Or if $\phi_s = 2\pi s_\lambda \sin \phi$, the fringe sweep rate will be doubled. Thus, by controlling ϕ_s the fringe pattern can be slowed down or speeded up as desired. Such *phase-shifting* or *lobe-sweeping* interferometers have been used by Little and Payne-Scott (1951) and by Brown, Palmer, and Thompson (1955). A lobe-sweeping antenna

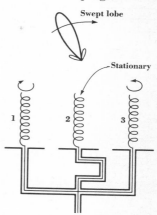

Fig. 6-30. Diagram of three-helix swept-lobe array. The outer two helices rotate in opposite directions.

has also been produced by rotating the helical elements of an antenna array (Kraus, 1958b). In this array, using three helical-beam antennas, as shown in the photograph of Fig. 6-29, a continuously swept lobe was obtained by rotating helix 1 clockwise and helix 3 counterclockwise while helix 2, at the center, remained stationary, as indicated in the diagram of Fig. 6-30. All helices (Kraus, 1950) were wound in the same sense so as to be responsive to the same hand of circular polarization. The helices were arranged on an east-west line and helices 1 and 3 rotated continuously. In this type of operation a small lobe appears an hour or so east of the meridian, then grows in amplitude and sweeps westward, reaching maximum amplitude at the meridian. After sweeping an hour or so west of the meridian, the lobe decreases to a small amplitude, and simultaneously a new small lobe appears east of the meridian, and the process is repeated, giving a continuously sweeping lobe (east to west) which crosses the meridian n times per minute for a helix speed of n revolutions per minute. By using more elements (helices) the beam width of the swept lobe can be made arbitrarily small.

Among still other types of interferometers mention should be made of the correlation interferometer of Brown and Twiss (1954) (Twiss and Little, 1959; Brown, 1964). In this system the outputs of two spaced receiving elements are detected and amplified independently with separate receivers. The low-frequency outputs of the receivers are then combined in a correlator, whose output is a function of the fringe visibility.

To summarize this discussion of interferometers (Secs. 6-10 to 6-13), the pattern and spatial-frequency characteristics of many of the types discussed are presented in Fig. 6-31.

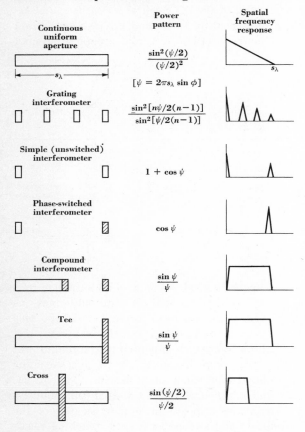

Fig. 6-31. Spatial-frequency characteristics and power-pattern expressions for a continuous uniform aperture and various interferometer arrangements. The switched portions of the interferometers are shaded. The width of the narrow interferometer elements is neglected in the pattern expressions.

6-14   Lunar Occultation and the Earth-Moon System as a High-resolution Interferometer   When the moon passes in front of a radio source, its disk occults the source, and a record, as in Fig. 6-32, may be obtained.   From the duration of the occultation and its midpoint two possible positions of the source with respect to the moon's position can be deduced from the geometry (Fig. 6-32a).   The ambiguity can be resolved, for example, if an approximate source position by other means is available.

The moon's position is accurately known, and it drifts across the sky at a sufficiently slow rate (about ½ sec of arc per second of time) so that position determinations of the order of 1 sec of arc are readily possible.†

Hazard, Mackey, and Shimmins (1963) used this lunar-occultation technique to determine the position of the quasi-stellar radio source 3C 273

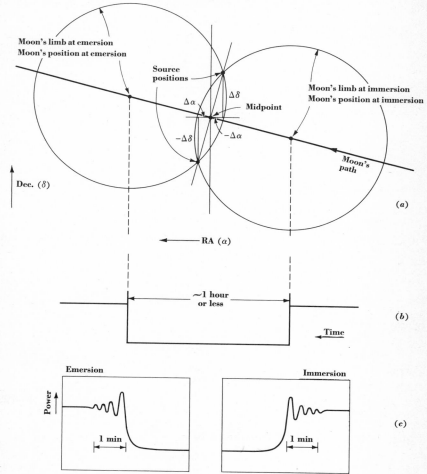

Fig. 6-32. (*a*) Geometry of moon occultation; (*b*) occultation record; and (*c*) occultation record at immersion (disappearance) and emersion (reappearance) to increased time scale, showing Fresnel diffraction patterns.

† The drift in right ascension averages about 51 min per day. The drift in declination ranges up to about ±5° per day. These drifts combine to give the drift of about ½ sec of arc per second of time indicated above.

to an accuracy of 1 sec of arc and also to deduce the source size and structure to a similar precision. The source distribution is obtained from an analysis of the Fresnel diffraction fringes of the record, as discussed below.

An analysis of lunar occultations may be facilitated by assuming that the moon acts as a straight edge. The well-known optical analysis of Fresnel diffraction by a straight edge may then be applied. Thus, from Fig. 6-33 the relative electric field at P (earth) due to a plane wave incident

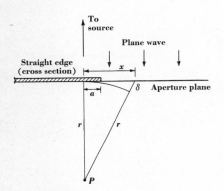

Fig. 6-33.   Occultation geometry.

on the aperture plane (moon's orbit) partly occulted by a straight edge (moon) perpendicular to the page is

$$E = \int_a^\infty e^{j\beta\delta} \, dx \tag{6-121}$$

where $\delta = x^2/2r$
$\quad\ r = $ earth-moon distance
$\quad\ \beta = 2\pi/\lambda$
Letting $k^2 = 2/r\lambda$ and $u = kx$, we have

$$E = \frac{1}{k} \int_{ka}^\infty e^{j\pi u^2/2} \, du \tag{6-122}$$

Equation (6-122) may be rearranged to

$$E = \frac{1}{k} \left(\int_0^\infty e^{j\pi u^2/2} \, du - \int_0^{ka} e^{j\pi u^2/2} \, du \right) \tag{6-123}$$

The integrals in (6-123) have the form of the complex Fresnel integral, so that (6-123) may be written (Born and Wolf, 1964; Schelkunoff, 1948)

$$E = \frac{1}{k} \{ \tfrac{1}{2} + j\tfrac{1}{2} - [C(ka) + jS(ka)] \} \tag{6-124}$$

where $C(ka) = \displaystyle\int_0^{ka} \cos(\pi u^2/2)\, du =$ Fresnel cosine integral

$S(ka) = \displaystyle\int_0^{ka} \sin(\pi u^2/2)\, du =$ Fresnel sine integral

The relative power response is then

$$P(ka) = EE^* = \tfrac{1}{2}\{[\tfrac{1}{2} - C(ka)]^2 + [\tfrac{1}{2} - S(ka)]^2\} \qquad (6\text{-}125)$$

It is assumed that the earth-based antenna at P tracks the source, so that the effect of its pattern does not enter into (6-125).

The occultation curve $P(ka)$ as a function of ka is illustrated by Fig. 6-34 for a point source. It is to be noted from (6-125) that when $ka = -\infty$ which corresponds to no edge present (moon remote), $P(ka) = 1$; when $ka = 0$, $P(ka) = \tfrac{1}{4}$; and when $ka = +\infty$, which corresponds to complete obscuration of the source by the straight edge (or moon), $P(ka) = 0$.

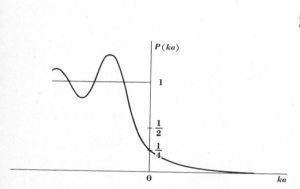

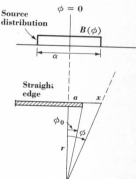

Fig. 6-34. Occultation curve for a point source.

Fig. 6-35. Occultation of source of width α.

From Fig. 6-35, $x/r = \tan\phi$ and for small ϕ, $x \simeq r\phi$. Let the equivalent interferometer power pattern of the earth-moon system be $P(\phi)$. In general, the occultation curve is the convolution of the source distribution $B(\phi)$ and the pattern (see Fig. 6-36), or

$$S(\phi_0) = \int_{-\alpha/2}^{+\alpha/2} B(\phi)\tilde{P}(\phi_0 - \phi)\, d\phi \qquad (6\text{-}126)$$

where

$$\tilde{P}(\phi_0 - \phi) = \tfrac{1}{2}(\{\tfrac{1}{2} - C[kr(\phi_0 - \phi)]\}^2 + \{\tfrac{1}{2} - S[kr(\phi_0 - \phi)]\}^2)$$

Scheuer (1962) has shown that differentiation of the occultation curve $S(\phi_0)$ yields an approximation to the true source brightness distribution and further that a restored, closer approximation to the true source bright-

ness distribution is given by the convolution of the differentiated occultation curve and the differentiated interferometer pattern. Scheuer states
that the resolution attainable and also the positional accuracy of the lunar-
occultation method are not a function of the frequency except insofar as
the frequency is related to the signal-to-noise ratio and receiver stability.

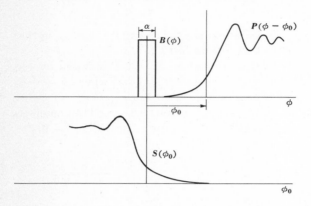

Fig. 6-36. The occultation curve is the convolution of
the source distribution and the occultation pattern for
a point source.

6-15   Radio-telescope Antenna Types and Steerability   Radio-
telescope antennas constitute a variety of types meeting the special requirements of radio astronomy.   Many are like antennas used for other purposes, but some are unique.   To provide sensitivity sufficient for the very
small flux densities of cosmic radio sources antennas of large aperture are
generally required.   Large apertures are also needed for adequate resolution.   For observing large regions of the sky some degree of steerability is
needed.   To reduce cost the steerability may be restricted to the declination coordinate, with observations made on the meridian.   Such *meridian-
transit telescopes* use the earth's rotation for scanning in right ascension.

An arbitrary but convenient way of classifying radio-telescope antennas is to divide them into three groups on the basis of their degree of
mechanical steerability as follows:

(1)   *Completely steerable types* which can be moved in two coordinates.

(2)   *Partially steerable types* movable in one coordinate, usually declination (meridian-transit telescopes belong to this group).

(3)   *Fixed types.*   Although the main antenna is mechanically stationary in this last group, the beam may be steered to a greater
or lesser extent by movement of the feed in a reflector type or by
phase changes in an array type.

The antennas of large radio telescopes are commonly parabolic or spherical reflectors or arrays of large numbers of elemental antenna types, such as dipoles or helical antennas. Examples of various kinds of radio-telescope antennas will be presented. A general discussion of radio telescopes is given by Findlay (1964). Data summarizing the aperture areas and beam widths of large radio telescopes are presented by Ko (1964a).

Examples of completely steerable parabolic-reflector radio telescopes are the Australian 210-ft parabolic-reflector antenna at Parkes, which appears in Fig. 6-37, (Bowen and Minnett, 1962, 1963) and the parabolic-reflector antennas at the Jodrell Bank radio observatory, Manchester, England, in Fig. 6-38. In Fig. 6-38 the 250-ft-diameter (Mark I) antenna is at the right (Brown and Lovell, 1958) and a newer 130- by 80-ft (Mark II) antenna at the left (Lovell, 1964; McElheny, 1964). A closeup view of the Mark II appears in Fig. 6-39.

Fig. 6-37. The 210-ft-diameter steerable parabolic-dish radio telescope of the CSIRO at Parkes, Australia. (*Photograph courtesy of Dr. J. L. Pawsey.*)

A crossed-grating interferometer with completely steerable parabolic-reflector elements at Stanford University is illustrated in Fig. 6-40 (Bracewell, 1957; Bracewell and Swarup, 1961). This antenna has 16 parabolic reflectors 10 ft in diameter in each arm to provide a 3 min of arc beam at

Fig. 6-38. Steerable parabolic dish antennas at the University of Manchester, England, Jodrell Bank radio observatory. The 250-ft dish antenna (Mark I) is at the right, and a newer, 130- by 80-ft antenna (Mark II) is at the left. (*Photograph by J. D. Kraus, Jr.*)

Fig. 6-39. Closeup of the Mark II 130- by 80-ft steerable dish antenna at Jodrell Bank. (*Photograph by J. D. Kraus, Jr.*)

9 cm wavelength.   Crossed-grating telescopes of this type are also used in Australia (Christiansen and Mathewson, 1958).

An example of a partially steerable (meridian-transit) array antenna is presented in Fig. 6-41.   This antenna, built in 1952 at the Ohio State University radio observatory, consists of an array of 96 helical-beam antennas, each of 11 turns, mounted on a tiltable steel ground plane 160 ft

Fig. 6-40. The 32-element crossed-grating interferometer at Stanford University. *Photograph courtesy of Dr. R. N. Bracewell.*)

ong (east-west) by 22 ft wide. At a wavelength of 1.2 m the beam width measured 1° in right ascension by 8° in declination (Kraus, 1953, 1956).

Serious consideration was given to enlarging the aperture of this antenna by increasing the east-west length and by adding other parallel sections to the south, in the manner of venetian blinds. However, the transmission-line losses and relatively narrow frequency response of this arrangement were unattractive, and effort was turned to the design and construction of a large fixed parabola with an associated tiltable flat reflector. Construction on this radio telescope began in 1956, with completion of the

Fig. 6-41. The 96-helix array at the Ohio State University radio observatory.

structure shown in Fig. 6-42 by 1962. The telescope is situated at the Ohio State–Ohio Wesleyan Radio Observatory, Delaware, Ohio, and is a partially steerable reflector antenna of the meridian-transit type. The telescope has two reflecting surfaces, a tiltable flat reflector 260 ft long by 100 ft in slant height, and a fixed standing parabola 360 ft long by 70 ft high (Kraus, 1955, 1963; Kraus, Nash, and Ko, 1961).

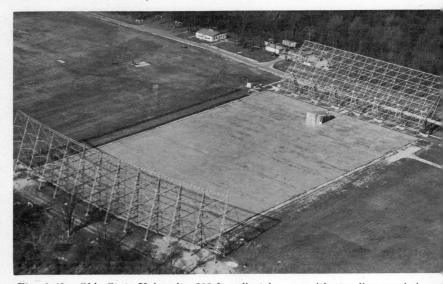

Fig. 6-42. Ohio State University 260-ft radio telescope with standing parabola an
tiltable flat reflector. The two reflectors are joined by a conducting ground plane, 3 acre
in extent. The feed horns are situated in the radome structure in front of the midpoir
of the tiltable flat reflector. The receiver laboratory is immediately under this radome
(*Photograph by Tom Root.*)

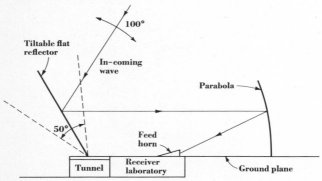

Fig. 6-43. Elevation cross section through standing-parabola
tiltable-flat-reflector radio telescope of the Ohio State University.

The principle of operation is indicated in the elevation cross section
of Fig. 6-43. Incoming waves are deflected by the flat reflector into the
parabola, which brings the waves to a focus at ground level near the base
of the flat reflector. By moving the flat reflector through 50° the antenna
beam is tilted through a 100° range in declination. The antenna may be

operated in two modes. In one mode, illustrated in Fig. 6-44*a*, the feed-horn axis is aligned with the center of the parabola and the ground plane is incidental. In the second mode (Fig. 6-44*c*) the horn axis is coincident with the conducting ground plane, which joins the parabola and flat reflector. The ground plane serves as a guiding boundary surface. In this mode polarization must be vertical, and the feed horn required is one-

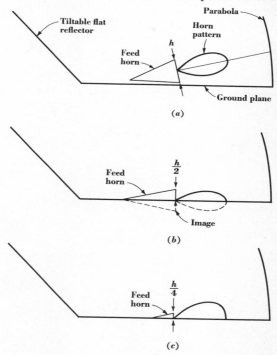

Fig. 6-44. Arrangement for feeding antenna without ground plane, as at (*a*), requires 4 times horn height as arrangement at (*c*) using ground plane. Diagram (*b*) illustrates the fact that half of the horn in (*a*) produces too sharp a pattern when used with the ground plane and must be reduced in size, as at (*c*).

fourth the height and one-half the length of the horn required in the first mode. This difference in horn size may be inferred with the aid of the three diagrams in Fig. 6-44. If a horn of the first mode (Fig. 6-44*a*) is placed with its axis coincident with the ground plane, the lower half is an image and may be discarded, as in Fig. 6-44*b*. However, the beam width of the horn is too narrow (by a factor of 2), so that its dimensions must be halved, as in Fig. 6-44*c*. Although the ground plane serves no primary function in the first mode of operation, its presence tends to reduce the antenna temperature by shielding minor lobes from direct ground pickup. Original plans called for a fixed parabola 720 ft long, or twice the present parabola length. The present antenna has been constructed as the central part of a 720-ft antenna and, if funds were available, could be enlarged.

An attractive feature of the design is that receivers may be situated in a spacious, stable, easily accessible laboratory directly under the focal point. Also there are no weight restrictions on the equipment placed at the feed point.

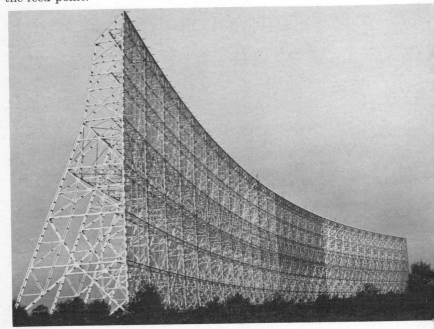

Fig. 6-45. The 1,000-ft curved reflector of Nancay, France, radio telescope. (*Photo-graph courtesy of Dr. J. L. Steinberg.*)

A larger antenna of the same design has been built at Nancay, France (Blum, Boischot, and Lequeux, 1963; Roret, 1962; Findlay, 1964). The antenna, shown in the two views of Figs. 6-45 and 6-46, has a tiltable reflector 200 m long by 40 m in slant height and a curved reflector 300 m long by 35 m high. The curved reflector is spherical, which permits the tracking of sources for about 1 hr before and after meridian transit by movement of the feed.

Another example of a partially steerable radio telescope is the aperture-synthesis 178-Mc interferometer at Cambridge, England. Figure 6-47 shows the fixed element of the interferometer, which is 1,450 ft long (east-west) by 65 ft wide, while the smaller movable section, mounted on

Fig. 6-46. Tiltable flat reflector of Nancay, France, radio telescope. (*Photograph courtesy of Dr. J. L. Steinberg.*)

Fig. 6-47. The 178-Mc tiltable parabolic-reflector antenna at Cambridge University, England. (*Photograph courtesy of Dr. Martin Ryle.*)

Fig. 6-48. Movable antenna mounted on tracks used with larger antenna of Fig. 6-in aperture-synthesis observations at Cambridge University, England. (*Photogra courtesy of Dr. Martin Ryle.*)

tracks, is illustrated in Fig. 6-48.   This section can be moved 1,000 north-south (Ryle and Hewish, 1960).

   The 300-ft-diameter parabolic reflector at the National Radio A tronomy Observatory, Green Bank, West Virginia, is also a partial steerable meridian-transit type (Findlay, 1963, 1964).   This antenn shown in Fig. 6-49, has a focal length of 128.5 ft, or an f/D (focal leng to diameter) ratio of 0.428.   It is used on frequencies up to about 1,500 M

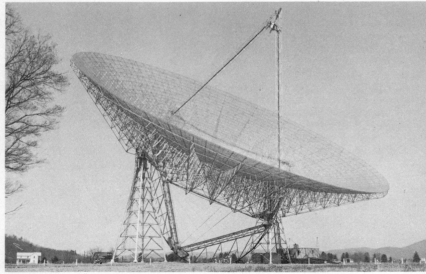

Fig. 6-49.   The 300-ft-diameter meridian-transit radio telescope at the National Rad Astronomy Observatory, Green Bank, W.Va. (*Photograph courtesy of Dr. J. W. Findla*

A partially steerable meridian-transit telescope at the Pulkova radio observatory, USSR, is shown in elevation cross section in Fig. 6-50. In his design a section of a paraboloid of revolution is approximated by 90 independently adjustable reflector elements, each 1.5 m wide by 3 m in slant height, deployed in an arc along the ground. For observing at the zenith the arc is circular, but for observing at the horizon the arc is parabolic, so that a change in declination requires adjustment of the angle and position of the reflector sections. At 3 cm wavelength a fan beam about 1 min of arc wide in right ascension is obtained (Khaikin and Kaidanovskii, 1959).

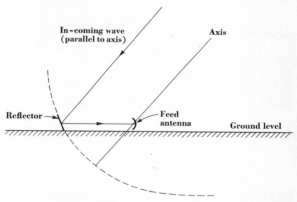

Fig. 6-50. Elevation cross section of Pulkova, USSR, radio telescope showing principle of operation.

The original Mills cross at Sydney, Australia (CSIRO) shown in Fig. 6-51 is an example of a mechanically fixed radio telescope (Mills, 1963). This instrument has east-west and north-south elements about 1,500 ft long. Each element has 500 half-wave dipoles and produces a fan beam about 50 by 0°.6 between half-power points at 85.5 Mc. By phase switching the two elements a pencil beam about 0°.8 between half-power points is obtained. The beam may be steered in declination by phase adjustments of the dipoles in the north-south element. A newer, cross telescope at the University of Sydney, Australia, has elements nearly 1 mile in length (Mills, Aitchison, Little, and McAdam, 1963). Both elements consist of cylindrical parabolic reflectors about 40 ft wide. The beam is steered in declination by a combination of a mechanical tilting of the east-west element on its long axis and an electrical phase shifting of the feeds of the mechanically fixed north-south element. Furthermore, the feed is arranged to provide simultaneous beams at several adjacent declinations. At 408 Mc the beam width between half-power points is about 3 min of

arc. A circular, as contrasted to a cross-type, array has been described by
Wild (1961), and an 80-Mc system using one hundred 40-ft-diameter steer
able dishes deployed around a circle 2 miles in diameter is being constructe
for solar observations (Sheridan, 1963).

Fig. 6-51. Original Mills-cross antenna of the
CSIRO, Australia. (*Photograph courtesy of B. Y.
Mills.*)

Fig. 6-52. Fixed cylindrical parabola radio telescope of the University of Illinois. (*Photograph
courtesy of Dr. G. W. Swenson, Jr.*)

The 400 ft (east-west) by 600 ft (north-south) cylindrical parabolic reflector at the University of Illinois is an example of a completely stationary meridian-transit antenna (Swenson and Lo, 1961). This telescope, shown in Fig. 6-52, has a parabolic reflector built in a depression in the ground. The tower-supported focal line is horizontal and runs north-south. Feeding is by a stationary linear array extending along the focal line. Beam steering in declination is accomplished by phase adjustments of the feed array.

Fig. 6-53. The 30-ft-diameter millimeter wave antenna with Cassegrain feed of North American Aviation Inc., Columbus, Ohio. (*Photograph courtesy of Dr. J. J. Myers.*)

A precision 30-ft-diameter millimeter-wavelength radio telescope built and operated by North American Aviation Inc., of Columbus, is illustrated in Fig. 6-53. This antenna uses a Cassegrain feed system; i.e., a mirror at the prime focus transfers the focal point to the apex of the parabola, where the receivers are situated. At a wavelength of 3 mm the beam width of this telescope approaches 1 min of arc.

A spherical reflector may be used in place of a parabolic one if the necessary corrections are made for spherical aberration (Ashmead and Pippard, 1946). One arrangement is to incorporate this correction into a line feed (Spencer, Sletten, and Walsh, 1949; Love, 1962). The 1,000-ft-diameter spherical reflector at Arecibo, Puerto Rico, uses this principle (Findlay, 1964; Gordon and Lalonde, 1961). This antenna, operated by Cornell University, is constructed of wire mesh anchored in a large de-

pression in the ground, as suggested in Fig. 6-54. The line feed swing
from a truss that rotates on the reflector axis to give beam steering any
where within 20° from the zenith. The truss and its support system ar
suspended by cables from three towers located around the periphery of th
reflector. At 430 Mc the half-power beam width is 10 min of arc.

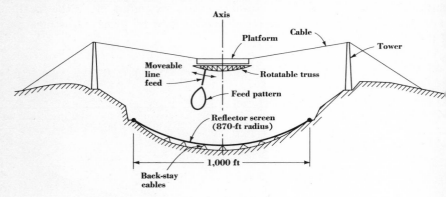

Fig. 6-54. Elevation cross section of fixed 1,000-ft-diameter spherical reflector of radi
telescope at Arecibo, Puerto Rico.

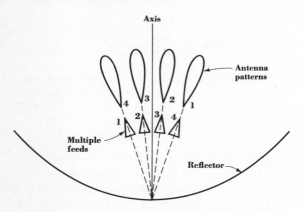

Fig. 6-55. Antenna with multiple feeds for producing
multiple beams.

Multibeaming, which was mentioned in connection with the nev
University of Sydney, Australia, cross antenna, can be accomplished i
reflector antennas by multiple feeds, as suggested in Fig. 6-55. Some bean
deterioration and decrease in gain result from off-axis operation in ;
parabola, but more information is accumulated per unit of time. Th

multiple feeds may be for the same frequency or different frequencies. In array antennas the multibeaming is accomplished in the feed system. At Nancay, France, Blum (1961) has constructed an array of 8 small paraboloids along a north-south line which produces 15 simultaneous beams along the meridian. Each of the 8 reflectors has its own radio-frequency amplifier and 15 output connections. These 120 outputs (= 8 × 15) are connected by cables of suitable length (and phase delay) to 15 receiver-recorders.

One approach to large aperture per unit cost in radio telescopes is to limit the mechanical steerability. Another approach is to subdivide the single aperture A_1 into n smaller apertures of area A such that $nA = A_1$, so that the total aperture is the same. This scheme requires that each subaperture have its own radio-frequency amplifier of substantially the same sensitivity as used for the larger single aperture. The aperture nA per unit cost is then given by

$$\frac{nA}{n(V_a + V_r)} \tag{6-127}$$

where V_a = cost of individual aperture
V_r = cost of individual radio-frequency amplifiers and associated connections

For parabolic reflectors the cost may be related to the aperture by

$$V_a = kA^\alpha \tag{6-128}$$

where k and α are constants, with the cost index α appearing to be in the range 1.2 to 1.5. Introducing (6-128) into (6-127) and maximizing yields the optimum individual aperture size, i.e., the aperture yielding the largest total aperture per unit cost. This is given by Drake (1964) as

$$A(\text{optimum}) = \left[\frac{V_r}{(\alpha - 1)k} \right]^{1/\alpha} \tag{6-129}$$

The variation of the aperture-cost ratio with aperture size is illustrated in Fig. 6-56, with the peak corresponding to the optimum aperture value. The result depends on the individual antenna and amplifier costs but does not depend on the number n. However, the minimum detectable flux density of the aggregation does depend on the total aperture nA(optimum) and, hence, on the number of units. Drake points out that on the basis of typical costs, arrays of several small antennas would appear to be more economical than a single large aperture.

It would appear that the aperture per unit cost of radio telescopes can be increased by restricting the steerability of single large units, as described earlier, or without sacrifice of steerability by subdividing the aperture into a number of smaller steerable units, as discussed in the preceding paragraph.

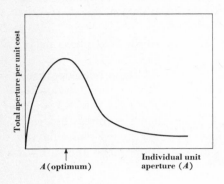

Fig. 6-56. Total aperture per unit cost as a function of individual-unit aperture size for a multiple-unit antenna.

6-16   Efficiency Considerations   Let a plane wave of total flux density S' be incident on an antenna, as in Fig. 6-57.   The power W delivered by the antenna to the receiver is then

$$W = S'A_e \tag{6-130}$$

where A_e is the *effective aperture* of the antenna, or

$$A_e = \frac{W}{S'} \tag{6-131}$$

A completely polarized point source is assumed with the antenna matched to the wave.

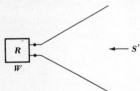

Fig. 6-57.   Wave of flux density S' incident on antenna.

The effective aperture is a unique quantity for any antenna.   The *physical aperture*, on the other hand, may or may not be.   For example, the physical aperture A_p of a horn antenna may quite properly be taken as equal to the area of the mouth of the horn.   But in the case of a stub antenna (monopole) projecting from an aircraft fuselage the situation is not so simple because antenna current may flow over the entire surface of the aircraft.   Thus, it is not clear in this case whether the physical aperture should be taken equal to the cross section of the stub antenna or of the entire aircraft or some other value.   However, in the case of antennas that

re large in terms of wavelength the physical aperture can usually be
eadily defined.

A preliminary discussion of effective aperture was given in Sec. 6-2, in
which the effect of ohmic losses was neglected (assumed zero). In the
more general case, where ohmic losses are not negligible, we may distinguish
etween the actual effective aperture (including the effect of ohmic losses)
nd an effective aperture based entirely on the pattern (losses neglected).
'hus, we may write

$$A_e = k_o A_{ep} \qquad (6\text{-}132)$$

where A_e = actual effective aperture, m²
$\qquad k_o$ = ohmic-loss factor, dimensionless $(0 \leq k_o \leq 1)$
$\qquad A_{ep}$ = effective aperture as determined entirely by pattern, m²
Using these symbols, the directivity of an antenna is given by

$$D = \frac{4\pi}{\Omega_A} = \frac{4\pi}{\lambda^2} A_{ep} \qquad (6\text{-}133)$$

'om which

$$A_{ep}\Omega_A = \lambda^2 \qquad (6\text{-}134)$$

nd

$$A_e\Omega_A = k_o\lambda^2 \qquad (6\text{-}135)$$

here Ω_A = antenna beam solid angle, rad²
$\qquad \lambda$ = wavelength, m
The *aperture efficiency* ϵ_{ap} is defined as

$$\epsilon_{ap} = \frac{A_e}{A_p} \qquad (6\text{-}136)$$

o that the ratio of the aperture and beam efficiencies is

$$\frac{\epsilon_{ap}}{\epsilon_M} = \frac{A_e\Omega_A}{A_p\Omega_M} = \frac{k_o\lambda^2}{A_p\Omega_M} \qquad (6\text{-}137)$$

instead of the simpler relation given in (6-27), where it was assumed that
, = 1.

The directivity of an antenna depends only on the radiation pattern,
o that the *directivity* D is given, as in (6-133), by

$$D = \frac{4\pi}{\lambda^2} A_{ep} \qquad (6\text{-}138)$$

'he *gain* G is then

$$G = Dk_o = k_o \frac{4\pi}{\lambda^2} A_{ep} \qquad (6\text{-}139)$$

'he *maximum directivity* D_m will be defined as the directivity obtainable from an antenna (assumed to be large) if the field is uniform over t aperture, i.e., the physical aperture. Hence,

$$D_m = \frac{4\pi}{\lambda^2} A_p \qquad (6\text{-}14$$

In designing an antenna the engineer may have a certain *desi directivity* D_d he wishes to achieve. In general, this will be less than I since, to reduce side lobes, some taper will probably be introduced ir the aperture distribution. Thus we may write

$$D_d = \frac{4\pi}{\lambda^2} A_p k_u = D_m k_u \qquad (6\text{-}14$$

where D_d = design directivity

The factor k_u is called the *utilization factor*. It is the ratio of the directivi chosen by design to that obtainable with a uniform aperture distrib tion, or

$$k_u = \frac{D_d}{D_m} \qquad 0 \le k_u \le 1 \qquad (6\text{-}14$$

After designing and building the antenna and measuring its performan it will probably be found, however, that the actual directivity D is l. than the design directivity D_d. It is possible, but unlikely, that D excee D_d. The actual directivity may then be expressed

$$D = D_d k_a = D_m k_u k_a \qquad (6\text{-}14$$

where k_a is the *achievement factor*, which is a measure of how well t designer achieved his objective. Thus,

$$k_a = \frac{D}{D_d} \qquad 0 \le k_a \le 1, \text{ usually} \qquad (6\text{-}14$$

The gain G of the antenna can now be written

$$G = D_m k_o k_u k_a = \frac{4\pi}{\lambda^2} A_p k_o k_u k_a = \frac{4\pi}{\lambda^2} A_p \epsilon_{ap} \qquad (6\text{-}14$$

where A_p = physical aperture
k_o = ohmic efficiency factor
k_u = utilization factor
k_a = achievement factor
ϵ_{ap} = aperture efficiency

We may also write

$$A_{ep} = k_u k_a A_p \qquad (6\text{-}1$$
$$A_e = k_o k_u k_a A_p \qquad (6\text{-}14$$
$$\epsilon_{ap} = \frac{A_e}{A_p} = k_o k_u k_a \qquad (6\text{-}1$$

where A_{ep} = effective aperture (as determined entirely by pattern)

A_e = actual effective aperture

A basic definition for the directivity of an antenna is that the directivity is equal to the ratio of the maximum to the average radiation intensity from the antenna (assumed transmitting) (Sec. 6-2). By reciprocity the directivity will be the same in the receiving case. Hence,

$$D = \frac{U_m}{U_{\text{avg}}} \qquad (6\text{-}149)$$

where U_m = maximum radiation intensity, watts rad^{-2}

U_{avg} = average radiation intensity, watts rad^{-2}

The average value may be expressed as the integral of the radiation intensity $U(\theta,\phi)$ over a solid angle of 4π divided by 4π. Thus

$$D = \frac{U_m}{\dfrac{1}{4\pi} \displaystyle\iint_{4\pi} U(\theta,\phi)\, d\Omega} = \frac{4\pi U_m}{W} \qquad (6\text{-}150)$$

where W is the total power radiated. From (6-150) the effective power (power which would need to be radiated if the antenna were isotropic) is

$$DW = 4\pi U_m \qquad (6\text{-}151)$$

Let us now consider an aperture under two conditions. Under condition 1 the field distribution is assumed to be uniform, so that the directivity is D_m. The power radiated is W'. Under condition 2 the field has its actual distribution, and the directivity and power radiated are D and W, respectively. If the effective powers are equal in the two conditions,

$$DW = D_m W' \qquad (6\text{-}152)$$

and

$$D = D_m \frac{W'}{W} = \frac{4\pi}{\lambda^2} A_p \frac{\dfrac{E_{\text{avg}} E_{\text{avg}}^*}{Z} A_p}{\displaystyle\iint_{A_p} \frac{E(x,y)E^*(x,y)}{Z}\, dx\, dy} \qquad (6\text{-}153)$$

In (6-153) the surface of integration has been collapsed over the antenna, with part of the surface coinciding with the aperture. Further, it is assumed that all the power radiated flows out through the aperture. In (6-153) Z is the intrinsic impedance of the medium (ohms per square) and E_{avg} is the average field across the aperture, as given by

$$E_{\text{avg}} = \frac{1}{A_p} \iint_{A_p} E(x,y)\, dx\, dy \qquad (6\text{-}154)$$

where $E(x,y)$ is the field at any point (x,y) of the aperture.   Rearranging (6-153), we have for the actual directivity

$$D = \frac{4\pi}{\lambda^2} A_p \frac{1}{\dfrac{1}{A_p} \displaystyle\iint_{A_p} \left[\dfrac{E(x,y)}{E_{\text{avg}}}\right]\left[\dfrac{E(x,y)}{E_{\text{avg}}}\right]^* dx\, dy} \tag{6-155}$$

This relation has been given by Bracewell (1961*a*).   Following Bracewell' discussion and also elaborating on it by introducing the utilization factor we can write for the design directivity

$$D_d = \frac{4\pi}{\lambda^2} A_p \frac{1}{\dfrac{1}{A_p} \displaystyle\iint_{A_p} \left[\dfrac{E'(x,y)}{E'_{\text{avg}}}\right]\left[\dfrac{E'(x,y)}{E'_{\text{avg}}}\right]^* dx\, dy} \tag{6-156}$$

where the primes indicate the design field values.   The right-hand facto in (6-156) may be recognized as the utilization factor k_u.   Multiplying and dividing (6-155) by k_u, as given in (6-156), yields

$$D = \frac{4\pi}{\lambda^2} A_p \frac{1}{\dfrac{1}{A_p} \displaystyle\iint_{A_p} \left[\dfrac{E'(x,y)}{E'_{\text{avg}}}\right]\left[\dfrac{E'(x,y)}{E'_{\text{avg}}}\right]^* dx\, dy}$$

$$\times \frac{\dfrac{1}{A_p} \displaystyle\iint_{A_p} \left[\dfrac{E'(x,y)}{E'_{\text{avg}}}\right]\left[\dfrac{E'(x,y)}{E'_{\text{avg}}}\right]^* dx\, dy}{\dfrac{1}{A_p} \displaystyle\iint_{A_p} \left[\dfrac{E(x,y)}{E_{\text{avg}}}\right]\left[\dfrac{E(x,y)}{E_{\text{avg}}}\right]^* dx\, dy} \tag{6-157}$$

We also have from (6-143)

$$D = D_m\, k_u\, k_a \tag{6-158}$$

Thus, the last factor in (6-157) is the achievement factor, a result giver by Bracewell.   Further, as done by Bracewell, let

$$E(x,y) = E_{\text{avg}} + \delta E_{\text{avg}} \tag{6-159}$$

or

$$\frac{E(x,y)}{E_{\text{avg}}} = 1 + \delta \tag{6-160}$$

where δ is the *complex deviation* (*factor*) of the field from its average value Thus, the denominator of the last factor in (6-157) may be written as

$$\frac{1}{A_p} \iint_{A_p} (1 + \delta)(1 + \delta)^*\, dx\, dy \tag{6-161}$$

Since the average of δ or δ^* over the aperture is zero, (6-161) simplifies to

$$1 + \frac{1}{A_p} \iint\limits_{A_p} \delta\delta^* \, dx \, dy = 1 + \text{var } \delta \tag{6-162}$$

where var δ is used to signify the *variance of* δ or average of $\delta\delta^*$ ($= |\delta|^2$) over the aperture. Thus, (6-157) may be written more concisely as

$$D = \frac{4\pi}{\lambda^2} A_p \frac{1}{1 + \text{var } \delta'} \frac{1 + \text{var } \delta'}{1 + \text{var } \delta} \tag{6-163}$$

where δ' = design value
δ = actual value

We also have that the utilization factor

$$k_u = \frac{1}{1 + \text{var } \delta'} \tag{6-164}$$

and the achievement factor

$$k_a = \frac{1 + \text{var } \delta'}{1 + \text{var } \delta} \tag{6-165}$$

Turning now to the *beam efficiency* ϵ_M, this was defined in Sec. 6-3 as the ratio of the solid angle of the main beam Ω_M to the total beam solid angle Ω_A. Thus,

$$\epsilon_M = \frac{\Omega_M}{\Omega_A} = \frac{\displaystyle\iint_{\substack{\text{main} \\ \text{lobe}}} P \, d\Omega}{\displaystyle\iint_{4\pi} P \, d\Omega} \qquad (0 \le \epsilon_M \le 1) \tag{6-166}$$

where P = antenna power pattern ($= EE^* = |E|^2$)

Let us consider next the effect of the aperture field distribution on the beam and aperture efficiencies. We take first the simplest case of a one-dimensional distribution, i.e., a rectangular aperture (L_λ by L'_λ) with a uniform distribution in the y direction (aperture L'_λ) and a distribution in the x direction (aperture L_λ) as in Fig. 6-58, given by

$$E(x) = K_1 + K_2 \left[1 - \left(\frac{2x}{L_\lambda}\right)^2 \right]^n \tag{6-167}$$

where $E(x)$ = field distribution
K_1 = const (see Fig. 6-58)
K_2 = const (see Fig. 6-58)
$L_\lambda = L/\lambda$ = aperture width, wavelengths
n = integer ($= 1, 2, 3, \ldots$)

If $K_2 = 0$, the distribution is uniform. If $K_1 = 0$, the distribution is parabolic for $n = 1$, and more severely tapered toward the edges for larger values of n, as indicated in Fig. 6-58. The beam and aperture efficiency of a one-dimensional aperture with $n = 1$ has been calculated by Nash (1964) as a function of the ratio $K_1/(K_1 + K_2)$, with the result shown in Fig. 6-59. For an abscissa value of 0 ($K_1 = 0$) the distribution tapers to zero at the edge (maximum taper). As the abscissa value increases, the taper decreases until at an abscissa value of 1 ($K_2 = 0$) there is no taper; i.e., the distribution is uniform. The curves of Fig. 6-59 show that the beam efficiency tends to increase with an increase in taper but the aperture efficiency decreases. *Maximum aperture efficiency occurs for a uniform aperture distribution, but maximum beam efficiency occurs for a highly tapered*

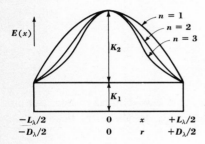

Fig. 6-58. Various shapes of aperture distribution.

distribution. In most cases a taper is used that is intermediate between the two extremes of Fig. 6-59 [$K_1/(K_1 + K_2) = 0$ or 1], and a compromise is reached between large beam and aperture efficiencies.

For a two-dimensional aperture distribution, i.e., a rectangular aperture (L_λ by L'_λ) with the same type of distribution in both x and y directions, the beam and aperture efficiencies as a function of taper have been calculated by Nash (1964). A field distribution as given by (6-167) with $n = 1$ is assumed (parabolic distribution when $K_1 = 0$). Thus for this case

$$E(x,y) = \left\{ K_1 + K_2 \left[1 - \left(\frac{2x}{L_\lambda} \right)^2 \right] \right\} \left\{ K_1 + K_2 \left[1 - \left(\frac{2y}{L'_\lambda} \right)^2 \right] \right\} \qquad (6\text{-}168)$$

Nash's data, which also show the effect of random phase errors across the aperture, are presented in Fig. 6-60. There is a family of four curves for the aperture efficiency (solid) and another family of four curves for the beam efficiency (dashed) for random displacements (or errors) of 0, 0.04λ, 0.06λ, and 0.08λ rms deviation. To obtain these curves the aperture and beam efficiencies calculated for the smooth distribution (6-168) were multiplied by a gain-degradation factor (Ruze, 1952) given by

$$k_g = e^{-(2\pi\delta/\lambda)^2} \qquad (6\text{-}169)$$

here δ is the rms phase front displacement from planar over the aperture. t is assumed that the correlation intervals of the deviations are greater han the wavelength. The curves of Fig. 6-60 indicate that the controlling ffect of the taper on the efficiencies (beam and aperture) tends to decrease s the phase error increases. The efficiencies are also reduced by the resence of the phase error, since such errors tend to scatter radiation into he side-lobe regions. Thus, the phase errors constitute a primary limita-ion on the antenna efficiency.

A circular aperture of diameter D_λ with a distribution as given by 6-167), where x is replaced by r and L_λ by D_λ, has also been investigated y Nash, with the results shown in Fig. 6-61. Two families of four curves ach are given for the beam and aperture efficiencies for four conditions

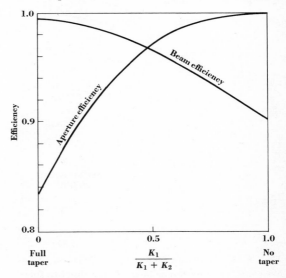

Fig. 6-59. Beam and aperture efficiencies for a one-dimensional aperture as a function of taper. (*After Nash*, 1964.) The aperture efficiency is a maximum with no taper, while the beam efficiency is a maximum with full taper.

f random phase error. It is assumed that $n = 1$ (K_2 part of distribution arabolic). The curves of Fig. 6-61 (circular aperture) are seen to be very imilar to those of Fig. 6-60 for the rectangular aperture. For reflector ntennas with an rms surface deviation δ' it should be noted that the phase ront deviation δ in (6-169) is approximately twice as large; i.e., $\delta = 2\delta'$.

Another problem to be considered in reflector antennas is the efficiency vith which the primary, or feed, antenna illuminates the reflector. This nay be defined as the *feed efficiency* ϵ_f, where

$$\epsilon_f = \frac{\displaystyle\iint_{\Omega_R} P_f \, d\Omega}{\displaystyle\iint_{4\pi} P_f \, d\Omega} \tag{6-170}$$

where P_f = power pattern of feed

Ω_R = solid angle subtended by reflector as viewed from feed point

If the first null of the feed-antenna pattern coincides with the edge of the reflector, the feed efficiency given by (6-170) would be identical with the beam efficiency of the feed.[†]

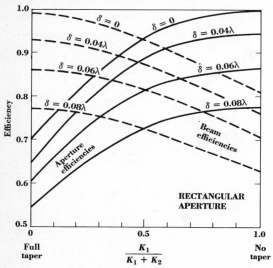

Fig. 6-60. Aperture efficiency (solid) and beam efficiency (dashed) of a rectangular aperture as a function of taper and phase error. (*After Nash*, 1964.)

In the general situation with a reflector antenna, the beam efficiency of the system may be expressed as

$$\epsilon_M = e^{-(4\pi\delta'/\lambda)^2} \frac{\displaystyle\iint_{\substack{\text{main}\\\text{lobe}}} P \, d\Omega \iint_{\Omega_R} P_f \, d\Omega}{\displaystyle\iint_{4\pi} P \, d\Omega \iint_{4\pi} P_f \, d\Omega} \tag{6-171}$$

[†] Common practice is to taper the feed illumination by 10 dB or so at the edges of the reflector (see Silver, 1949, chaps. 12 and 13; also Kraus, 1950, Sec. 12-6.)

where δ' = rms surface error of reflector

P = power pattern of reflector due to aperture distribution produced by the feed assuming no phase error

P_f = power pattern of feed

Surface leakage is neglected. If it is appreciable, another factor would be required.

An experimental procedure for determining the beam efficiency ϵ_M of a large radio telescope antenna is as follows. The main-beam solid angle Ω_M is evaluated from

$$\Omega_M = k_p\theta_{\mathrm{HP}}\phi_{\mathrm{HP}} \tag{6-172}$$

where k_p = factor between about 1.0 for a uniform aperture distribution and 1.13 for a Gaussian power pattern (Ko, 1964b)

θ_{HP} = half-power beam width in θ plane, rad^2

ϕ_{HP} = half-power beam width in ϕ plane, rad^2

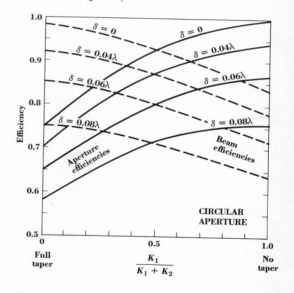

Fig. 6-61. Aperture efficiency (solid) and beam efficiency (dashed) of a circular aperture as a function of taper and phase error. (*After Nash*, 1964.)

The half-power beam widths are measured, while k_p may be calculated or estimated from the pattern shape. The half-power beam width in right ascension in degrees is given by the observed half-power beam width of a drift profile in minutes of time multiplied by cos $\delta/4$, where δ is the declination of the observed point source. That is,

$$\text{HPBW(deg)} = \frac{\text{HPBW(min) } \cos \delta}{4} \tag{6-173}$$

The total beam solid angle Ω_A is obtained from the relation

$$\Omega_A = k_o \lambda^2 / A_e \tag{6-174}$$

where the effective aperture A_e is determined by observing a point source of known flux density.† That is,

$$A_e = \frac{2kT_A}{S} \tag{6-175}$$

Combining relations, the beam efficiency is then given by

$$\epsilon_M = \frac{\Omega_M}{\Omega_A} = \frac{k_p \theta_{\text{HP}} \phi_{\text{HP}} 2kT_A}{k_o \lambda^2 S} \tag{6-176}$$

where k = Boltzmann's constant (1.38×10^{-23} joule °K^{-1})

T_A = antenna temperature due to radio source (measured value corrected for cable loss), °K

S = flux density of radio source, watts m^{-2} cps^{-1}

λ = wavelength, m

k_o = antenna ohmic-loss factor, dimensionless

and where k_p, θ_{HP}, and ϕ_{HP} are as defined in (6-172).

A procedure for determining the aperture efficiency ϵ_{ap} of a large radio-telescope antenna is to observe a radio source of known flux density to find A_e by (6-175), from which

$$\epsilon_{\text{ap}} = \frac{A_e}{A_p} \tag{6-177}$$

where A_p = physical aperture

It is often of interest to the antenna designer, however, to break ϵ_{ap} up into a number of factors, as in (6-148), in order to account as completely as possible for all the causes of efficiency degradation. Thus, from knowledge of the antenna structure and its conductivity the ohmic-loss factor k_o may be determined. The utilization factor k_u can be calculated from design considerations. The achievement factor k_a could be calculated from the relation in (6-157) if the actual fields across the aperture were accurately known; however, these are rarely measured. As an alternative the achievement factor can be separated into many factors involving the random surface error, feed efficiency, aperture blocking, feed displacement normal to axis (squint or coma), feed displacement parallel to axis (astigmatism), etc. It is assumed that each of these subfactors can be independently calculated or estimated.

† The flux density, in turn, is obtained from (6-175) using an antenna of accurately known effective aperture. A horn antenna is often used for this, since its effective aperture can be readily calculated (Schelkunoff and Friis, 1952).

The aperture efficiency ϵ_{ap} may then be expressed as

$$\epsilon_{\text{ap}} = \frac{A_e}{A_p} = k_o k_u k_a = k_o k_u k_1 k_2 \cdot \cdot \cdot \cdot \; k_N$$

$$= k_o k_u \prod_{n=1}^{N} k_n \tag{6-178}$$

where k_o = ohmic-loss factor
k_u = utilization factor (by design)
k_a = achievement factor
$k_1 = e^{-(4\pi\delta'/\lambda)^2}$ = random-surface-error factor
$k_2 = \epsilon_f$ = feed-efficiency factor
k_3 = aperture-blocking factor
k_4 = squint factor
k_5 = astigmatism factor
k_6 = surface-leakage factor
  etc.

A close agreement between ϵ_{ap} as calculated by (6-178) and as measured by (6-177) does not necessarily mean that the designer has taken all factors properly into account (he could have overestimated one and underestimated another), but it does provide him with some confidence as to his understanding of the factors. However, if there is significant disagreement between the two methods, the designer knows his analysis is incorrect or incomplete in one or more respects. This comparative method has been used by Nash (1961) in analyzing the performance characteristics of the Ohio State University 260-ft radio telescope. A discussion of tolerances in large antennas is given by Bracewell (1961a).

6-17 Telescope Resolution and Sensitivity As indicated in (6-20), the number of sources N_r which a radio telescope can resolve may be stated as

$$N_r = \frac{4\pi}{\Omega_M} \tag{6-179}$$

Since $\Omega_A > \Omega_M$, a more conservative value might be

$$N_r = \frac{4\pi}{\Omega_A} = \frac{4\pi}{\lambda^2} A_e = D \tag{6-180}$$

where A_e = effective aperture
D = directivity

In practice, however, N_r, as given by (6-180), is found to be too large and a number perhaps 10 percent as large is more realistic. However, (6-180) does provide an order of magnitude upper limit.

A determination (Kraus, 1958*a*) of the number of sources N_d which a radio telescope can detect requires a number of assumptions. First let us assume that N_d is proportional to the telescope range cubed, or

$$N_d = K_1 R^3 \tag{6-181}$$

where K_1 = const
 R = maximum range ($= R_0 \sqrt{S_0/\Delta S_{min}}$)
 ΔS_{min} = minimum detectable flux density
 R_0 = range of standard source
 S_0 = flux density of standard source
Introducing the value of ΔS_{min} from (3-123)

$$N_d = K_2 A_e^{3/2} \lambda^{1.5(n+f)-0.75} \tag{6-182}$$

where K_2 = const
 A_e = effective aperture, m^2
 λ = wavelength, m
 n = radio-source spectral index
 f = receiver noise-figure index (the system temperature is assumed proportional to the noise figure F, which in turn is proportional to λ^{-f}) ($f = 1$ if F is proportional to frequency)
In (6-182) the receiver bandwidth is taken as proportional to the frequency. Restricting attention to nonthermal sources, we may take $n = \frac{2}{3}$. For vacuum-tube amplifiers a reasonable assumption is that the noise figure is proportional to the square root of the frequency ($f = +\frac{1}{2}$). Introducing $n = \frac{2}{3}$ and $f = \frac{1}{2}$ in (6-182), we have for vacuum-tube amplifiers that

$$N_d = K_2 A_e^{3/2} \lambda \tag{6-183}$$

For the more advanced solid-state amplifiers the noise figure may be regarded as independent of the frequency ($f = 0$), so that we have for this case

$$N_d = K_2 A_e^{3/2} \lambda^{1/4} \tag{6-184}$$

From (6-180), (6-183), and (6-184) the number of sources which a radio telescope can resolve and detect is shown in Fig. 6-62 as a function of the frequency for two aperture sizes corresponding to circular apertures 150 and 500 ft in diameter (46 and 152 m). N_d values are also given for both vacuum tube (VT) and solid state (SS) amplifiers. It is assumed that the antenna ohmic losses are negligible and that the antenna aperture efficiency $\epsilon_{ap} = 0.5$. The constant K_2 has been taken empirically as 0.1. Refrigeration of a solid-state amplifier would displace the N_d line upward.

The maximum number of sources are both detected and resolved where $N_r = N_d$. At lower frequencies $N_r < N_d$, and the telescope is *resolution-limited*. At higher frequencies $N_d < N_r$ and the telescope is *detection- or sensitivity-limited*. Because of the assumptions involved, too much significance should not be attached to a graph such as Fig. 6-62. In practice only 10 percent or so of the values given may be achieved. However, the graph does illustrate that there is in principle a frequency at which a maximum number of sources may be detected and resolved and that at the lower frequencies it may be uneconomical to improve the

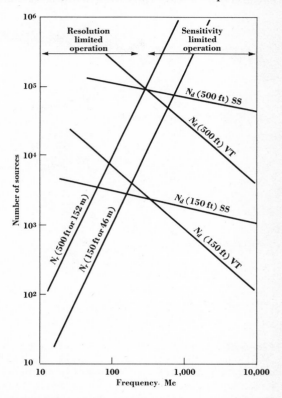

Fig. 6-62. Number of sources a radio telescope can resolve N_r and number it can detect N_d as a function of the frequency. The number it can detect is given for both solid-state (SS) and vacuum-tube (VT) receivers.

receiver, since N_r is already less than N_d. On the other hand, an increase in resolution (without changing the aperture area) by rearranging the aperture into some other configuration, such as a Mills cross, would be worthwhile.

Radio-telescope records illustrating sensitivity-limited and resolution-limited operation are presented in Fig. 6-63. There are five profiles through the center of the Andromeda galaxy (M 31). The top profile (Fig. 6-63a) is the result of a single drift scan at 1,415 Mc with the Ohio State University 260-ft radio telescope operating with 10 sec integration time. The next profile (Fig. 6-63b) is the average of four scans taken on different days, the rms system noise reduced by a factor of 2 ($= \sqrt{4}$). By increasing the effective integration time from 10 to 30 sec an additional noise reduction

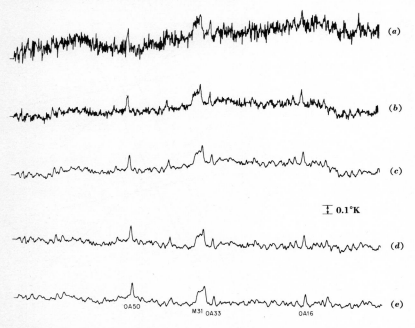

$\updownarrow$ 0.1°K

OA50   M31 OA33   OA16

Fig. 6-63. Five drift profiles through the nucleus of the Andromeda galaxy (M 31) made with the Ohio State University 260-ft radio telescope, illustrating sensitivity-limited and resolution-limited conditions of operation. The top profile (*a*) is the result of one scan; (*b*) the average of four scans; (*c*) the same as profile (*b*) but with a threefold increase in integration time; (*d*) the same as profile (*c*) but with long-term drift removed; and (*e*) the average of eleven scans with increased integration time and drift removed as in (*d*). Profiles (*a*) and (*b*) are sensitivity-limited, while profile (*e*) is resolution- or confusion-limited. The rms telescope noise temperature T_{min} is about 0.01°K for (*c*) and (*d*). The maximum temperature for the largest sources on the profile (M 31 and OA 50) is about 0.15°K. The profiles are 2.5 hr long in right ascension.

of 1.7 ($= \sqrt{3}$) is achieved in the third profile (Fig. 6-63c). The fourth profile (Fig. 6-63d) is the same as the previous one except that long-term drift has been removed. The last profile (Fig. 6-63e) is the average of 11 scans (with 30 sec integration and drift removed), which should result in a further reduction of the rms noise by a factor of 1.65 ($= \sqrt{11\frac{1}{4}}$). Although the smallest-amplitude noise is reduced in (*e*) as compared to (*d*), there is relatively little difference with respect to most of the structure, even the finer details. Thus, with 11 scans the results are at the resolution or confusion limit of the telescope. This means that a further increase in the telescope sensitivity would reveal no more significant detail at this frequency, although the same results could be achieved with fewer scans. Even with four scans the results are close to the resolution limit. Accordingly, profiles (a) and (b) are examples of sensitivity-limited operation, while profile (e) is an example of resolution-limited operation in which the rms system noise has been reduced below the sky background roughness or confusion level. Profiles (c) and (d) are intermediate.

Although a profile such as (e) in Fig. 6-63 will not be altered appreciably if still more scans are averaged, it does not mean that all the detail is significant. The sources, OA 16, OA 33, M 31, and OA 50 are sufficiently strong to be highly significant. Some of the slightly weaker sources may also be significant and represent actual sources. However, the significance of the smallest peaks is much more uncertain. At this level some of the sources may be real, but others may be spurious, representing single or blended side-lobe responses of the antenna to intense sources in other parts of the sky. Hence, this level is called the *resolution* or *confusion level*. A source has a high probability of being real if its limiting level is at least 5 times the square root of the sum of the squares of the rms system and confusion noise levels. Observations at different wavelengths and/or with telescopes of different design are also desirable in establishing the reality of any source.

The profiles in Fig. 6-63 were computer-plotted, using digital data obtained at 1,415 Mc with the Ohio State University 260-ft radio telescope. The basic digital-integration time was 10 sec, the receiver bandwidth 8 Mc, and the system temperature about 100°K. The increase in effective integration time was accomplished by convolving profile (b) with a 30-sec-long square pulse. To remove the long-term drift the tenth-order polynomial of best fit to (c) was subtracted from the profile, resulting in the profile at (d). The averaging, convolving, and drift removal were performed by programs with an IBM 7094 computer (Kraus, Dixon, and Fisher, 1966).

References

ARSAC, J.: Application of Mathematical Theories of Approximation to Aerial Smoothing in Radio Astronomy, *Australian J. Phys.*, vol. 10, pp. 16–28, 1957.

ARSAC, J.: Correction approchée de l'effect de lobe en radio astronomie, *Opt. Acta*, vol. 6, pp. 103–116, April 1959a.

ARSAC, J.: Essai de détermination des très faibles diamètres apparents en radio astronomie, *Opt. Acta*, vol. 6, pp. 77–98, January, 1959b.

ASHMEAD, J., and A. B. PIPPARD: The Use of Spherical Reflectors as Microwave Scanning Aerials, *J. Inst. Elec. Engrs. London*, vol. 93, pt. IIIA, pp. 627–632, March–May, 1946.

BARBER, N. F.: Compound Interferometers, *Proc. IRE*, vol. 46, p. 1951, December, 1958.

BLUM, E. J.: Sensibilité des radiotelescopes et récepteurs à correlation, *Ann. Astrophys.*, vol. 22, pp. 140–163, 1959.

BLUM, E. J.: Le réseau nord-sud à lobes multiples, *Ann. Astrophys.*, vol. 24, p. 359, 1961.

BLUM, E. J., A. BOISCHOT, and J. LEQUEUX: Radio Astronomy in France, *Proc. IRE Australia*, vol. 24, pp. 208–213, February, 1963.

BOOKER, H. G., and P. C. CLEMMOW: The Concept of an Angular Spectrum of Plane Waves and Its Relation to That of Polar Diagram and Aperture Distribution, *Proc. Inst. Elec. Engrs. London*, ser. 3, vol. 97, pp. 11–17, January, 1950.

BOOKER, H. G., J. A. RATCLIFFE, and D. H. SHINN: Diffraction from an Irregular Screen with Applications to Ionospheric Problems, *Phil. Trans. Roy. Soc. London*, *Ser. A*, vol. 247, p. 579, 1950.

BOLTON, J. G.: The Australian 210-foot Reflector and Its Research Program, *Proc. IRE Australia*, vol. 24, pp. 106–112, February, 1963.

BORN, M., and E. WOLF: "Principles of Optics," The Macmillan Company, New York, 1964.

BOWEN, E. G., and H. C. MINNETT: The Australian 210-foot Radio Telescope, *J. Brit. Inst. Radio Eng.*, vol 23, pp. 49–54, January, 1962.

BOWEN, E. G., and H. C. MINNETT: The Australian 210-foot Radio Telescope, *Proc. IRE Australia*, vol. 24, pp. 98–105, February, 1963.

BRACEWELL, R. N., and J. A. ROBERTS: Aerial Smoothing in Radio Astronomy, *Australian J. Phys.*, vol. 7, pp. 615–640, 1954.

BRACEWELL, R. N.: "Antenna Problems in Radio Astronomy," *IRE Nat. Conv. Record*, vol. 5, pt. 1, *Antennas Propagation*, pp. 68–71, 1957.

BRACEWELL, R. N.: Radio Interferometry of Discrete Sources, *Proc. IRE*, 46, pp. 97–105, January, 1958.

BRACEWELL, R. N.: Tolerance Theory of Large Antennas, *IRE Trans. Antennas Propagation*, vol. AP-9, pp. 49–58, January, 1961a.

BRACEWELL, R. N.: Interferometry and the Spectral Sensitivity Island Diagram, *IRE Trans. Antennas Propagation*, vol. AP-9, pp. 59–67, January, 1961b.

BRACEWELL, R. N., and G. SWARUP: The Stanford Microwave Spectroheliograph Antenna: A Microsteradian Pencil Beam Interferometer, *IRE Trans. Antennas Propagation*, vol. AP-9, pp. 22–30, January, 1961.

BRACEWELL, R. N.: Radio Astronomy Techniques, in S. Flügge (ed.), "Handbuch der Physik," vol. 54, pp. 42–129, Springer-Verlag OHG, Berlin, 1962a.

BRACEWELL, R. N.: Electrical Scanning and Low Pass Filtering in Radio Astronomy, *Australian J. Phys.*, vol. 15, pp. 447–449, 1962b.

BRACEWELL, R. N.: Proposal Leading to Future Large Radio Telescopes, *Proc. Nat. Acad. Sci., U.S.*, vol. 49, pp. 766–777, June, 1963.

BRACEWELL, R. N.: "The Fourier Transform and Its Applications," McGraw-Hill Book Company, New York, 1965. Detailed discussion of convolution, autocorrelation, sampling theorem, etc.

BROWN, R. H., and R. Q. TWISS: A New Type of Interferometer for Use in Radio Astronomy, *Phil. Mag.*, vol. 45, p. 663, 1954.

BROWN, R. H., H. P. PALMER, and A. R. THOMPSON: A Rotating Lobe Interferometer and Its Application to Radio Astronomy, *Phil. Mag.*, vol. 46, p. 857, 1955.

BROWN, R. H.: The Stellar Interferometer at Narrabri Observatory, *Sky & Telescope*, vol. 28, pp. 64–69, August, 1964.

BROWN, R. H., and A. C. B. LOVELL: "The Exploration of Space by Radio," John Wiley & Sons, Inc., New York, 1958.

BRUNDAGE, W. D., and J. D. KRAUS: Preliminary Results of a Hydrogen-line Survey of M 31 with the Ohio State University Radio Telescope, *Astron. J.*, vol. 70, p. 669, November, 1965.

CAMPBELL, G. A., and R. M. FOSTER: Fourier Integrals for Practical Applications, *Bell Telephone System Tech. J.*, *Publ. Monograph* B-584, 1931.

CHRISTIANSEN, W. N., and D. S. MATHEWSON: Scanning the Sun with a Highly Directional Array, *Proc. IRE*, vol. 46, pp. 127–131, January, 1958.

CHRISTIANSEN, W. N., and J. A. WARBURTON: The Distribution of the Radio Brightness over the Solar Disk at a Wavelength of 21 cm, *Australian J. Phys.*, vol. 6, pp. 190–202, June, 1953.

COSTAIN, C. H., and F. G. SMITH: The Radio Telescope for 7.9 Meters Wavelength at the Mullard Observatory, *Monthly Notices Roy. Astron. Soc.*, vol. 121, pp. 405–412, 1960.

COVINGTON, A.: A Compound Interferometer, *J. Roy. Astron. Soc. Canada*, vol. 54, pp. 17–68, 1960.

COVINGTON, A. E., and G. A. HARVEY: Resolving Power of 3 Antenna Patterns Derived from the Same Aperture, *Can. J. Phys.*, vol. 37, pp. 1216–1229, 1959.

DRAKE, F. D.: Optimum Size of Radio Astronomy Antennas, *Proc. IEEE*, vol. 52, pp. 108–109, January, 1964.

FEDERER, C. A.: Some Current Programs at Arecibo, *Sky & Telescope*, vol. 28, pp. 4–8, July, 1964, and pp. 73–77, August, 1964.

FINDLAY, J. W.: The 300-foot Radio Telescope at Green Bank, *Sky & Telescope*, vol. 25, pp. 68–75, February, 1963.

FINDLAY, J. W.: Radio Telescopes, *IEEE Trans. Military Electron.*, vol. MIL-8, pp. 187–198, July–October, 1964; also *IEEE Trans. Antennas Propagation*, vol. AP-12, pp. 853–864, December, 1964.

GORDON, W. E., and L. M. LaLONDE: The Design and Capabilities of an Ionospheric Radar Probe, *IRE Trans. Antennas Propagation*, vol. AP-9, pp. 17–22, January, 1961.

HAZARD, C., M. B. MACKEY, and A. J. SHIMMINS: Investigation of the Radio Source 3C 273 by the Method of Lunar Occultations, *Nature*, vol. 197, pp. 1037–1039, Mar. 16, 1963.

JENNISON, R. C.: "Fourier Transforms and Convolutions for the Experimentalist," Pergamon Press, New York, 1961.

KHAIKIN, S. E., and N. L. KAIDANOVSKII: A New Radio Telescope of High Resolving Power, in R. N. Bracewell (ed.), "Paris Symposium on Radio Astronomy," pp. 166–170, Stanford University Press, Stanford, Calif., 1959.

KO, H. C.: Radio Telescope Antennas, chap. 4 in H. C. Hansen (ed.), "Microwave Scanning Antennas," Academic Press Inc., New York, 1964a.

KO, H. C.: Radio Telescope Antenna Parameters, *IEEE Trans. Military Electron.*, vol. MIL-8, pp. 225–232, July–October, 1964b.

KRAUS, J. D.: "Antennas," McGraw-Hill Book Company, New York, 1950.

KRAUS, J. D.: The Ohio State Radio Telescope, *Sky & Telescope*, vol. 12, pp. 157–159, April, 1953.

KRAUS, J. D.: "Radio Telescopes," *Sci. Am.*, vol. 192, pp. 33–43, March, 1955.

KRAUS, J. D.: The Radio Sky, *Sci. Am.*, vol. 195, pp. 32–37, July, 1956.

KRAUS, J. D.: Radio Telescope Antennas of Large Aperture, *Proc. IRE*, vol. 46, pp. 92–97, January, 1958a.

KRAUS, J. D.: Planetary and Solar Radio Emission at 11 Meters Wavelength, *Proc. IRE*, vol. 46, pp. 266–274, January, 1958b.

KRAUS, J. D., R. T. NASH, and H. C. KO: Some Characteristics of the Ohio State University 360-foot Radio Telescope, *IRE Trans. Antennas Propagation*, vol. AP-9, pp. 4–8, January, 1961.

KRAUS, J. D.: The Large Radio Telescope of the Ohio State University, *Sky & Telescope*, vol. 26, pp. 12–16, July, 1963.

KRAUS, J. D., R. S. DIXON, and R. O. FISHER: A High Sensitivity Map of the M 31 Region at 1,415 Mc, *Astrophys, J.*, vol. 144, May, 1966.

LITTLE, A. G., and R. PAYNE-SCOTT: The Position and Movement on the Solar Disk of Sources of Radiation at a Frequency of 97 mcs, *Australian J. Sci. Res. A.*, vol. 4, pp. 489–507, December, 1951.

LO, Y. T.: A Probabilistic Approach to the Design of Large Antenna Arrays, *IEEE Trans. Antennas Propagation*, vol. AP-11, pp. 95–96, 1963.

LOVE, A. W.: Spherical Reflecting Antennas with Corrected Line Sources, *IRE Trans. Antennas Propagation*, vol. AP-10, pp. 529–539, September, 1962.

LOVELL, B.: Jodrell Bank Mark II Radio Telescope, *Nature*, vol. 203, pp. 11–13, July 4, 1964.

MAHER, T. M., and D. K. CHENG: Random Removal of Radiators from Large Linear Arrays, *IEEE Trans. Antennas Propagation*, vol. AP-11, pp. 106–112, 1963.

McCREADY, L. L., J. L. PAWSEY, and R. SCOTT-PAYNE: Solar Radiation at Radio Frequencies and Its Relation to Sunspots, *Proc. Roy. Soc. London, Ser. A.*, vol. 190, p. 357, 1947.

McELHENY, V. K.: Jodrell Bank Observatory, *Science*, vol. 144, pp. 520–523, 1964.

MILLS, B. Y.: Cross-type Radio Telescopes, *Proc. IRE Australia*, vol. 24, pp. 132–140, February, 1963.

MILLS, B. Y., R. E. AITCHISON, A. G. LITTLE, and W. B. McADAM: The Sydney University Cross-type Radio Telescope, *Proc. IRE Australia*, vol. 24, pp. 156–165, February, 1963.

MILLS, B. Y., and A. G. LITTLE: A High Resolution Aerial System of a New Type, *Australian J. Phys.*, vol. 6, pp. 272–278, 1953.

MILLS, B. Y., A. G. LITTLE, K. V. SHERIDAN, and O. B. SLEE: A High Resolution Radio Telescope for Use at 3.5 Meters, *Proc. IRE*, vol. 46, pp. 65–84, 1958.

NASH, R. T.: A Multi-reflector Meridian Transit Radio Telescope Antenna for the Observation of Waves of Extra-terrestrial Origin, Ph.D. dissertation, Ohio State University, 1961.

NASH, R. T.: Beam Efficiency Limitations of Large Antennas, *IEEE Trans. Military Electron.*, vol. MIL-8, pp. 252–257, July–October, 1964.

PAPOULIS, A.: "The Fourier Integral and Its Applications," McGraw-Hill Book Company, New York, 1962.

PAWSEY, J. L., and R. N. BRACEWELL: "Radio Astronomy," Oxford University Press, Fair Lawn, N.J., 1955.

RHODES, D. R.: The Optimum Line Source for the Best Mean Square Approximation to a Given Radiation Pattern, *IEEE Trans. Antennas Propagation*, vol. AP-11, pp. 440–446, July, 1963.

RORET, J.: Construction du plus grande radio telescope du monde à Nancay pour l'Observatoire de Paris, *Ann. Inst. Tech. Batiment Trav. Publ.*, No. 169, suppl., January, 1962.

RUZE, J.: Physical Limitations on Antennas, *Mass. Inst. Technol. Res. Lab. Electron Tech. Rept.* 248, Oct. 30, 1952.

RYLE, M.: A New Radio Interferometer and Its Application to the Observation of Weak Radio Stars, *Proc. Roy. Soc. London Ser. A*, vol. 211, pp. 351–375, 1952.

RYLE, M.: The New Cambridge Radio Telescope, *Nature*, vol. 194, pp. 517–518, May 12, 1962.

RYLE, M., and A. HEWISH: The Synthesis of Large Radio Telescopes, *Monthly Notices Roy. Astron. Soc.*, vol. 120, pp. 220–230, 1960.

RYLE, M., A. HEWISH, and J. R. SHAKESHAFT: The Synthesis of Large Radio Telescopes by the Use of Radio Interferometers, *IRE Trans. Antennas Propagation*, vol. AP-7, suppl., pp. S120–124, December, 1959.

SCHELKUNOFF, S. A.: "Applied Mathematics for Engineers and Scientists," D. Van Nostrand Company, Inc., Princeton, N.J., 1948.

SCHELKUNOFF, S. A., and H. T. FRIIS: "Antennas: Theory and Practice," John Wiley & Sons, Inc., New York, 1952.

SCHEUER, P. A. G.: On the Use of Lunar Occultations for Investigating the Angular Structure of Radio Sources, *Australian J. Phys.*, vol. 15, pp. 333–343, September, 1962.

SCOTT, P. F., M. RYLE, and A. HEWISH: First Results of Radio Star Observations Using the Method of Aperture Synthesis, *Monthly Notices Roy. Astron. Soc.*, vol. 122, pp. 95–111, 1961.

SHERIDAN, K. V.: Techniques for the Investigation of Solar Radio Bursts at Metre Wavelengths, *Proc. IRE Australia*, vol. 24, pp. 174–184, February, 1963.

SILVER, S. (ed.): "Microwave Antenna Theory and Design," McGraw-Hill Book Company, New York, 1949.

SNEDDON, I. N.: "Fourier Transforms," McGraw-Hill Book Company, New York, 1951.

SPENCER, R. C., C. J. SLETTEN, and J. E. WALSH: Correction of Spherical Aberration by a Phased Line Source, *Proc. Natl. Electron. Conf.*, vol. 5, pp. 320–333, 1949.

STONE, J. S.: U.S. Pats. 1,643,323 and 1,715,433.

SWENSON, G. W., JR., and Y. T. LO: The University of Illinois Radio Telescope, *IRE Trans. Antennas Propagation*, vol. AP-9, pp. 9–16, January, 1961.

TWISS, R. Q., and A. G. LITTLE: The Detection of Time-correlated Photons by a Coincidence Counter, *Australian J. Phys.*, vol. 12, pp. 77–93, 1959.

WILD, J. P.: Solar Radio Interferometry, *Rend. Scuola Intern. Fis. Enrico Fermi*, vol. 12, pp. 281–295, 1960.

WILD, J. P.: Circular Aerial Arrays for Radio Astronomy, *Proc. Roy. Soc. London, Ser. A*, vol. 262, pp. 84–99, 1961.

Problems

6-1. Starting with (6-4), show that the directivity D of an antenna may be written

$$D = \frac{\dfrac{E(\theta,\phi)_{max}\, E^*(\theta,\phi)_{max}}{Z} r^2}{\dfrac{1}{4\pi} \displaystyle\iint_{4\pi} \dfrac{E(\theta,\phi)\, E^*(\theta,\phi)}{Z} r^2\, d\Omega}$$

6-2. For symmetrical pencil-beam patterns (function only of θ) show that the main-beam solid angle Ω_M is given by

1.13 θ_{HP}^2 for a Gaussian pattern
0.988 θ_{HP}^2 for a $(\sin x)/x$ pattern
1.008 θ_{HP}^2 for a Bessel pattern

where θ_{HP} is the half-power beam width. The Gaussian power pattern may be expressed
$$P(\theta) = e^{-(u\theta)^2}$$
the $(\sin x)/x$ by

$$P(\theta) = \left[\frac{\sin(u\sin\theta)}{u\sin\theta}\right]^2$$

and the Bessel pattern by

$$P(\theta) = \left[\frac{2J_1(u \sin \theta)}{u \sin \theta} \right]^2$$

6-3. Show that the half-power beam width is $50°.8/L_\lambda$ for a long uniform array.

6-4. Show that the *average* directive gain over the minor lobes of a highly directive antenna is nearly equal to the stray factor.

6-5. Calculate the aperture efficiency and directivity of an antenna with rectangular aperture x_1y_1 with a uniform field distribution in the y direction and a cosine field distribution in the x direction (zero at edges, maximum at center) if $x_1 = 20\lambda$ and $y_1 = 10\lambda$.   *Ans.* $\epsilon_{ap} = 0.81$, $D = 2,040$

6-6. Repeat Prob. 6-5 for the case where the aperture field has a cosine distribution in both x and y directions.   *Ans.* $\epsilon_{ap} = 0.66$, $D = 1,650$

6-7. A radio-telescope antenna of 500 m² effective aperture is directed at the zenith. Calculate the antenna temperature assuming that the sky temperature is uniform and equal to 10°K. Take the ground temperature equal to 300°K and assume that half the minor-lobe beam area is in the back direction. The wavelength is 20 cm, and the beam efficiency is 0.7.   *Ans.* 53.5°K

6-8. A simple two-element 200-Mc interferometer has an east-west spacing between elements of 500 m. What is the time between fringes of the interferometer pattern for a point source at the meridian with a declination of 50°, and with a declination of 10°?   *Ans.* 64 and 42 sec

6-9. In (6-110) the power-pattern expression can also be written

$$P(\phi) = \frac{\sin (\psi/4)}{\psi/4} \frac{\sin \psi}{4 \sin (\psi/4)}$$

Show that the two expressions are equivalent.

6-10. (*a*) Calculate the power pattern of a continuous uniform array of aperture s_λ; (*b*) repeat for the case where the aperture is divided into two halves, operated as a simple interferometer; (*c*) repeat as in (*b*) but with the two halves operated as a phase-switched interferometer.

6-11. The positions of the moon at the start and end of the lunar occultation of a radio source are

	RA	Dec (N.)
Start	03ʰ34ᵐ05ˢ	15°56′55″
End	03 35 45	16 04 37

If the moon's diameter was 33′04″ during the occultation, find the two possible positions of the radio source.

6-12. Calculate and construct a graph of the lunar occultation curve for a uniformly bright source 5 sec of arc wide in the direction of the moon's path. The wavelength is 1 m. Take the earth-moon distance as 3.8×10^8 m. The source position coincides with the moon's path. Compare with the occultation curve for a point source. *Ans.* Relative amplitudes of first maximum, first minimum, and second maximum are 1.37:0.78:1.20 for a point source and 1.28:0.92:1.05 for a 5 sec of arc uniform source

6-13. Differentiate the result of Prob. 6-12, and perform its convolution with the differentiated "interferometer" pattern. Compare the result with the distribution given in Prob. 6-12.

6-14. (*a*) Calculate the beam efficiency of a circular aperture of diameter D_λ with the distribution of (6-167) (replace x by r and L_λ by D_λ) for $K_1 = 0$ and $n = 1$. Assume zero phase errors.   *Ans.* 0.983

    (*b*) Repeat for $n = 2$.   *Ans.* 0.996

    (*c*) Repeat for $K_2 = 0$ (distribution uniform)   *Ans.* 0.838

6-15. Show that the visibility function observed with a simple interferometer of spacing s_λ for a uniform source of width α with a symmetrical uniform bright center of width $\beta = \alpha/6$ is

$$V(s_\lambda) = \frac{\sin (\pi s_\lambda \alpha) + 3 \sin (\pi s_\lambda \alpha/6)}{3\pi s_\lambda \alpha/2}$$

if the center brightness is 4 times the side brightness (see Fig. 6-19*b*).

6-16. Show that the visibility function observed with a simple interferometer of spacing s_λ for two equal uniform sources of width $\alpha/6$ spaced between centers by $5\alpha/6$ is

$$V(s_\lambda) = \frac{\sin \pi s_\lambda \alpha}{\pi s_\lambda \alpha} - \frac{2}{3} \frac{\sin (2\pi s_\lambda \alpha/3)}{2\pi s_\lambda \alpha/3}$$

6-17. Calculate the reflector diameter for a multiunit parabolic antenna array designed to have the maximum aperture per unit cost at 3 cm wavelength as based on the following assumptions: unit amplifier cost (maser): $100,000; 85-ft-diameter reflector antenna cost: $400,000. Assume the cost index $\alpha = 1.35$. *Ans.* 75 ft

6-18. For the phase-switched interferometer of Fig. P6-18 show that the relative power pattern is given by

$$P(\phi) = \sum_{k=1}^{8} \cos (2k - 1)\psi' \simeq \frac{\sin 16\psi'}{16\psi'}$$

where $\psi' = 2\pi a_\lambda \sin \phi$, where a_λ is the basic spacing unit. The overall array size is 15 a_λ, and the individual (nonisotropic) elements are spaced at multiples of a_λ from the left-hand (0) element, as indicated by the integers 2, 4, 6, etc. *Hint:* See Barber (1958). Note that the above result is equivalent to that of (6-110) for an interferometer with a continuous array of length $6a_\lambda$ replacing the four-element array in Fig. P6-18.

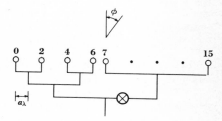

Fig. P6-18. Interferometer arrangement.

6-19. Show that the output of a simple interferometer of spacing s and bandwidth $\nu_0 \pm \Delta\nu/2$ for a point source is given by

$$S_0 \Delta\nu \left[1 + \frac{\sin \dfrac{2\pi s}{c} \dfrac{\Delta\nu}{2} \phi}{\dfrac{2\pi s}{c} \dfrac{\Delta\nu}{2} \phi} \cos \frac{2\pi s}{c} \nu_0\phi \right]$$

where c = velocity of light. Note that this expression can be obtained by integrating a relation similar to (6-75) with frequency replacing angle as the variable of integration.

6-20. Show that for an interferometer with bandwidth $\nu_0 \pm \Delta\nu/2$ (Prob. 6-19), the condition $\Delta\nu/\nu_0 \ll 1/n$, where n = fringe order, must hold in order that the fringe amplitude not be decreased.

6-21. Show that when the spacing of an interferometer approaches zero, $S(\phi_0, s_\lambda)$ approaches S_0, the source flux density.

6-22. A uniform linear array has 16 isotropic in-phase point sources with a spacing of one-half wavelength. Calculate exactly (*a*) the half-power beam width, (*b*) the level of the first side lobe, (c) the beam solid angle, (d) the beam efficiency, (e) the directivity, and (f) the effective aperture.

Ans. (a) $6°22'$, (b) -13.35 dB, (c) $\pi/4$ rad^2, (d) 0.88, (e) 16, (f) $1.27\lambda^2$

6-23. A square unidirectional aperture (x_1y_1) is 10 wavelengths on a side and has a design distribution for the electric field which is uniform in the x direction but triangular in the y direction with maximum at the center and zero at the edges. Design phase is constant across the aperture. However, in the actual aperture distribution there is a plus-and-minus-30° sinusoidal phase variation in the x direction with several phase cycles per wavelength. Calculate (a) the design directivity, (b) the utilization factor, (c) the actual directivity, (d) the achievement factor, (e) the effective aperture, and (f) the aperture efficiency.

Ans. (a) 940, (b) $\frac{3}{4}$, (c) 818, (d) 0.87, (e) $65.2\lambda^2$, (f) 0.65

6-24. Apply the Fourier-transform method to obtain the far-field pattern of an array of two equal in-phase isotropic point sources with a separation d. Reduce the expression to its simplest trigonometric form.

6-25. An array consists of two isotropic point sources, one at the origin and one at a distance of one-half wavelength in the x direction. If the source at the origin has twice the amplitude (field) of the other source, find the position of the phase center of the array.

Ans. On the x axis at $x = 0.167, 0.147, 0.044,$ and 0.0 wavelength for $\theta = 0, 30, 60,$ and 90°, respectively, where θ is measured from the y axis

6-26. A uniform linear array has 24 isotropic point sources with a spacing of one-fourth wavelength. If the phase difference $\delta = -\pi/2$ (ordinary end-fire condition) calculate exactly the six quantities listed in Prob. 6-22.

Ans. 44°, -13.3 dB, 0.17π, 0.89, 24, $1.9\lambda^2$

6-27. Calculate the antenna temperature of the array of Prob. 6-26 if it is located over a flat nonreflecting ground and directed at the zenith. The sky brightness temperature is 5°K between the zenith and 45° from the zenith, 50°K between 45° from the zenith and the horizon, and 300°K for the ground (below the horizon). The antenna is 99 percent efficient ($k_o = 0.99$) and is at a physical temperature of 300°K.

Ans. 13.5°K

6-28. Verify the half-power beam width and first side-lobe level data given for various apertures in Appendix 6.

6-29. For the following aperture distributions show that the far-field patterns are as given:

(a) $E(\phi) = \dfrac{2}{3}\dfrac{\sin\psi}{\psi} + \dfrac{1}{3}\dfrac{\sin\psi/2}{\psi/2}$ where $\psi = \pi L_\lambda \sin\phi$

      Stepped

(b) $E(\phi) = \dfrac{2J_1(\psi)}{\psi}$ where $\psi = 2\pi L_\lambda \sin\phi$

      Circular

(c) $E(\phi) = \dfrac{\sin^2\psi}{\psi^2}$ where $\psi = \dfrac{\pi}{2}L_\lambda \sin\phi$

      Triangular

(d) $E(\phi) = \dfrac{\sin\psi}{\psi} + j\left(\dfrac{\cos\psi}{\psi} - \dfrac{\sin\psi}{\psi^2}\right)$ where $\psi = \pi L_\lambda \sin\phi$

      Triangular
      asymmetric

6-30. Show that the directivity of an antenna may be expressed as

$$D = \frac{4\pi}{\lambda^2} \frac{\displaystyle\iint_{A_p} E(x,y)\ dx\ dy \iint_{A_p} E^*(x,y)\ dx\ dy}{\displaystyle\iint_{A_p} E(x,y)\ E^*(x,y)\ dx\ dy}$$

here $E(x,y)$ is the aperture field distribution.

6-31. A circular parabolic dish antenna has an effective aperture of 100 m². If ne 45° sector of the parabola is removed, find the new effective aperture. The rest of ie antenna, including the feed, is unchanged. *Ans.* 76.6 m²

6-32. (a) On a contour map of a radio source obtained with a radio telescope iow that the flux density of the source is given approximately by

$$S = \frac{2k}{A\,\Omega_M} \iint_{\text{source}} \Delta T_A(\theta,\phi)\ d\Omega \tag{1}$$

here $\Delta T_A(\theta,\phi)$ is the antenna temperature due to the source and where integration is irried out over the observed source extent. Ω_M may be determined from (6-172) or ? in Prob. 6-33. Discuss cases where Ω'_M might be more appropriate than Ω_M. It is ssumed that $\Delta T(\text{max})$ is measured at the antenna terminals.

(b) Show that for a point source

$$S = \frac{2k\,\Delta T_A(\text{max})}{A_e} \tag{2}$$

(c) Show that an equivalent expression to (1) is given by

$$S = \frac{2k}{A_e} \sum \Delta T_A(\theta,\phi) \tag{3}$$

here the $\Delta T_A(\theta,\phi)$ values are summed over the observed source extent at points spaced y the HPBW in each direction (α and δ) so as to form a rectangular grid.

6-33. Show that by definition the main-beam area can be obtained from the easured contour map of an isolated point source as

$$\Omega_M = \frac{\iint \Delta T_A(\theta,\phi)\ d\Omega}{\Delta T_A(\text{max})}$$

'ote that the contour map of $\Delta T_A(\theta,\phi)$ is the true antenna pattern and that the inte-ation is over this pattern, not including minor lobes if these are present.

7

Radio-Telescope
Receivers by *Martti E. Tiuri*†

7-1 General Principles of Radio-Telescope Receivers

7-1a. Introduction. The function of a radio-telescope receiver is
detect and measure the radio emission of celestial sources. In most cas
the emission consists of incoherent radiation whose statistical properti
do not differ from the noise originating in the receiver or from the bac
ground radiation coupled to the receiver by the antenna. The power lev
of the signal in radio-telescope receivers is usually quite small, of the ord
of 10^{-15} to 10^{-20} watt. The power received from the background (Fig. 7-1
may be much higher than this, so that both high sensitivity and high st
bility of the receiver are important requirements. However, there a
cases involving, for example, solar bursts or Jovian radiation, where th
radiation is relatively strong, and other receiver characteristics, such a
the ability to detect the signal spectrum as a function of time, becom
important.

7-1b. Receiver Types. Radio-telescope receivers are basically sim
lar in construction to receivers used in other branches of radio science an
engineering. The most common type is the superheterodyne receive
Figure 7-2 gives the block diagram of a typical *superheterodyne receive*
The signal power, having a center frequency ν_{RF}, is coupled to the receiv
by an antenna and is first amplified in a radio-frequency (RF) amplifie
with a gain of the order of 10 to 30 dB. The next stage is a mixer, wher
the weak signal is mixed with a strong local-oscillator signal at a frequenc
ν_0 producing an output signal on an intermediate frequency (IF), the I
signal power being directly proportional to the RF signal power. Th
IF signal is then amplified with a gain of the order of 60 to 90 dB. Th
largest part of the gain in a superheterodyne receiver is obtained in this I
amplifier, which also usually determines the predetector bandwidth of th
receiver. The IF amplifier is followed by a detector, which is normally

† Institute of Technology, Helsinki, Finland.
‡ The power is related to the antenna temperature as in (7-3).

236

square-law device in radio-telescope receivers (d-c output voltage proportional to the input-voltage amplitude squared). This means that the output d-c voltage of the detector is directly proportional to the output noise power of the predetection section of the receiver. Final stages may

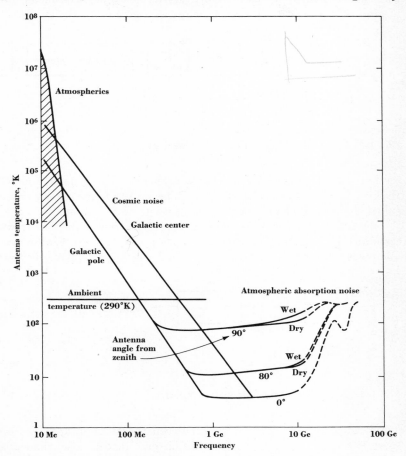

Fig. 7-1. Antenna sky noise temperature as a function of frequency and antenna angle. A beam angle (HPBW) of less than a few degrees and 100 percent beam efficiency are assumed. (*After Kraus and Ko*, 1957, *cosmic noise below 1 Gc; Penzias and Wilson*, 1965, *and Dicke et al.*, 1965, *cosmic noise above 1 Gc; Croom*, 1964, *atmospheric noise; and CCIR*, 1964, *atmospherics*).

consist of a low-pass amplifier or integrator and a data-recording system, such as an analog recorder or a digital output system. The integrator integrates the observed signal power for a predetermined length of time. The actual value used, commonly of the order of seconds, is usually a compromise between too short a period, for which the output noise is excessive, and too long a period, causing excessive smoothing and loss of information.

In superheterodyne receivers (Fig. 7-2) the section after the mixer is the same for all frequencies. Only the RF amplifier, the mixer, and the local oscillator must be designed separately for each frequency range. The

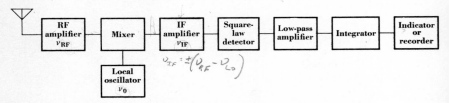

Fig. 7-2. A superheterodyne radio-telescope receiver.

section before the detector is usually called the high-frequency part of the receiver or the predetection section. The section following the detector is called the low-frequency part or the postdetection section.

Figure 7-3 shows a *two-channel superheterodyne receiver*. It was the standard receiver on microwave frequencies prior to the advent of low-noise microwave amplifiers, and it is still widely used, particularly in the milli-

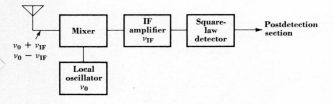

Fig. 7-3. Two-channel superheterodyne receiver.

meter range. If no filter is used between the antenna and the receiver, there is no RF selectivity before the mixer. In this case, the signal frequency

$$\nu_{RF} = \nu_0 + \nu_{IF} \qquad (7\text{-}1)$$

and the image frequency

$$\nu'_{RF} = \nu_0 - \nu_{IF} \qquad (7\text{-}2)$$

can be received. These two frequencies are usually equally effective in giving intermediate frequency power, and, hence, the receiver has two input channels separated in frequency by $2\nu_{IF}$. In continuum measurements signals in both channels may be practically equal in power and statistically independent, resulting in a receiver which has a sensitivity of about twice that of the one-channel receiver.

The block diagram of the so-called *direct receiver* is shown in Fig. 7-4. This type is frequently used when very broad-band (of the order of 1 octave) measurements are made. The RF section might consist of a traveling-wave tube or a tunnel-diode amplifier.

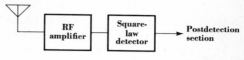

Fig. 7-4. Direct receiver.

Figure 7-5 shows the block diagram of a *video receiver*. The first active stage of this kind of receiver is the detector. Radio-frequency selectivity can be achieved by a suitable filter. The video receiver is mostly used in the millimeter region, where standard superheterodyne receivers are more difficult to construct.

Fig. 7-5. Video receiver.

From a radio-astronomy point of view, receivers can be divided in two groups, *continuum receivers* and *spectral-line receivers*. In the former the *exact* frequency of operation is not critical, but in the latter the precise frequency of reception may be of paramount importance and may also need to be variable (tuning done by changing the local-oscillator frequency). An alternative scheme for spectral observations is to use a *multichannel receiver*, as in Fig. 7-6. This type can be a normal superheterodyne receiver

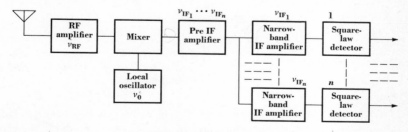

Fig. 7-6. Multichannel superheterodyne receiver.

where a part of the IF amplifier and the rest of the receiver consist of several narrow-band stages in parallel. Multichannel receivers are usually preferred in radio-astronomy work because it is easier to construct several equally sensitive channels than to keep the sensitivity constant when sweeping the local-oscillator frequency.

7-1c. System Noise. As has been discussed in Chap. 3, the noise power per unit bandwidth from an antenna is given by kT_A, where k is Boltzmann's constant and T_A is the effective noise temperature of the antenna radiation resistance. The noise power from the antenna is

$$W_{NA} = kT_A \, \Delta\nu \tag{7-3}$$

where W_{NA} = antenna noise power, watts

k = Boltzmann's constant ($= 1.38 \times 10^{-23}$ joule °K⁻¹)

T_A = antenna temperature, °K

$\Delta\nu$ = bandwidth, cps

The receiver also contributes noise due to the thermal noise in the receiver components, shot noise in the tubes or transistors, etc. In addition, losses in the transmission line (coaxial line or wave guide) between the antenna and receiver will add noise. Thus, the total or system noise power at the antenna terminals is

$$W_{\text{sys}} = W_{NA} + W_{NR} = k(T_A + T_{RT}) \, \Delta\nu \tag{7-4}$$

and the total or system noise temperature referred to the antenna terminals (Fig. 7-7) is

$$T_{\text{sys}} = T_A + T_{RT} \tag{7-4a}$$

where T_A = antenna temperature, °K

T_{RT} = receiver noise temperature (including transmission line), °K

W_{NR} = receiver noise power referred to the antenna terminals, watts

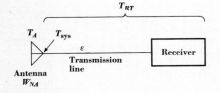

Fig. 7-7. The antenna, transmission line, and receiver contribute to the system temperature.

The system noise temperature T_{sys} of radio telescopes varies from ten or so degrees Kelvin to thousands of degrees Kelvin, depending on the frequency and type of antenna and receiver. The signal temperature ΔT may be a small part of 1 °K. Hence, the receiver must be able to detect small differences in the total noise. The basic theory of the noise temperature of receivers is discussed in more detail in Sec. 7-2.

7-1d. Total-power Receiver and Its Sensitivity. Any receiver which measures the total noise power from the antenna and from the receiver is called a *total-power receiver*. This is in distinction to receivers which measure, for example, the difference in powers from the antenna and a reference (see Secs. 7-1*f* and following). The characteristics and sensitivity of the total-power receiver will be analyzed in this section.

A block diagram of the total-power receiver is presented in Fig. 7-8. It is assumed that the amplifiers of the receiver are linear and have constant gain and that the bandpass characteristics are rectangular. The detector is assumed to be of the square-law type. The system noise temperature is taken to be T_{sys}, and ΔT is the signal noise temperature or change in

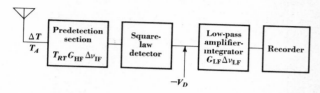

Fig. 7-8.   Total-power receiver.

antenna temperature to be measured. Figure 7-9 shows the voltage wave forms and power spectra at different parts of the receiver. The incident power consists of broad-band noise. The RF amplifiers and mixer accept only the frequency components around the signal frequency ν_{RF}, while the mixer converts this spectrum to the IF frequency ν_{IF}. The predetection part of the receiver is assumed to have a rectangular passband of width $\Delta\nu_{\text{HF}}$, determined effectively by the bandwidth of the IF amplifier. The filtered voltage V_{IF} at the IF amplifier output resembles a randomly modulated carrier wave of frequency ν_{IF} (Fig. 7-9). The amplitude v of the envelope of the wave form has a Rayleigh distribution, as shown in Fig. 7-10 (Rice, 1945):

$$P(v) = \frac{v}{V_{\text{eff}}}\, e^{-(v^2/2V^2_{\text{eff}})} \qquad (7\text{-}5)$$

The most probable value of the envelope amplitude v is the rms value V_{eff} of the noise. The bandwidth $\Delta\nu_{\text{HF}}$ allows the envelope to have a rise time of approximately $1/\Delta\nu_{\text{HF}}$ sec. This means that after the square-law detector there are about $\Delta\nu_{\text{HF}}$ independent noise pulses per second. In practical receivers, $\Delta\nu_{\text{HF}}$ varies from a few kilocycles per second to tens of megacycles per second.

The IF amplifier output power W_{HF} is

$$W_{\text{HF}} = G_{\text{HF}}k(T_{\text{sys}} + \Delta T)\,\Delta\nu_{\text{HF}} \qquad \text{watts} \qquad (7\text{-}6)$$

where G_{HF} = power gain of predetection section

This power is fed to the square-law detector.   The detector has an output voltage V_{det}, which varies with the input voltage v, according to

$$V_{det} = \alpha v^2 \tag{7-7}$$

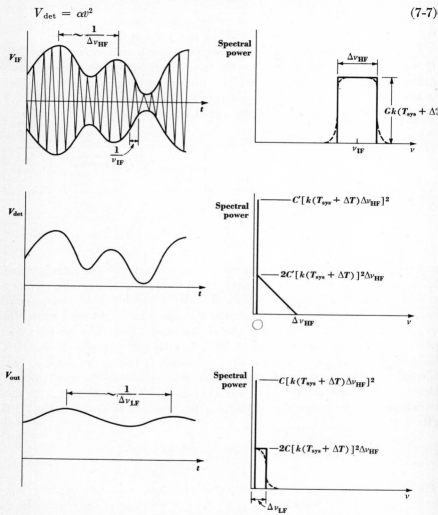

Fig. 7-9.   Voltage wave forms and spectral power (watts cps^{-1}) spectra at different stages in a radio-telescope receiver.

where α is a constant.   Hence, the detector d-c output voltage, $V_D + \Delta V$, is directly proportional to the input power, or

$$V_D + \Delta V = \beta G_{HF}kT_{sys}\,\Delta\nu_{HF} + \beta G_{HF}k\Delta T\,\Delta\nu_{HF} \tag{7-8}$$

where β is a constant. In most square-law detectors V_{eff}, and hence V_D, must be kept small (about 0.1 volt) in order to ensure proper square-law operation. Because $\Delta V \ll V_D$ ($\Delta T \ll T_{\text{sys}}$) high amplification of ΔV is needed in order to get an output indication on the recorder. To make the amplification easier the voltage V_D due to the system noise T_{sys} is now canceled by a d-c voltage $-V_D$, leaving ΔV as the output signal voltage from the detector. The corresponding signal power is equal to

$$W' = C'(k \, \Delta T \, \Delta\nu_{\text{HF}})^2 \qquad \text{watts} \tag{7-9}$$

where C' is a constant, watt^{-1}.

In addition to ΔV, a noise voltage also exists at the detector output. The low-frequency components of this voltage from d-c to $\Delta\nu_{\text{HF}}$ cps are due to different IF noise-voltage components in the frequency range from

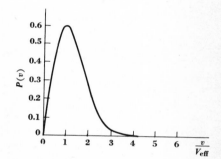

Fig. 7-10. Probability distribution of the amplitude of the envelope of the predetection output noise voltage. (*After Rice*, 1945.)

$\nu_{\text{IF}} - \Delta\nu_{\text{HF}}/2$ to $\nu_{\text{IF}} + \Delta\nu_{\text{HF}}/2$ beating with each other in the detector. The resulting low-frequency (LF) power spectrum is triangularly shaped, because the number of IF noise-voltage components giving a certain noise component at frequency ν_{LF} is proportional to $\Delta\nu_{\text{HF}} - \nu_{\text{LF}}$ (ν_{LF} varies from zero to $\Delta\nu_{\text{HF}}$). The maximum LF power density close to zero frequency is equal to (Tiuri, 1964)

$$W_{\text{LF}_{\max}} = 2C'(kT_{\text{sys}})^2 \, \Delta\nu_{\text{HF}} \tag{7-10}$$

assuming $\Delta T \ll T_{\text{sys}}$.

The detector output voltage is fed to the recorder through a low-pass filter amplifier in order to reduce fluctuations. The effect of a low-pass filter is often obtained by a long-time-constant (order of seconds) RC integrator circuit so that the effective bandwidth $\Delta\nu_{\text{LF}}$ of the low-pass filter is much smaller than the bandwidth of predetection section $\Delta\nu_{\text{HF}}$. If the low-pass amplifier filter has a rectangular passband from zero to $\Delta\nu_{\text{LF}}$ and a power gain G_{LF}, then the fluctuating noise power output is equal to

$$W_{\text{LF}} = G_{\text{LF}} 2C'(kT_{\text{sys}})^2 \, \Delta\nu_{\text{HF}} \, \Delta\nu_{\text{LF}} \tag{7-11}$$

The corresponding signal power due to the noise temperature ΔT is from (7-9) equal to

$$W = G_{LF}C'(k\, \Delta T\, \Delta \nu_{HF})^2 \tag{7-12}$$

The sensitivity or the minimum detectable signal ΔT_{min} of the radiometer is defined to be the signal noise temperature ΔT which produces a receiver d-c output power W equal to the noise output power W_{LF}. Hence from (7-11) and (7-12)

$$\Delta T_{min} = T_{sys}\sqrt{\frac{2\Delta\nu_{LF}}{\Delta\nu_{HF}}} \tag{7-13}$$

A smaller ΔT_{min} corresponds to a higher sensitivity.

In deriving (7-13) rectangular bandpass characteristics were assumed for the pre- and postdetection sections. For actual filter characteristics the following equivalent bandwidths can be calculated (Tiuri, 1964).

$$\Delta\nu_{HF} = \frac{\left[\int_0^\infty G_{HF}(\nu)\, d\nu\right]^2}{\int_0^\infty [G_{HF}(\nu)]^2\, d\nu} \tag{7-14}$$

$$\Delta\nu_{LF} = \frac{\int_0^\infty G_{LF}(\nu)\, d\nu}{G_{LF}(0)} \tag{7-15}$$

where $G_{HF}(\nu)$ = predetection power gain [= (voltage gain)^2] as function of frequency

$G_{LF}(\nu)$ = postdetection power gain

$G_{LF}(0)$ = postdetection d-c power gain

An ideal integrator with an integration time t_{LF} can be used as a low-pass filter. An ideal integrator has the following gain function $G(\nu)$

$$G(\nu) = \frac{\sin^2\,(\tfrac{1}{2}\,\omega t_{LF})}{(\tfrac{1}{2}\,\omega t_{LF})^2} \tag{7-16}$$

The corresponding postdetection bandwidth $\Delta\nu_{LF}$ is, from (7-15), equal to

$$\Delta\nu_{LF} = \frac{1}{2t_{LF}} \tag{7-17}$$

Hence,

$$\Delta T_{min} = \frac{T_{sys}}{\sqrt{\Delta\nu_{HF}\, t_{LF}}} \tag{7-18}$$

The equivalent ideal integration time of any type of smoothing filter can be calculated from (7-15) and (7-17) as

$$t_{LF} = \frac{G_{LF}(0)}{2\int_0^\infty G_{LF}(\nu)\, d\nu} \tag{7-19}$$

Table 7-1 shows some predetection equivalent bandwidths $\Delta\nu_{\mathrm{HF}}$ calculated from (7-14) for certain IF amplifier circuits. For single-tuned circuits, $\Delta\nu_{\mathrm{HF}}$ is more than 1.5 times the half-power bandwidth. For staggered-tuned circuits or double-tuned circuits it is close to the 3-dB bandwidth.

Table 7-1
Predetection equivalent bandwidth
$\Delta\nu_{\mathrm{HF}}$ ***(7-14) as compared to***
half-power bandwidth $\Delta\nu$

Type of filter	$G(\nu)$	$\Delta\nu_{\mathrm{HF}}/\Delta\nu$
n cascaded single-tuned stages	$\left[1 + \left(\dfrac{\Delta\omega}{\pi\,\Delta\nu_s}\right)^2\right]^{-n}$	
$\quad n = 1$		3.14
$\quad n = 2$		1.96
$\quad n = 3$		1.76
$\quad n = 5$		1.62
$\quad n = \infty$ (Gaussian)	$2^{-[\Delta\omega/(\pi\Delta\nu)]^2}$	1.50
m cascaded $2n$-pole Butterworth filters	$\left[1 + \left(\dfrac{\Delta\omega}{\pi\,\Delta\nu_s}\right)^{2n}\right]^{-m}$	
$\quad m = 1,\, n = 2$		1.48
$\quad m = 1,\, n = 3$		1.26
$\quad m = 1,\, n = \infty$		1.00
$\quad m = 2,\, n = 2$		1.30

$\Delta\omega = \omega - \omega_0$, where ω_0 = angular center frequency
$\Delta\nu_s$ = 3-dB bandwidth of a single section
$\Delta\nu$ = 3-dB bandwidth of the amplifier

Table 7-2 shows some postdetection equivalent bandwidths and the equivalent integration times calculated from (7-15) and (7-19).

Equation (7-18) can also be deduced as follows. As stated earlier, there are $\Delta\nu_{\mathrm{HF}}$ effectively independent noise-pulse contributions per second producing the detector output. These are averaged in the postdetection integrator over the integration time t_{LF}. During this time there are $\Delta\nu_{\mathrm{HF}}\,t_{\mathrm{LF}}$ independent contributions and, therefore, the *root-mean-square deviation (standard deviation)* is $1/\sqrt{\Delta\nu_{\mathrm{HF}}\,t_{\mathrm{LF}}}$ times the average d-c output.

Equations (7-8) to (7-19) assume that a square-law detector is used. The square-law detector is used in radio-telescope receivers because it makes it possible to have an output calibration independent of the detector power level with fixed predetection gain. The system noise temperature may vary slowly because of variations in background noise, and the detector input power will change correspondingly. From (7-8), when the postdetection gain is constant, the receiver output d-c voltage will be

$$V_R = \gamma T_{\mathrm{sys}} + \gamma\,\Delta T \tag{7-20}$$

where γ is a constant. Hence a certain ΔT will cause the same deflection in the recorder independent of T_{sys}, as illustrated in Fig. 7-11.

Table 7-2
Postdetection equivalent bandwidth
$\Delta\nu_{\text{LF}}$ **(7-15) as compared to half-power**
bandwidth $\Delta\nu$ **and equivalent**
integration time t_{LF} **(7-19)**

Type of filter	$G_{\text{LF}}(\nu)$	$\Delta\nu_{\text{LF}}$	t_{LF}
Pure integrator, integrating time t_{LF}	$\dfrac{\sin^2\left(\frac{1}{2}\omega t_{\text{LF}}\right)}{\left(\frac{1}{2}\omega t_{\text{LF}}\right)^2}$	$\dfrac{1}{2t_{\text{LF}}}$	t_{LF}
Ideal low-pass filter	1	$\Delta\nu$	$1/2\Delta\nu$
n RC filters in independent cascade, time constant t_{RC}	$(1 + \omega^2 t_{RC}^2)^{-n}$		
$n = 1$		$1.57\Delta\nu$	$2t_{RC}$
$n = 2$		$1.22\Delta\nu$	$4t_{RC}$
$n = \infty$ (Gaussian)		$1.06\Delta\nu$	$1/2.12\Delta\nu$
Second-order filter (ω_0 = undamped natural frequency of the filter; ζ = damping constant)	$\dfrac{\omega_0^4}{(\omega_0^2 - \omega^2)^2 + (2\zeta\omega\omega_0)^2}$	$\dfrac{\omega_0}{8\zeta}$	$\dfrac{4\zeta}{\omega_0}$

$\Delta\nu$ = 3-dB bandwidth

With detector laws differing from square law, γ in (7-20) depends on T_{sys}. It has been shown that the sensitivity of the receiver is highest when a square-law detector is used (Kelly, Lyons, and Root, 1963).

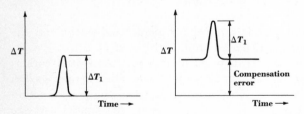

Fig. 7-11. Effect of d-c compensation error with square-law detector.

The minimum detectable signal temperature ΔT_{min} is equivalent to an output signal-to-noise ratio of unity. To increase the probability that the temperature change is significant ΔT should be higher than ΔT_{min}. A common criterion is that the weakest significant source should produce a deflection equal to the peak-to-peak value of the noise fluctuations. The probability distribution of the fluctuations depends on filter and detector characteristics. With a square-law detector and $\Delta\nu_{\text{HF}} \gg \Delta\nu_{\text{LF}}$, the instantaneous value of the low-frequency fluctuations has a Gaussian distribution (Bell, 1960; Kac and Siegert, 1947). In this case the probability of the fluctuation amplitude exceeding 4 times the rms value is only 3×10^{-5}, which means that this event may occur once every $3.3 \times 10^4 \ t_{LF}$ seconds. For example, when $t_{LF} = 5$ sec, this interval is about 2 days.

Example 1. Find the sensitivity of a radiometer when the system noise temperature is 150°K, the IF amplifier has five single-tuned stages with total 3-dB bandwidth of 5 Mc, and the integrator is a simple RC circuit with time constant of 3 sec.

Solution. From Table 7-1:
$$\Delta \nu_{HF} = 1.62 \times \Delta \nu = 8.1 \ \text{Mc}$$
From Table 7-2:
$$t_{LF} = 2t_{RC} = 6 \ \text{sec}$$
From (7-18):
$$\Delta T_{\min} = \frac{150}{\sqrt{8.1 \times 10^6 \times 6}} = 0.02°\text{K}$$

Example 2. The radiometer receiver of Example 1 delivers an output voltage of 1 volt to the recorder when $\Delta T = 1°$K. The recorder input impedance is 1,000 ohms. Find (approximately) the required HF and LF power gains, when the rms voltage at the detector input is 0.1 volt and the equivalent detector input impedance is 5,000 ohms.

Solution. The receiver input power is (for $\Delta \nu_N$ see Sec. 7-2e)
$$W_{NA} + W_{NR} = kT_{\text{sys}} \Delta \nu_N = 1.38 \times 10^{-23} \times 150 \times 1.11 \times 5 \times 10^6$$
$$= 1.15 \times 10^{-14} \ \text{watt}$$
Detector input power is
$$W_d = \frac{V^2}{R} = \frac{0.1^2}{5 \times 10^3} = 2 \times 10^{-6} \ \text{watt}$$
HF power gain is
$$G_{HF} = \frac{W_d}{W_{NA} + W_{NR}} = 1.8 \times 10^8 = 82 \ \text{dB}$$
Actual gain of the amplifiers must be higher due to the losses in the mixer, etc.

The detector output d-c voltage V_D can be assumed to be close to 0.1 volt. Output d-c voltage due to signal temperature ΔT is then $\Delta V = (\Delta T / T_{\text{sys}}) V_D = \frac{1}{150} \times 0.1$ volt = 0.67 mv. Assuming a source impedance of 5,000 ohms, the signal power is $W_D = 0.9 \times 10^{-10}$ watt. The required signal power to the recorder is $W_R = 10^{-3}$ watt. Hence, the low-frequency gain is $G_{LF} = W_R / W_D = 0.9 \times 10^7 = 69.5$ dB. The actual gain must be higher due to the losses in integration, etc.

7-1e. Sensitivity Reduction Due to Receiver Instability.

In the analysis of sensitivity in the preceding section it was assumed that the receiver gain was constant. In practice gain variations are unavoidable, especially with the very high gains that are necessary in radio-telescope receivers (see Example 2 just preceding). Short- and long-period gain variations can occur that are due, for example, to supply-voltage variations and to ambient-temperature fluctuations. Gain variations of a few percent are common in radio receivers, and only by carefully stabilizing all supply voltages and the operating temperature can gain stabilities of the order of to 0.1 percent per hour be achieved (Steinberg, 1952; Yaroshenko, 1964).

The detector cannot distinguish an increase in signal power from an increase due to higher predetection gain. According to (7-8), an increase in G_{HF} by a factor f will result in a signal-temperature increase

$$\Delta T = f T_{sys} \tag{7-21}$$

For the case of the receiver in Example 1 (just preceding), where $T_{sys} = 150°K$, an increase of gain by 1 percent gives a change in temperature at $\Delta T = 1.5°K$, which is more than 70 times the sensitivity ($0.02°K$) the receiver would have if it were completely stable.

Output fluctuations due to gain variations are independent of the fluctuations resulting from system noise. Hence, the actual sensitivity of the total-power receiver can be stated as

$$\Delta T_{min} = T_{sys} \sqrt{\frac{1}{\Delta \nu_{HF}\, t_{LF}} + \left(\frac{\Delta G}{G_{HF}}\right)^2} \tag{7-22}$$

where G_{HF} = average predetection power gain

ΔG = effective value of the detected receiver power-gain variations

Variation in postdetection gain will have no effect on the receiver sensitivity when d-c compensation (canceling of V_D) in the detector is used. These variations will change the calibration only.

Variations of the receiver noise temperature and of the bandwidth of the receiver decrease the sensitivity in the same way as gain instability.

Several methods for gain stabilization in a total-power receiver have been proposed and tested (Colvin, 1961; Cooper, 1961). Automatic-gain-control (AGC) schemes are not very useful in practice because an AGC is apt to change the receiver noise figure and the receiver bandwidth and so introduce spurious output fluctuations.

7-1f. Dicke Receiver. Measurements show that gain fluctuations decrease rapidly as their frequency increases (Steinberg, 1952; Yaroshenko, 1964). Figure 7-12 shows some results of gain-variation measurements concerning audio-frequency amplifiers with a total gain of 120 dB (Yaroshenko, 1964). The variation of gain with frequency can be expressed as

$$\frac{G(\nu)}{G} = K\nu^{-\alpha}$$

where K is a constant (dependent on the stability of supply voltages) and α is between 2 and 2.5. Figure 7-12 indicates that d-c supply voltages produce a gain stability that is about 1000 times better than unregulated supply voltages. High-frequency amplifiers probably have considerably higher gain variations than audio amplifiers, but the spectral density of gain variations can be assumed to go down according to the same law and will be negligible above a few tens of cycles.

The effect of gain variations can be reduced if the receiver input is continuously switched between the antenna and a comparison noise source at a frequency high enough so that the gain has no time to change during one cycle. Figure 7-13 shows the block diagram of such a switched receiver,

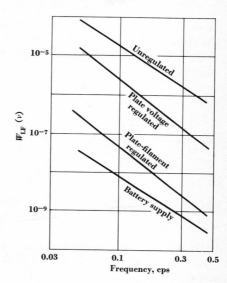

Fig. 7-12. Dependence of the gain-variation spectrum on different supply voltages (audio amplifier with 120 dB gain). (*After Yaroshenko*, 1964.)

introduced by *Dicke* (1946). The antenna and the comparison load are connected alternately to the receiver half of the time at the switching frequency ν_M, which ranges in practice from about 10 to 1,000 cycles. Receivers commonly employ semiconductor diode switches or ferrite

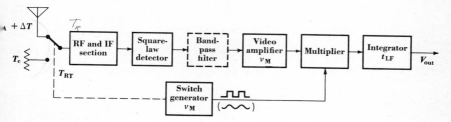

Fig. 7-13. Switched or Dicke receiver (bandpass filter optional).

switches having an attenuation of 0.2 to 0.3 dB in "on" position and 20 to 40 dB in "off" position.

The noise power from the antenna is equal to

$$W_{NA} = kT_A \, \Delta\nu_{HF} \tag{7-23}$$

The noise power from the comparison load is equal to

$$W_{NC} = kT_C \, \Delta\nu_{\mathrm{HF}} \tag{7-24}$$

If W_{NC} equals W_{NA} ($T_C = T_A$), then the signal power

$$W_{NS} = k \, \Delta T \, \Delta\nu_{\mathrm{HF}} \tag{7-25}$$

is square-wave amplitude modulated at the switching frequency ν_M because the signal can enter the receiver only during alternate half cycles. This amplitude-modulated signal is detected and fed to the multiplier (also called a phase-sensitive detector or a synchronous demodulator), as shown in Fig. 7-13. The multiplier switches the receiver detector output in synchronism with the antenna switch (see Fig. 7-14). Hence, the output

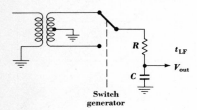

Switch
generator

Fig. 7-14. Multiplier and integrator stages of a Dicke receiver.

voltage from the integrator will be directly proportional to the signal temperature ΔT, and there will be no d-c output if $T_C = T_A$ with no signal present. This assumes that the gain and noise temperatures of the receiver remain the same in both switch positions.

When the receiver is connected to the antenna, the d-c output voltage is

$$V_A = C_1 G_{\mathrm{HF}} k (T_A + T_{RT}) \, \Delta\nu_{\mathrm{HF}} \tag{7-26}$$

When it is connected to the comparison load it is

$$V_C = C_1 G_{\mathrm{HF}} k (T_C + T_{RT}) \, \Delta\nu_{\mathrm{HF}} \tag{7-27}$$

Hence, the output voltage from the multiplier is

$$V = C_2(V_A - V_C) = C_1 C_2 G_{\mathrm{HF}} k (T_A - T_C) \, \Delta\nu_{\mathrm{HF}} \tag{7-28}$$

Gain variations ΔG will now give an output indication

$$\Delta T_G = (T_A - T_C) \frac{\Delta G}{G_{\mathrm{HF}}} \tag{7-29}$$

If $T_C = T_A$, output-voltage fluctuations due to gain instability disappear, and the receiver sensitivity is determined by the system noise alone.

In Dicke receivers the signal is connected to the receiver half of the time only (for square-wave modulation). Hence, the sensitivity of the Dicke receiver is

$$\Delta T_{\min} = 2 \frac{T_{\mathrm{sys}}}{\sqrt{\Delta\nu_{\mathrm{HF}} \, t_{\mathrm{LF}}}} \tag{7-30}$$

or one-half of the theoretical sensitivity of the total-power system discussed in Sec. 7-1*d*. In (7-30) square-wave modulation and square-wave multiplication are assumed.

It is advantageous to introduce the required postdetection gain before the multiplier stage, in which case a video amplifier can be used instead of a d-c amplifier. The bandwidth of the video amplifier must be from ν_M to about 10 times ν_M in order to include all the important harmonics of the square-wave signal.

The amplifier will require fewer stages and the danger of overloading will be avoided if the video amplifier is made narrow band (ν_M plus or minus a few cycles), which can be done by using a narrow bandpass filter, such as shown in Fig. 7-13 by dashed lines.† This means some reduction in the sensitivity, because only the first harmonic of the square-wave signal reaches the multiplier. The amplitude of the first harmonic is $4/\pi$ times the amplitude of the square wave, and its effective value is $4/\pi\sqrt{2}$, hence, $\Delta T_{\min}$ in (7-30) must be multiplied by $(\pi\sqrt{2})/4$. The sensitivity equation for a Dicke receiver with square-wave input modulation and narrow-band filter is then given by

$$\Delta T_{\min} = \frac{\pi}{\sqrt{2}} \frac{T_{\text{sys}}}{\sqrt{\Delta\nu_{\text{HF}}\, t_{\text{LF}}}} \tag{7-31}$$

Because the reduction in the sensitivity is only a little more than 10 percent [as compared to (7-30)], this modification of the Dicke receiver is usually preferred. In this case the wave form from the switching generator to the multiplier can be sinusoidal. If the input power modulation is sinusoidal instead of square-wave, the sensitivity is further reduced by the factor $4/\pi$.

In the normal Dicke receiver the signal power is observed only half of the time. This may be considered to be an inefficient use of the signal power collected by a radio-telescope antenna. Full efficiency in observing time can be achieved by switching the telescope antenna between two receivers. Both receivers can be of the Dicke type, and the outputs of both can be added (or averaged), thereby increasing the sensitivity by $\sqrt{2}$. By using separate digital outputs for the two receivers the addition can be performed after the observations have been made, or the combining operation can be effected electronically during the actual observation. One method for this has been presented by Graham (1958) and is shown in Fig. 7-15. By adding the two independent observations the sensitivity is improved by a factor of $\sqrt{2}$ as compared to a single receiver.

In the sensitivity equations, (7-30) and (7-31), a balanced receiver ($T_C = T_A$) is assumed. There are several different methods of achieving this. One method is to use a second antenna, which is directed toward cold

---

† This is equivalent to an audio amplifier tuned to ν_M.

regions in the sky, so that its noise temperature is less than that of the telescope antenna.   The balance is then introduced by feeding additional noise to the comparison load from an adjustable noise source through a directional coupler to make T_C equal to T_A.   In hydrogen-line measurements the signal antenna can also serve as the reference load by switching the receiving frequency during the reference period outside the hydrogen-

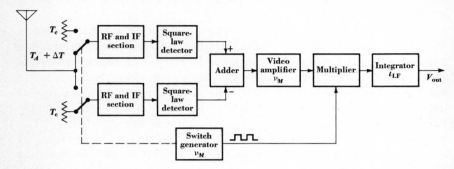

Fig. 7-15.   Graham's receiver.

line part of the spectrum.   An input switch is not required if switching is accomplished by such side tuning of the local oscillator.

A second method is to use an unsymmetrical square wave of suitable form for modulating and multiplying in order to balance the noise powers in both positions of the switch.   This will cause some reduction in sensi-

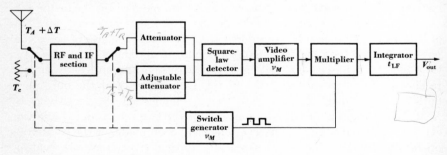

Fig. 7-16.   Dicke receiver using gain modulation for balancing.

tivity, because the modulation of the signal is not so effective as in the symmetrical case.

A useful method for obtaining the balanced operation in Dicke receivers is by means of gain modulation (Orhaug and Waltman, 1962).   Figure 7-16 shows the block diagram of a gain-modulation receiver.   Two stable, passive attenuators are switched alternately in synchronism with the input switch. The HF gain of the receiver with the attenuator is G_A when the signal antenna is connected and G_C when the comparison load is connected. If

$$(T_A + T_{RT})G_A = (T_C + T_{RT})G_C \tag{7-32}$$

then the receiver is balanced. In a low-noise receiver only a small difference between T_A and T_C can be balanced without introducing excessive gain modulation.

The Dicke receiver suffers from gain instability when the signal is present, especially when the signal is relatively strong. Instabilities will distort the shape of the signal curve and reduce the accuracy of the results. Several methods for gain stabilization in Dicke receivers have been presented (Colvin, 1961).

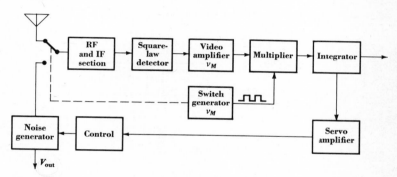

Fig. 7-17. Null-balancing Dicke receiver of Machin, Ryle, and Vonberg.

A receiver where the balanced condition is realized continuously is the *null-balancing* Dicke receiver (Machin, Ryle, and Vonberg, 1952). Its block diagram is shown in Fig. 7-17. The comparison load has an adjustable noise source. The output noise power of the source is controlled by the receiver integrator output, so that the output is always zero. The actual output signal of the radio-telescope receiver is then the controlling signal. On very high and ultrahigh frequencies the comparison source can be a noise diode whose anode current is directly proportional to the noise power. The anode current can then be used as the receiver output. At microwave frequencies the comparison source could be a fixed-output noise generator (discharge tube, cold load, etc.) in connection with a current-controlled attenuator. The null-balancing method can also be combined with the gain-modulation method by using a servo-controlled current-dependent attenuator in the comparison attenuator on the intermediate frequency.

Example 1.   A Dicke receiver has an antenna noise temperature of 25°K and a system noise temperature of 150°K.   The comparison-load temperature is 300°K. Find the value of the attenuator for balancing the receiver by using the gain-modulation method.

Solution.   From $(T_A + T_{RT})G_A = (T_C + T_{RT})G_C$, with $T_{RT} = 125°K$, we get $G_A/G_C = 2.83$.   Hence, a 4.5-dB attenuator must be switched into the IF amplifier when the comparison load is connected to the receiver.

7-1g. Interferometer Receiver.   A radio interferometer consists of two (or more) antennas separated from each other by a distance of several wavelengths (see Chap. 6).   For highest sensitivity each may be equipped with its own preamplifier or the entire predetection section of the receiver (see Fig. 7-18).   It is assumed that the antennas and receivers are identical and that both antennas are pointed in the same direction.   A source will induce equal signal voltages to the receivers with a certain phase difference depending on the source direction.

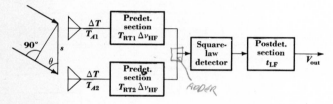

Fig. 7-18.   Simple interferometer receiver.

In such an interferometer receiver the IF output voltages are added and fed to a square-law detector.   Referring to Sec. 6-10, the IF output signal voltages are in the same phase and will produce the maximum d-c signal from the detector when

$$s \cos \theta = n\lambda \tag{7-33}$$

and minimum output when

$$s \cos \theta = \frac{n\lambda}{2}$$

where s = antenna spacing
     θ = direction angle measured from base line
     λ = wavelength
     n = integer

Designating the two antennas and associated amplifiers 1 and 2, the signal power from antenna 1 due to the discrete source is equal to $k \, \Delta T \, \Delta\nu$. In addition to this, antenna 1 adds a noise power $kT_{A1} \, \Delta\nu$, consisting of background noise, noise from side lobes, etc.   Corresponding noise powers from antenna 2 to amplifier 2 are $k \, \Delta T \, \Delta \nu$ and $kT_{A2} \, \Delta \nu$, respectively. Finally, there are receiver noise powers $kT_{RT1} \, \Delta \nu$ and $kT_{RT2} \, \Delta \nu$. Noise powers $kT_{A1} \, \Delta \nu$, $kT_{A2} \, \Delta \nu$, $kT_{RT1} \, \Delta \nu$, and $kT_{RT2} \, \Delta \nu$ can be assumed to be independent. Hence, when no discrete source is present, the detector input noise power is equal to $G_{HF}k(T_{s1} + T_{s2}) \, \Delta \nu_{HF}$, where G_{HF} is the predetection power gain; $T_{s1} = T_{A1} + T_{RT1}$ is the system noise temperature of the receiver with front end 1 and T_{s2} the system noise temperature of the receiver with front end 2.

Comparing this to the total-power receiver (7-11), the fluctuating noise power output from the integrator is equal to

$$W_{LF} = G_{LF}2C'(kT_{s1} + kT_{s2})^2 \, \Delta \nu_{HF} \, \Delta \nu_{LF} \tag{7-34}$$

When a weak discrete source is present, noise voltages proportional to $\sqrt{k} \, \Delta T \, \Delta \nu_{HF}$ are obtained from both IF amplifiers. If these voltages are in phase, a d-c voltage proportional to $4k \, \Delta T \, \Delta \nu_{HF}$ is obtained from the square-law detector. If the phase angle between the IF signal voltages is equal to ϕ, the corresponding d-c voltage is $2k \, \Delta T \, \Delta \nu_{HF}(1 + \cos \phi)$. Comparing this to the total-power receiver (7-12), the signal output power from the integrator is equal to

$$W = G_{LF}C'[2k \, \Delta T(1 + \cos \phi) \, \Delta \nu_{HF}]^2 \tag{7-35}$$

The sensitivity of the simple interferometer is obtained by putting $W = W_{LF}$, giving

$$\Delta T_{min} = \frac{T_{s1} + T_{s2}}{2(1 + \cos \phi)\sqrt{\Delta \nu_{HF} \, t_{LF}}} \tag{7-36}$$

or, assuming $T_{s1} = T_{s2} = T_{sys}$,

$$\Delta T_{min} = \frac{T_{sys}}{(1 + \cos \phi)\sqrt{\Delta \nu_{HF} \, t_{LF}}} \tag{7-37}$$

The maximum sensitivity is twice the sensitivity of the total-power receiver with the same T_{sys}. This is achieved by using two antennas and two receivers.

The simple interferometer suffers from the same drawbacks as the total-power receiver. Synchronous switching in both receivers could be used to stabilize against gain variations, but a correlation method is usually preferred. This is considered in the next section.

7-1h. Correlation Receiver.

In a correlation receiver the IF output voltages from the two separate receivers, as in an interferometer, are multiplied instead of added and detected. The block diagram of the correlation receiver is shown in Fig. 7-19.

When uncorrelated noise voltages due to the noise temperatures T_{s1} and T_{s2} are multiplied, the product has an average value zero (zero d-c output voltage from the multiplier). As in the detector of the total-power receiver the uncorrelated noise-voltage components from both receivers beat with each other in the multiplier, resulting in a low-frequency, fluctuating noise-voltage output. Referring to (7-11) for the total-power receiver, the fluctuating noise power from the integrator will be

$$W_{\text{LF}} = G_{\text{LF}}C'kT_{s1}kT_{s2}\,\Delta\nu_{\text{HF}}\,\Delta\nu_{\text{LF}} \tag{7-38}$$

A discrete source produces correlated signal-noise powers $k\,\Delta T\,\Delta\nu_{\text{HF}}$ at the receiver inputs and corresponding IF output voltages with amplitudes

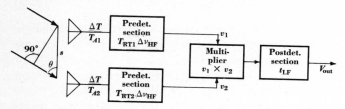

Fig. 7-19.   Interferometer with correlation receiver.

proportional to $\sqrt{k\,\Delta T\,\Delta\nu_{\text{HF}}}$. If both are in phase, the multiplier output d-c voltage is proportional to $k\,\Delta T\,\Delta\nu_{\text{HF}}$. If the phase angle between them is ϕ, the d-c output voltage is equal to $k\,\Delta T\,\Delta\nu_{\text{HF}}\cos\phi$. The signal output power from the integrator is, from (7-12),

$$W = G_{\text{LF}}C'(k\,\Delta T\,\Delta\nu_{\text{HF}})^2\cos^2\phi \tag{7-39}$$

The sensitivity of the correlation receiver is obtained by putting $W = W_{\text{LF}}$ from (7-38) and (7-39), or

$$\Delta T_{\min} = \frac{1}{\cos\phi}\sqrt{\frac{T_{s1}T_{s2}\,\Delta\nu_{\text{LF}}}{\Delta\nu_{\text{HF}}}} \tag{7-40}$$

or if $T_{s1} = T_{s2} = T_{\text{sys}}$ and $t_{\text{LF}} = 1/2\Delta\nu_{\text{LF}}$

$$\Delta T_{\min} = \frac{1}{\sqrt{2}\cos\phi}\frac{T_{\text{sys}}}{\sqrt{\Delta\nu_{\text{HF}}\,t_{\text{LF}}}} \tag{7-41}$$

The sensitivity of a correlation receiver is hence $2\sqrt{2}$ times better than the sensitivity of a Dicke receiver with the same system noise temperature and one antenna.

Because only correlated noise voltages give a d-c output, voltage-gain instabilities will not affect the sensitivity of the correlation receiver. Gain variation will change only the calibration of the receiver. However, random phase variations in the amplifiers of the predetection sections are undesirable (Fujimoto, 1964). For the same reason scintillations in the ionosphere will reduce the sensitivity. One of the advantages of the correlation receiver is that there is no switch and, hence, no extra losses between the antenna and the receiver, which means that the noise temperature of the receiver will be lower.

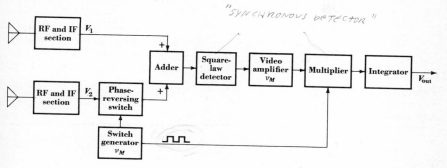

Fig. 7-20. Phase-switched receiver

The correlation principle is also applied in the phase-switched interferometer (Ryle, 1952) in Fig. 7-20. The IF signal of one receiver goes through a phase-reversing switch, which is operated at the frequency ν_M. If signals v_1 and v_2 are uncorrelated, the switching will have no effect on the square-law-detector output. When v_1 and v_2 contain correlated components, the detector output is different for $v_1 + v_2$ and for $v_1 - v_2$. This means that the detector output varies at the frequency ν_M because of the correlated

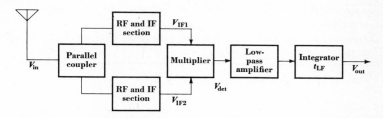

Fig. 7-21. Correlation receiver.

signal. Assuming that the desired signal is the only correlated signal, it is clear that the sensitivity of the phase-switching receiver is the same as the sensitivity of the simple Dicke receiver using a similar low-frequency section (O'Donnell, 1963).

The correlation technique can be used with one antenna by dividing the output signal from the antenna between two identical receivers (Fig. 7-21). In this case the antenna noise power ($T_A/2$) is also correlated in addition to signal noise. Hence, this modification is useful only when T_A

is small (Tiuri, 1964).   In practice it may be difficult to couple two receivers in parallel.   Care must be taken so that the input circuit noise from one receiver does not enter the other receiver, since this will cause an extra correlated signal.   The coupling can be arranged, for example, by using two circulators.

7-1i. Sensitivity Comparison of Radiometers.   A comparison of the sensitivities of the above receivers is presented in Table 7-3.

Table 7-3
*Sensitivity constants K_s of different
radio-telescope receivers*†

Receiver type	K_s
Total-power receiver (Fig. 7-8)	1
Dicke receiver (Fig. 7-13), square-wave modulation, square-wave multiplication	2
Dicke receiver (Fig. 7-13), square-wave modulation, narrow-band video amplifier (sine-wave multiplication)	$\dfrac{\pi}{\sqrt{2}} = 2.22$
Dicke receiver (Fig. 7-13), sine-wave power modulation, narrow-band video amplifier (sine-wave multiplication)	$2\sqrt{2} = 2.83$
Graham's receiver (Fig. 7-15), square-wave modulation, square-wave multiplication	$\sqrt{2} = 1.41$
Simple interferometer‡ (Fig. 7-18)	½
Correlation interferometer‡ (Fig. 7-19) (system noise temperature of one antenna and one receiver $= T_{\text{sys}}$)	$\dfrac{1}{\sqrt{2}} = 0.71$
Phase-switching interferometer (Fig. 7-20), square-wave switching and multiplication	2
Correlation receiver (Fig. 7-21) (antenna noise small in comparison to receiver noise)	$\sqrt{2} = 1.41$

† The constant K_s is defined by

$$\Delta T_{\min} = K_s \frac{T_{\text{sys}}}{\sqrt{\Delta \nu_{\text{HF}}\, t_{\text{LF}}}}$$

where T_{sys} is the system temperature, $\Delta \nu_{\text{HF}}$ is the predetection (high-frequency) equivalent bandwidth (see Table 7-1), t_{LF} is the postdetection (low-frequency) equivalent integration time (see Table 7-2), and $\Delta T_{\min}$ is the minimum detectable temperature (rms system temperature). An increase in sensitivity corresponds to a decrease in $\Delta T_{\min}$.
‡ Two identical antennas.

7-2   Noise Temperature of Receivers

7-2a. Noise Temperature and Noise Figure of a Linear Two-port.   In Sec. 7-1 it was shown that system noise temperature determines the sensitivity of the radio-telescope receiver.   In this section the receiver noise temperature is examined in more detail.

In order to define exactly the noise temperature of the receiver it is useful first to consider its linear two-port network model. A two-port is a device with an input and an output. The predetection section of the receiver is normally linear with respect to the signal. Hence, the receiver ahead of the detector can be considered to consist of cascaded linear two-ports.

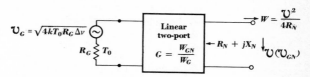

Fig. 7-22. An ideal signal generator and a linear two-port. ($\mho_{GN}$ is due to the noise from the generator).

Figure 7-22 shows a linear two-port with a signal generator feeding it. The signal generator has a source impedance R_G at a reference temperature of $T_0 = 290°$K. The signal generator is assumed to be ideal; i.e., it will generate no other noise than the thermal noise in the resistance R_G. The corresponding noise at frequency ν has an rms emf $\mho_G$ given by

$$\bar{\mho}_G{}^2 = 4R_G k T_0 \, \Delta\nu \tag{7-42}$$

$\Delta\nu$ is a narrow bandwidth extending from ν to $\nu + \Delta\nu$.

The *exchangeable power* (Haus and Adler, 1959) from a two-port is defined as

$$W = \frac{\mho^2}{4R_N} = \frac{g^2}{4G_N} \qquad R_N \neq 0; \; G_N \neq 0 \tag{7-43}$$

where $\mho$ = open-circuit voltage
  g = short-circuit current
$R_N + jX_N$ = output impedance
$G_N + jB_N$ = output admittance

The exchangeable power from a two-port having a positive R_N is equal to the maximum power obtained from the two-port by varying the load impedance. If R_N is negative, W becomes negative.

The exchangeable noise power from the signal generator in Fig. 7-22 is

$$W_G = \frac{\bar{\mho}_G{}^2}{4R_G} = kT_0 \, \Delta\nu \tag{7-44}$$

The exchangeable output noise power from the two-port network due to the generator noise is W_{GN}. The *exchangeable power gain G* of the two-port is defined as

$$G = \frac{W_{GN}}{W_G} \tag{7-45}$$

G is independent of the actual load impedance of the two-port.   It is negative when R_N is negative.   In many cases the individual stages of the receiver are matched, and the exchangeable power gain of a section is the actual power gain of the same section.

The exchangeable output noise power of the two-port due to the noise sources of the two-port is W_N.   The same exchangeable noise power is obtained if the two-port is generating no noise and the temperature of the generator resistance has the value T such that

$$GkT \, \Delta\nu = W_N \tag{7-46}$$

The temperature T is called the (spot) *noise temperature* of the two-port network.   Thus,

$$T = \frac{W_N}{Gk \, \Delta\nu} \tag{7-47}$$

If T and G are known, (7-46) can be used to find the output noise power due to the noise sources inside the two-port.   The (spot) *noise figure F* (Friis, 1944; IRE, 1960) of the two-port is defined by

$$F = \frac{W_{GN} + W_N}{W_{GN}} \tag{7-48}$$

From (7-44) to (7-46),

$$F = 1 + \frac{T}{T_0} \tag{7-49}$$

and hence

$$T = (F - 1)T_0 \tag{7-50}$$

where T = two-port noise temperature
   T_0 = ambient temperature (290°K)
If the two-port contributes no noise, the noise temperature is zero ($T = 0$), and the noise figure is unity.   For the low-noise two-port F is very close to unity, and it is common practice to specify the noise temperature instead of the noise figure.

The noise temperature and noise figure can be calculated by (7-47) and (7-48) for different types of two-ports, such as for transistor or vacuum-tube amplifier stages, for mixers, etc. (Bell, 1960; Bennett, 1960; Rhein-felder, 1964; van der Ziel, 1954).   It can also be measured by using a noise generator (see Sec. 7-4b).

In some cases it is simpler to calculate the noise temperature by using noise powers obtained in the actual load impedance of the two-port.   In this case T is found so that

$$W'_{GN} = W'_N \tag{7-51}$$

where W'_{GN} = load noise power due to generator at temperature T
    W'_N = load noise power due to noise sources of two-port
Equation (7-51) gives the correct value for T because both noise powers will decrease by the same ratio if the actual load impedance departs from the output impedance of the two-port.

The noise figure is often expressed in decibels; i.e.,

$$F_{dB} = 10 \log F \tag{7-52}$$

Example. Find the noise temperature of an amplifier stage having a noise figure of 1 dB.

Solution. The noise figure F is antilog $(F_{dB}/10) = 1.26$. Hence, the noise temperature is

$$T = (1.26 - 1)290 = 75°K$$

7-2b. Noise Temperature of an Attenuator. A two-port commonly used in receivers is a passive attenuator. For example, a transmission line between an antenna and a receiver always has some losses and can be considered to be an attenuator. If such an attenuator is connected in series with a matched generator, the exchangeable output noise power from the attenuator due to the generator at temperature T_0 is

$$W_{GN} = \epsilon k T_0 \, \Delta \nu = \frac{1}{L} k T_0 \, \Delta \nu \tag{7-53}$$

where ϵ is the transmission-efficiency coefficient of the attenuator. It is a dimensionless quantity with values between 0 (complete attenuation) and unity (no attenuation) $(0 \le \epsilon \le 1.)$ L is the loss factor, or attenuation, of the attenuator and is equal to the reciprocal of ϵ $(\infty \ge L \ge 1)$. The exchangeable power gain of the attenuator is hence

$$G = \epsilon \tag{7-54}$$

The exchangeable noise power from the attenuator due to the attenuator is

$$W_N = (1 - \epsilon) \, k T_{LP} \, \Delta \nu \tag{7-55}$$

where T_{LP} is the physical temperature of the attenuator. The noise temperature T_T of the attenuator is given by [see (3-121)]

$$T_T = \left(\frac{1}{\epsilon} - 1\right) T_{LP} = (L - 1) T_{LP} \tag{7-55a}$$

7-2c. Noise Temperature of Linear Two-ports in Series Connection. As stated earlier, the receiver can be considered as a series connection of linear two-ports. Figure 7-23 shows the receiver divided into separate two-ports. The noise temperatures of the two-ports are T_1,

$T_2, \ldots, T_n$; noise figures, $F_1, F_2, \ldots, F_n$; and exchangeable power gains $G_1, G_2, \ldots, G_n$.

The exchangeable output power from the whole system is

$$W_{GN} + W_N = G_1 G_2 \cdots G_n k T_0 \, \Delta\nu + G_1 G_2 G_3 \cdots G_n k T_1 \, \Delta\nu +$$
$$+ G_2 G_3 \cdots G_n k T_2 \, \Delta\nu + \cdots + G_n k T_n \, \Delta\nu \qquad (7\text{-}56)$$

It follows that the noise temperature of the series connection is

$$T = T_1 + \frac{T_2}{G_1} + \frac{T_3}{G_1 G_2} + \cdots + \frac{T_n}{G_1 \cdots G_{n-1}} \qquad (7\text{-}57)$$

Equation (7-57) indicates that if the exchangeable power gain of each two-port is large in comparison with unity, the noise temperature of the series

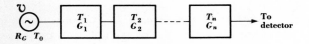

Fig. 7-23.   Linear two-ports in cascade connection.

connection is effectively determined by the first few two-ports (assuming that none of the noise temperatures is extremely high).

The corresponding equation for the noise figure is

$$F = F_1 + \frac{F_2 - 1}{G_1} + \frac{F_2 - 1}{G_1 G_2} + \cdots + \frac{F_n - 1}{G_1 \cdots G_{n-1}} \qquad (7\text{-}58)$$

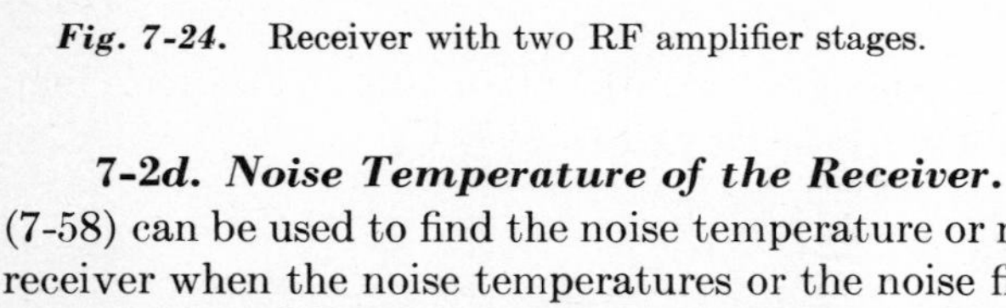

```
┌──────────┐   ┌──────────┐   ┌────────┐   ┌──────────┐
│  1st RF  │   │  2d RF   │   │ Mixer  │   │   IF     │
│ amplifier│   │ amplifier│   │ G₃ T₃  │   │ amplifier│
│  G₁ T₁   │   │  G₂ T₂   │   │        │   │          │
└──────────┘   └──────────┘   └────────┘   └──────────┘
                                  │
                             ┌────────┐
                             │ Local  │
                             │oscillator│
                             └────────┘
```

Fig. 7-24.   Receiver with two RF amplifier stages.

7-2d. Noise Temperature of the Receiver.   Equations (7-57) and (7-58) can be used to find the noise temperature or noise figure of the actual receiver when the noise temperatures or the noise figures and the available power gains of different receiver stages are known.

A typical receiver may have two RF amplifier stages followed by a mixer and an IF amplifier, as shown in Fig. 7-24.   If the available power gain $G_1 G_2$ of the RF amplifiers is high enough, the noise temperature of the receiver is given by

$$T_R = T_1 + \frac{T_2}{G_1} + \frac{T_3}{G_1 G_2} \tag{7-59}$$

where T_3 is the noise temperature of the mixer-IF-amplifier combination.

The noise temperature T_R of the receiver given by the manufacturer does not include the noise-temperature contribution from the transmission line. This extra noise is dependent on the losses in the transmission line and can be calculated separately. From (7-57) the noise temperature of the receiver-transmission-line combination is

$$T_{RT} = T_T + \frac{T_R}{G} = T_T + \frac{T_R}{\epsilon} = T_T + L T_R \tag{7-60}$$

Here the transmission-line "gain" G is less than unity ($= \epsilon$). Introducing (7-55a) yields

$$T_{RT} = (L - 1) T_{LP} + L T_R \tag{7-60a}$$

The extra noise temperature added to T_R due to the transmission line is then

$$T_{RT} - T_R = (T_{LP} + T_R)(L - 1) \tag{7-61}$$

Example 1. Find the noise temperature of a receiver in which the noise temperature of the first RF amplifier is 45°K and its gain 13 dB and the noise temperature of the second RF amplifier is 140°K and its gain 13 dB. The rest of the receiver has the noise temperature of 800°K.

Solution. The temperature contribution from the first stage is 45°K. The contribution from the second stage is

$$T_2' = \frac{T_2}{G_1} = {}^{140}\!/_{20} = 7°\text{K}$$

The contribution from the third stage is

$$T_3' = \frac{T_3}{G_1 G_2} = \frac{800}{20 \times 20} = 2°\text{K}$$

The noise temperature of the entire receiver is

$$T_R = 45° + 7° + 2° = 54°\text{K}$$

Example 2. Find the noise-temperature increase for the receiver in Example 1 above when the antenna transmission line is connected. The transmission line has losses of 0.4 dB and a physical temperature of 290°K.

Solution. The transmission line loss $L =$ antilog ${}^4\!/_{100} = 1.1$. The increase in noise temperature is $(54° + 290°)(1.1 - 1) = 34°\text{K}$.

7-2e. Integrated Noise Temperature and Noise Equivalent Bandwidth. The definitions and equations given so far assume a constant noise-power density in a narrow bandwidth $\Delta\nu$ and give a spot noise temperature and a spot noise figure at the frequency ν. Actually the output noise power in (7-47) is a function of frequency. Thus $W_N = W_N(\nu)$. The exchangeable power gain is also a function of frequency, or $G = G(\nu)$.

The total output noise power W_{NT} is obtained by integrating over a fre quency range from zero to infinity:

$$W_{NT} = \int_0^\infty W_N(\nu)\ d\nu \qquad (7\text{-}6\text{?})$$

The total output noise power due to the generator is

$$W_{GNT} = \int_0^\infty G(\nu)kT_0\ d\nu \qquad (7\text{-}6\text{?})$$

The noise equivalent bandwidth $\Delta\nu_N$ of the two-port is defined by

$$\Delta\nu_N = \frac{1}{G_{\max}} \int_0^\infty G(\nu)\ d\nu \qquad (7\text{-}6\text{?})$$

where $G_{\max}$ is the maximum value of the exchangeable power gain. so-called *average noise temperature* of a linear two-port is defined as

$$T_{\mathrm{avg}} = \frac{W_{NT}}{G_{\max}k\ \Delta\nu_N} \qquad (7\text{-}6\text{?})$$

When several two-ports are connected in series and the noise band width of the nth two-port is small in comparison with other two-ports, can be concluded that the average noise temperature of the series connectic is

$$T_{\mathrm{avg}} = T_1 + \frac{T_2}{G_1} + \cdots + \frac{T_n}{G_1 \cdots G_{n-1}} \qquad (7\text{-}6\text{?})$$

If the gain before the nth stage is high,

$$T_{\mathrm{avg}} = T \qquad (7\text{-}6\text{?})$$

or the average noise temperature equals the spot noise temperature.

The section determining the predetection noise bandwidth of the superheterodyne receiver is normally the IF amplifier. If its noise band width is indicated by $\Delta\nu_N$, the output noise power due to the receiver given by

$$W_{NRT} = G_{\max}kT_{RT}\ \Delta\nu_N \qquad (7\text{-}6\text{?})$$

and the noise power due to the receiver (referred to the antenna terminals)

$$W_{NR} = kT_{RT}\ \Delta\nu_N \qquad (7\text{-}6\text{?})$$

The noise bandwidth of the IF amplifier is close to the 3-dB band width. Table 7-4 shows noise equivalent bandwidths $\Delta\nu_N$ for some I amplifier types as compared to the 3-dB bandwidth $\Delta\nu$.

Table 7-4
Noise equivalent bandwidths of IF amplifiers

Type of amplifier	$G(\nu)$	$\Delta\nu_N/\Delta\nu$
m cascaded single-tuned stages	$\left[1 + \left(\dfrac{\Delta\omega}{\pi\,\Delta\nu_s}\right)^2\right]^{-m}$	
$\quad m = 1$		1.57
$\quad m = 3$		1.16
$\quad m = 5$		1.11
$\quad m = \infty$ (Gaussian)		1.06
m cascaded $2n$-pole Butterworth filters	$\left[1 + \left(\dfrac{\Delta\omega}{\pi\,\Delta\nu_s}\right)^{2n}\right]^{-m}$	
$\quad m = 1, n = 2$		1.11
$\quad m = 1, n = 3$		1.05
$\quad m = 1, n = 5$		1.02
$\quad m = 1, n = \infty$ (ideal filter)		1.00
$\quad m = 2, n = 2$		1.04

$\Delta\nu_s$ = 3 dB bandwidth of a single amplifier stage.

7-2f. Noise Temperature of a Two-channel Receiver. A common receiver used in radio-astronomy observations is the superheterodyne type with a crystal-mixer input stage giving a two-channel receiver, as shown in Figs. 7-3 and 7-25.

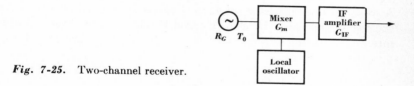

Fig. 7-25. Two-channel receiver.

The output noise power of the mixer is normally given by stating a so-called (broad-band) *noise-temperature ratio* of the mixer t_m (Strum, 1953; Messenger and McCoy, 1957). The noise-temperature ratio gives the ratio by which the actual mixer output noise power exceeds its thermal output noise power, assuming the mixer temperature and the signal-source temperature $T_0 = 290°K$. The conversion loss of the mixer is indicated by $L_m = 1/G_m$, and the noise temperature of the IF amplifier by T_{IF}.

The output noise power from the mixer is then

$$W_{GX} + W_X = kt_m T_0\,\Delta\nu \tag{7-70}$$

The available noise power from the signal generator through the mixer, counting only the signal-frequency channel, is

$$W_{GX} = G_m kT_0\,\Delta\nu \tag{7-71}$$

Hence, the mixer noise temperature is from (7-47),

$$T_m = \frac{W_X}{G_m k \, \Delta \nu} = \frac{k t_m T_0 \, \Delta \nu - G_m k T_0 \, \Delta \nu}{G_m k \, \Delta \nu}$$

$$= \left(\frac{t_m}{G_m} - 1 \right) T_0 = (L_m t_m - 1) T_0 \tag{7-72}$$

Comparing this with (7-50), it can be seen that the noise figure of the mixer is $L_m t_m$.

The noise temperature of the entire crystal-mixer receiver, taking into account only the signal-frequency channel, is then

$$T_{R1} = (L_m t_m - 1) T_0 + L_m T_{IF} \tag{7-73}$$

and the noise figure of the receiver is

$$F_R = L_m t_m + L_m(F_{IF} - 1) = L_m(t_m + F_{IF} - 1) \tag{7-74}$$

By definition, the noise figure of the receiver is always given for only one channel (IRE, 1960).

Actually there are two equal signal channels. These are the signal-frequency channel and the image-frequency channel. Hence, the effect of the noise contributed by the receiver is only one-half that of the one-channel receiver. The two-channel-receiver noise temperature T_{R2} is thus

$$T_{R2} = \tfrac{1}{2} T_{R1} = \tfrac{1}{2}(F_R - 1) T_0 \tag{7-75}$$

A so-called *two-channel noise figure* F_{R2} can be defined as a noise figure of the receiver measured by using a broad-band noise generator without taking into consideration the effect of the image channel (see Sec. 7-4b). The measurement gives

$$F_{R2} = \tfrac{1}{2}(F_R + 1) \tag{7-76}$$

Using this two-channel noise figure, the two-channel noise temperature is given by

$$T_{R2} = (F_{R2} - 1) T_0 \tag{7-77}$$

Example.   A 2-Gc crystal-mixer receiver has a 7-dB noise figure. Find the two channel noise temperature.

Solution.   The noise figure $F = 5$. From (7-75) $T_{R2} = \tfrac{1}{2}(5 - 1)290°\text{K} = 580°\text{K}$

7-3   Low-noise Amplifiers

7-3a. Principles of Negative-resistance Amplifiers.   Radio astronomy receivers commonly employ low-noise amplifiers as radio-frequency preamplifiers to achieve the required sensitivity. In this section the properties of some low-noise amplifiers are considered.

From Fig. 7-1 it can be seen that the antenna sky noise temperature is a few degrees Kelvin in the frequency range from 1 to 10 Gc. Hence, in this frequency range a low receiver noise temperature is essential. Below 100 Mc the sky noise temperature is so high that all receivers normally have a noise temperature well below the antenna noise temperature. But UHF receivers, and especially microwave superheterodyne receivers, require a low-noise preamplifier before the crystal mixer and IF amplifier. A good crystal-mixer receiver (Mackey, 1964) has a one-channel noise

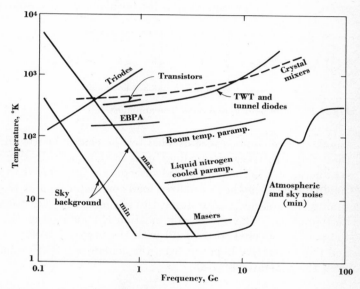

Fig. 7-26. Noise temperatures of low-noise amplifiers as of the year 1964 compared to sky noise levels.

temperature of 800 to 1000°K. From (7-57) it is evident that an RF amplification of 30 dB is needed to reduce the noise-temperature contribution of the crystal mixer to 1°K.

Negative-resistance amplifiers find wide application as low-noise devices. In these amplification is obtained by introducing a negative resistance into the signal circuit. Masers, parametric amplifiers, and tunnel-diode amplifiers are of this type. Figure 7-26 shows the noise temperatures of some low-noise amplifiers as of the year 1964. Parametric amplifiers have a noise temperature roughly proportional to the operating temperature. The actual noise temperature of the helium-cooled parametric amplifiers and masers is largely dependent on the losses of the input and output transmission lines.

Parametric amplifiers and masers are relatively narrow band devices (relative bandwidth 20 percent or less). When a broad-band amplifier (of the order of 1 octave) is required, tunnel-diode amplifiers (Chow, 1964) and traveling-wave tube (TWT) amplifiers (Pierce, 1950) must be used. In VHF receivers triode and transistor amplifiers (Rheinfelder, 1964) give a satisfactory noise temperature.

The principles of negative-resistance amplifiers are next considered. In a simple negative-resistance amplifier a negative resistance or conductance is coupled directly to the terminals of the signal source, as shown in Fig. 7-27. The parameter g_G is the normalized signal-source conductance

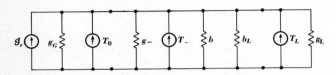

Fig. 7-27.   Simple negative-resistance amplifier.

(signal-source conductance divided by the characteristic admittance Y_0 of the connecting transmission line), g_- is the normalized value of the negative conductance, b is the normalized susceptance introduced with the negative conductance, g_L is the normalized load conductance, and b_L is a normalized tuning susceptance. The quantities g_-, b, and b_L are usually dependent on frequency. The power amplification G is defined as the ratio between the power delivered to the load with the negative conductance connected and the exchangeable power from the generator. The exchangeable power from the generator is (7-43)

$$W_G = \frac{|g_s|^2}{4G_G} \tag{7-78}$$

where g_s = short-circuit signal current
G_G = generator conductance
The power to the load with the negative conductance at the resonant frequency is equal to

$$W_L = \frac{|g_s|^2 g_L}{(g_G + g_L + g_-)^2} \tag{7-79}$$

The power gain is hence

$$G = \frac{4 g_G g_L}{(g_G + g_L + g_-)^2} \tag{7-80}$$

The gain is high only when $|g_-|$ is close to $g_G + g_L$. The disadvantage of the simple negative-resistance amplifier is that with reasonably high gain small changes in generator, load, or negative conductance will cause large change in gain. On microwave frequencies changes of effective generator impedance are unavoidable. If, for example, the standing-wave ratio in the antenna feed line is s and the line is several wavelengths long, the effective generator admittance at the receiver end depends strongly on frequency as it ranges from Y_0/s to sY_0, where Y_0 is the line admittance. The gain varies correspondingly, as suggested in Fig. 7-28. This renders the simple negative-resistance amplifier unsatisfactory in radio-telescope applications, but with the introduction of a circulator (see below) to isolate the amplifier its use becomes practical.

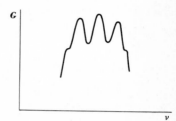

Fig. 7-28. Gain variation of the simple negative-resistance amplifier due to standing waves in the transmission line.

Example. A simple negative-resistance amplifier has an antenna impedance of 50 ohms and a load impedance of 50 ohms. Find the value of the parallel negative resistance giving 20 dB gain and the standing-wave ratio of the antenna transmission line giving 3 dB increase of gain.

Solution. From (7-80) with $g_G = g_L = 1$ and $G = 100$, the negative conductance is $g_- = -1.8$. The corresponding parallel resistance is $R_- = -50/1.8 = -27.8$ ohms. With a standing-wave ratio s such that $g_g = 1/s$ the gain is 3 dB higher, or 200 when $s = 1.06$.

Fig. 7-29. Negative conductance with equivalent noise generator.

$$\bar{g}^2 = 4kT_-\,|g_-|\,\Delta\nu$$

The noise temperature of a negative-resistance amplifier is dependent on the noisiness of the negative resistance. This can be expressed by attaching to the negative conductance an equivalent noise temperature T_-, so that the *mean square noise current* (see Fig. 7-29) is

$$\bar{g}^2 = 4kT_-|g_-|Y_0\,\Delta\nu \tag{7-81}$$

The noise temperature of a simple negative-resistance amplifier (Fig. 7-27) can be calculated using (7-47). The noise from the load conductance must be taken into account because it appears in the same circuit as the generator noise. The exchangeable noise power W_N due to the amplifier is thus

$$W_N = \frac{g_N{}^2}{4G_N} = \frac{k(T_-|g_-| + T_L g_L)\,\Delta\nu}{g_G + g_-} \tag{7-82}$$

where $\mathcal{I}_N$ = short-circuit noise current, amp

G_N = output conductance, mhos

k = Boltzmann's constant ($= 1.38 \times 10^{-23}$ joule °K^{-1})

T_- = noise temperature of negative conductance, °K

T_L = noise temperature of load, °K

g_G = generator conductance (normalized) ($= G_G/Y_0$), mhos

g_L = load conductance (normalized), mhos

g_- = negative conductance (normalized), mhos

$\Delta\nu$ = bandwidth, cps

The exchangeable power gain is, from (7-45),

$$G = \frac{g_G}{g_G + g_-} \tag{7-83}$$

The noise temperature is, from (7-47),

$$T = \frac{T_-|g_-| + T_L g_L}{g_G} \tag{7-84}$$

In the case of a high-gain amplifier, where $g_G = g_L = 1$, the negative conductance is approximately $g_- = -2$, and

$$T = T_L + 2T_- \tag{7-85}$$

The load conductance g_L must be made small in order to achieve a low noise temperature.

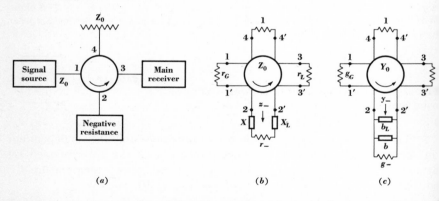

Fig. 7-30. (a) Reflection amplifier; (b) negative-resistance reflection amplifier; and (c) negative-conductance reflection amplifier.

Negative-resistance amplifiers can be constructed with the aid of a circulator, as shown in Fig. 7-30. The *circulator* is a nonreciprocal three-or four-port network which has such characteristics that an incident voltage wave entering port 1 comes out port 2, an incident voltage wave entering port 2 comes out port 3, etc. The circulation direction is shown by the arrow. Practical circulators use the nonreciprocal characteristics of satura-tion-magnetized ferrites (Lax and Button, 1962; Fay and Comstock, 1965). They have small losses (0.1 to 0.5 dB) from port to port in the forward direction and an isolation of 20 to 40 dB in the backward direction.

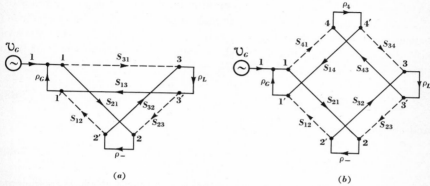

(a)   *(b)*

Fig. 7-31. (a) Flow graph of three-port circulator; (b) flow graph of four-port circulator.

The negative-resistance amplifier using a circulator is called a *reflection amplifier*. The voltage wave entering from the antenna to the negative resistance is reflected back from it to the circulator and to the main receiver. Amplification is obtained because the wave reflected from the negative resistance is stronger than the incident wave. The voltage-reflection coeffi-cient (ratio of the reflected voltage to the incident voltage) is equal to

$$\rho_- = \frac{z_- - 1}{z_- + 1} = \frac{1 - y_-}{1 + y_-} \tag{7-86}$$

where z_- is the normalized input impedance and y_- the normalized input admittance of the negative-resistance device. Assuming that the antenna impedance and the input impedance of the main receiver are matched, the power gain is

$$G = |\rho_-|^2 \tag{7-87}$$

For example, if the characteristic impedance of transmission lines and the circulator is $Z_0 = 50$ ohms and z_- is real $(= r_-)$, a 20-dB gain is obtained with $|\rho_-| = 10$ and $r_- = -0.82$ or $R_- = -41$ ohms.

If the positive impedances connected to the different ports of the reflection amplifier circulator are not matched, a feedback voltage wave will appear in the circuit and will change the gain periodically as a function of frequency depending on the phase of the feedback wave. The effect of mismatch can be studied using flow diagrams (Kuhn, 1963; Chow and Cassignal, 1962). Flow diagrams of three-port and four-port reflection amplifiers are shown in Fig. 7-31. Voltage waves can enter in the direction of the arrow only. ρ_G is the reflection coefficient measured at port 1 toward the generator, and ρ_L is the reflection coefficient measured at port 3 toward the main receiver. S_{ji} is a complex coefficient by which the voltage wave must be multiplied when entering from the node point i to the node point j. In an ideal three-port circulator $|S_{21}| = |S_{32}| = |S_{13}| = 1$ with equal phase and $|S_{31}| = |S_{23}| = |S_{12}| = 0$. Assuming an ideal three-port circulator, the flow diagram of Fig. 7-31a simplifies to Fig. 7-32a. This can be reduced further, as shown in Fig. 7-32b, c, and d, by using the reduction rules of flow diagrams. From Fig. 7-32d the voltage wave entering the load from port 3 is (assuming $S_{21} = S_{32} = S_{13} = 1$)

$$V_3 = \mathcal{U}_G \frac{\rho_-}{1 - \rho_G \rho_L \rho_-} \tag{7-88}$$

and the power gain is

$$G = (1 - |\rho_L|^2) \left| \frac{V_3}{\mathcal{U}_G} \right|^2 = (1 - |\rho_L|^2) \left| \frac{\rho_-}{1 - \rho_G \rho_L \rho_-} \right|^2 \tag{7-89}$$

Care must be taken to keep the antenna and load-reflection coefficients small enough to ensure stable operation. The load-reflection coefficient

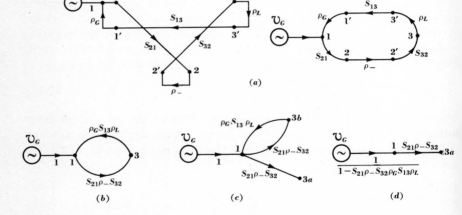

Fig. 7-32. Flow graph of reflection amplifier using an ideal circulator.

can be stabilized by using a ferrite isolator ahead of the load. A ferrite isolator (Lax and Button, 1962) is a nonreciprocal transmission-line device whose attenuation is small in the direction of the incident wave (0.1 to 1.0 dB) and high in the direction of the reflected wave (20 to 40 dB). A four-port circulator has a similar effect because the fourth port can be terminated with a carefully matched load.

Example. Find the maximum and minimum gain of a three-port reflection amplifier where $\rho_- = 10$, the standing-wave ratio of the antenna is 1.4, and the standing-wave ratio of the load is 1.4.

Solution. The reflection coefficients $|\rho_G|$ and $|\rho_L|$ are

$$|\rho_G| = |\rho_L| = \frac{s-1}{s+1} = \frac{0.4}{2.4} = 0.17$$

Maximum gain occurs at the frequency for which $\rho_G = \rho_L = 0.17$, so that

$$G_{max} = (1 - 0.03)\left(\frac{10}{1 - 0.28}\right)^2 = 187$$

Minimum gain occurs at the frequency for which $\rho_G = 0.17$ and $\rho_L = -0.17$, so that

$$G_{min} = (1 - 0.03)\left(\frac{10}{1 + 0.28}\right)^2 = 59$$

The amplifier can start to oscillate when $|\rho_G \rho_L \rho_-| = 1$ or $|\rho_G \rho_L| = 0.1$ in this example. If $|\rho_G| = |\rho_L|$, this corresponds to a standing-wave ratio of 1.9.

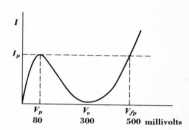

Fig. 7-33. Current-voltage (I-V) characteristic of a tunnel diode.

The noise temperature of the reflection amplifier can be calculated from (7-51). The load noise power due to the amplifier is (see Fig. 7-30)

$$W'_N = 4kT_-|g_-|\,\Delta\nu\,\frac{g_L}{(g_- + g_L)^2} \tag{7-90}$$

In a matched case $g_G = g_L = 1$. The load noise power due to the generator at a temperature T is

$$W'_{GN} = \left|\frac{1-g_-}{1+g_-}\right|^2 kT\,\Delta\nu \tag{7-91}$$

Hence, from (7-51)

$$T = \frac{4T_-|g_-|}{(1-g_-)^2} = \left(1 - \frac{1}{G}\right)T_- \tag{7-92}$$

In a high-gain amplifier $|g_-| \approx g_G = 1$, and

$$T \approx T_- \tag{7-93}$$

7-3b. Tunnel-diode Amplifiers. A tunnel diode is a semiconductor diode which has a static current-voltage (I-V) characteristic as shown in Fig. 7-33. With a bias voltage between V_p and V_v the diode has a negative dynamic conductance. The equivalent circuit of the diode is shown in

Fig. 7-34. Tunnel diodes can be used in negative-resistance amplifiers or frequencies where the input admittance Y_d has a negative real part. The real part of Y_d is zero at a resistive cutoff frequency

$$\nu_{ro} = \frac{G'_-}{2\pi C_d} \sqrt{1 - \frac{1}{R_d G'_-}} \tag{7-94}$$

The resistive cutoff frequency can be made larger than 100 Gc. At frequencies $\nu \ll \nu_{ro}$, R_d and L_s can be neglected. The susceptance of C_d is tuned out by an inductance. In actual amplifiers an impedance transformer is usually required to transform G'_- to the suitable value for amplification.

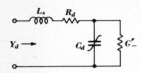

Fig. 7-34.   Equivalent circuit of a tunnel diode.

The predominant noise of a tunnel diode in the negative-conductance region is shot noise. The mean square value of the shot-noise current is (Chow, 1964)

$$\overline{g^2} = 2eI \, \Delta\nu \tag{7-95}$$

where I is the diode direct current and e is the charge of an electron. The corresponding noise temperature of the conductance G'_- is

$$T'_- = \frac{2eI}{4k|G'_-|} = \frac{IT_0}{V_1|G'_-|} \tag{7-96}$$

where $V_1 \approx 50 \times 10^{-3}$ volt. For the best tunnel diodes $T'_- \approx 1.7T_0 \approx$ 490°K. Hence, tunnel-diode amplifiers do not have very low noise temperatures, but their great simplicity makes them attractive (Seling, 1964).

7-3c. Parametric Amplifiers.

The parametric amplifier (paramp is a negative-resistance device which employs the properties of nonlinear reactances to achieve amplification (Blackwell and Kotzebue, 1961; Penfield and Rafuse, 1962).

There are several types of nonlinear elements which can be used for parametric amplification. At present, however, the most useful device is the variable-capacitance diode, or varactor.

The *varactor* is a *p-n* junction diode with voltage-dependent capacitance,

$$C(V) = \frac{C(O)}{(1 - V/\phi)^n} \tag{7-97}$$

Here $C(V)$ is the junction capacitance at (negative) bias voltage V, ϕ the contact potential, which depends on the semiconductor, and n an exponent which varies between $\frac{1}{2}$ and $\frac{1}{3}$.

Because of the finite resistivity of the semiconductor material, the varactor always possesses a series resistance. The equivalent circuit is shown in Fig. 7-35a. The cutoff frequency at bias voltage V is defined as

$$\nu_{CV} = \frac{1}{2\pi R_d C(V)} \tag{7-98}$$

where R_d = diode series resistance
$C(V)$ = junction capacitance from (7-97)
The Q of the varactor (diode) at frequency ν is given by

$$Q_d = \frac{1}{2\pi\nu C(V)R_d} = \frac{\nu_{CV}}{\nu} \tag{7-99}$$

where ν_{CV} is as given by (7-98).

The packaging of the diode will add series inductance L_s and parallel capacitance C_p. The varactor has series and parallel resonant frequencies given by

$$\nu_{d1} = \frac{1}{2\pi\sqrt{C(V)L_s}} \tag{7-100}$$

$$\nu_{d2} = \nu_{d1}\sqrt{1 + \frac{C(V)}{C_p}} \tag{7-101}$$

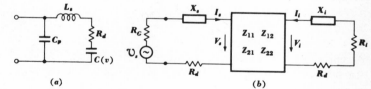

(a)   (b)

Fig. 7-35. (a) Varactor equivalent circuit; (b) equivalent circuit of a negative-resistance parametric amplifier ($\nu_i = \nu_p - \nu_s$). X_s and X_i are reactances for tuning.

If the varactor is "pumped" by a sinusoidal voltage (or current), it acts like a time-varying capacitance. Fourier-analyzing its value, as done by Blackwell and Kotzebue (1961), we have

$$C(t) = C_0(1 + 2\gamma \cos \omega_p t + \cdots) \tag{7-102}$$

where C_0 = d-c term of Fourier expansion
γ = capacitance modulation factor
ω_p = radian pump frequency
Higher harmonics in (7-102) are neglected.

If a signal frequency ω_s is also impressed on the varactor, mixing products of ω_p and ω_s will arise. In the *negative-resistance parametric amplifier* only voltages at the difference frequency $\omega_p - \omega_s = \omega_i$, called the idler frequency, are allowed to exist in addition to signal and pump voltages. If $\omega_s = \omega_i = \omega_p/2$, the amplifier is called *degenerate*.

The parametric amplifier can now be represented by the equivalent circuit of Fig. 7-35*b* and the relations (Blackwell and Kotzebue, 1961)

$$\begin{bmatrix} V_S \\ V_i^* \end{bmatrix} = \begin{bmatrix} Z_{11} & Z_{12} \\ Z_{21} & Z_{22} \end{bmatrix} \begin{bmatrix} I_s \\ I_i^* \end{bmatrix} \tag{7-103}$$

$$Z_{11} = \frac{1}{j\omega_s C} \qquad Z_{12} = \frac{\gamma}{j\omega_i C} \qquad Z_{21} = -\frac{\gamma}{j\omega_s C}$$

$$Z_{22} = -\frac{1}{j\omega_i C} \qquad C = C_0(1 - \gamma^2)$$

X_i and X_s are reactances which tune the idler and signal circuits to resonance. In practice the modulation factor γ has a maximum value of 0.25 to 0.30.

From (7-103) the input impedance Z_- of the amplifier becomes

$$Z_- = \frac{\gamma^2}{\omega_s \omega_i C^2 Z_i^*} \tag{7-104}$$

where Z_i^* is the complex conjugate of the impedance of the idler circuit. At resonance, Z_- becomes a negative real resistance, given by

$$R'_- = -\frac{\gamma^2}{\omega_s \omega_i C^2 (R_i + R_d)} = -\frac{\gamma^2 Q_{Li}}{\omega_s C} \tag{7-105}$$

R_i represents any loading of the idler circuit, external to the diode. In the degenerate case it is equal to the signal-source resistance R_G. Q_{Li} is the loaded Q of the idler circuit.

The effective negative resistance R_- in the signal circuit is

$$R_- = R_d + R'_- = \frac{1}{\omega_s C}\left(\frac{1}{Q_{ds}} - \gamma^2 Q_{Li}\right) \tag{7-106}$$

and the power gain of the reflection amplifier is

$$G = \left[\frac{R_- - R_G}{R_- + R_G}\right]^2 \tag{7-107}$$

An impedance transformer is usually needed to transform the actual R_- to a suitable value. In order to obtain sufficient gain R_- must be nearly equal to R_G, which means that any fluctuations of the negative resistance R'_- or the generator resistance R_G will have a marked effect on the gain. The negative resistance is proportional to the square of the capacitance modulation factor γ, which in turn is a function of pump-power level. For a low-noise amplifier with about 20 dB gain, the gain changes by about 0.1 dB for a 0.01-dB change in pump power. Hence, the power stability and also the frequency stability of the pump source, if the pump circuit is not broad-band, are important.

For a nondegenerate amplifier ($\omega_s \neq \omega_i$) with single-tuned signal and idler circuits, the gain-bandwidth product for large gain ($R_- \approx R_G$) is given by de Jager (1964) as

$$\Delta\nu\sqrt{G} = \frac{2}{\sqrt{2 + q^2 + 1/q}} \frac{\nu_s}{Q_M} \tag{7-108}$$

where $\Delta\nu$ = 3-dB bandwidth
$\quad G$ = power gain

$$q = \frac{\gamma}{\omega_s C_0(R_i + R_d)} \sqrt{\frac{d_i}{d_s}\left(\frac{\omega_s}{\omega_i}\right)^2}$$

$$d_s = \frac{1}{2} + \frac{1}{2}\omega_s{}^2 C_0 \left(\frac{dX_s}{d\omega}\right)_{\omega=\omega_s}$$

$$d_i = \frac{1}{2} + \frac{1}{2}\omega_i{}^2 C_0 \left(\frac{dX_i}{d\omega}\right)_{\omega=\omega_i}$$

$$Q_M = \frac{1}{\gamma} \sqrt{d_s d_i \frac{\omega_s}{\omega_i}}$$

d_i and d_s are the slope factors of the idler and signal circuits, respectively. In order to achieve maximum bandwidth Q_M should be as small as possible. The best gain-bandwidth product is achieved when the idler frequency lies between ν_{d1} and ν_{d2} (Heinlein and Mezger, 1962). If ν_i is raised above ν_{d1}, the bandwidth falls off rapidly, since d_s increases rapidly. Self-resonant frequencies for varactors in current use range from 6 to 14 Gc. With more complicated tuning arrangements than simple resonators in the signal and idler circuits, the bandwidth may be raised considerably (Matthaei, 1961), but tuning becomes more critical and requirements on pump stability more severe.

The noise of a parametric amplifier comes from thermal noise in the circuit resistances at the signal and idler frequencies. The noise at the idler frequency is transferred to the signal frequency by the mixing process, the power-transforming ratio being ω_s/ω_i. The noise temperature of the negative resistance R'_- is then (Blackwell and Kotzebue, 1961)

$$T'_- = \frac{\omega_s}{\omega_i} \frac{R_d T_d + R_i T_i}{R_d + R_i} \tag{7-109}$$

where T_d = physical temperature of diode
$\quad T_i$ = physical temperature of external idler loading

The lowest noise temperature occurs when $T_i R_i = 0$, which, in practice, means that $R_i = 0$. In this nondegenerate case

$$T'_- = \frac{\omega_s}{\omega_i} T_d \tag{7-110}$$

The effective noise temperature T_- of the effective negative resistance (Fig. 7-36) in Eq. (7-106) is, from (7-110),

$$T_- = T_d \frac{R_d + |R'_-| \omega_s/\omega_i}{|R_d + R'_-|} = T_d \frac{1 + (\omega_s/\omega_i)\gamma^2 Q_{ds} Q_{di}}{|1 - \gamma^2 Q_{ds} Q_{di}|} \tag{7-111}$$

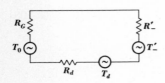

Fig. 7-36. Noise sources of a negative-resistance parametric amplifier.

The effective noise temperature ratio as a function of ω_s/ω_i is illustrated in Fig. 7-37 (Heinlein and Mezger, 1962). From (7-92) the noise temperature of a nondegenerate parametric reflection amplifier is

$$T = (1 - 1/G)T_- \approx T_- \tag{7-112}$$

The noise temperature is directly proportional to the physical temperature T_d of the amplifier and can therefore be lowered by refrigeration.

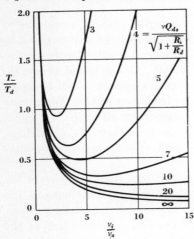

Fig. 7-37. Noise temperature ratio of a high-gain negative-resistance parametric amplifier as a function of idler-to-signal frequency ratio. (*After Heinlein and Mezger*, 1962.)

Example. A negative-resistance parametric amplifier at a signal frequency 2 Gc uses a varactor diode with $\nu_{CV} = 30$ Gc. The pump frequency is 12 Gc. Find the theoretical noise temperature when γ is 0.25 and $T_d = 78°$K.

Solution. The varactor diode Q at the signal frequency is $\nu_{VC}/\nu_s = 15$ and at the idler frequency ($\nu_i = 10$ Gc) is 3. From (7-111), $T_- = 68°$K.

In the almost-degenerate case, where $\omega_s \approx \omega_i$, $R_i = R_G$, and $T_i = T_0$ (Fig. 7-38), the noise temperature of the negative resistance is, from (7-109),

$$T'_- = \frac{R_d T_d + R_G T_0}{R_d + R_G} \tag{7-113}$$

and the effective noise temperature T_- in the high-gain case $(R_G + R_d \approx |R'_-|)$ is

$$T_- = 2\,\frac{R_d T_d}{R_G} + T_0 = \frac{2}{\gamma Q_{ds} - 1}\,T_d + T_0 \tag{7-114}$$

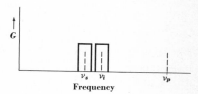

Fig. 7-38. Almost-degenerate parametric amplifier.

In (7-114) the first term is due to the signal and idler noise from the diode series resistance, and T_0 is the temperature ($= 290°K$) of the idler (and signal) external resistance R_G.

In a fully degenerate case ($\omega_s = \omega_i$) the signal and idler channels in Fig. 7-39 coincide, and the effective noise temperature of the amplifier becomes with high gain

$$T_{dg} = \frac{R_d T_d}{R_G} = \frac{T_d}{\gamma Q_{ds} - 1} \tag{7-115}$$

In the degenerate amplifier, output-noise contributions at the signal frequency and through the idler-signal conversion are correlated. This causes the fluctuation noise power in the radiometer after the square-law detector

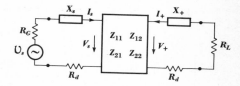

Fig. 7-39. Equivalent circuit of an up-converter ($\nu_+ = \nu_p + \nu_s$).

and the integrator to be twice that of uncorrelated case (de Jager and Robinson, 1961). Hence, $\Delta T_{\min}$ of the receiver using a degenerate parametric amplifier is $\sqrt{2}$ times that of the nondegenerate receiver with the same T_R and T_A.

Example. Find the noise temperature of a degenerate amplifier when the signal frequency is 2 Gc, the diode cutoff frequency is 30 Gc, γ is 0.25, and T_d is 78°K.

Solution. The diode Q at the signal frequency is 15. From (7-115), $T_{dg} = 28°K$.

A parametric amplifier in which the sum frequency $\omega_s + \omega_p = \omega_+$ is utilized is called an *up-converter*. The up-converter may be analyzed similarly to the negative-resistance amplifier (Blackwell and Kotzebue, 1961). Referring to Fig. 7-39, the matrix representation becomes

$$\begin{bmatrix} V_s \\ V_+ \end{bmatrix} = \begin{bmatrix} Z_{11} & Z_{12} \\ Z_{21} & Z_{22} \end{bmatrix} \begin{bmatrix} I_s \\ I_+ \end{bmatrix} \tag{7-116}$$

where

$$Z_{11} = \frac{1}{j\omega_s C_0} \qquad Z_{12} = \frac{\gamma}{j\omega_+ C_0}$$

$$Z_{21} = -\frac{\gamma}{j\omega_s C_0} \qquad Z_{22} = \frac{1}{j\omega_+ C_0}$$

The signal power is fed in at frequency ω_s and comes out at ω_+. No negative resistance is involved, and the amplifier is stable. The gain at resonance of the up-converter is given by

$$G = \frac{4R_G R_L}{[(R_G + R_d)(R_L + R_d)\,\omega_s C/\gamma + \gamma/\omega_+ C]^2} \tag{7-117}$$

Maximizing with respect to R_G and R_L yields for maximum gain

$$G_{\max} = \frac{\omega_+}{\omega_s}\frac{x}{(1 + \sqrt{1 + x})^2} \tag{7-118}$$

where $x = (\omega_s/\omega_+)(\gamma Q_{ds})^2$. The gain is always less than ω_+/ω_s, and when $\omega_+/\omega_s \to \infty$, the gain approaches a limiting value

$$G'_{\max} = \tfrac{1}{4}(\gamma Q_{ds})^2 \tag{7-119}$$

This type of amplifier is suitable only at relatively low signal frequencies, such as at the UHF band with output in the EHF band, as shown in Fig. 7-40. The up-converter places much less stringent requirements on the stability of the pump source than the negative-resistance amplifier.

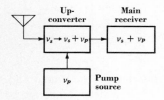

Fig. 7-40. A receiver using an up-converter as a preamplifier.

The minimum noise temperature of the up-converter amplifier is given by Blackwell and Kotzebue (1961) as

$$T_{\min} = 2T_d\left[\frac{1}{\gamma Q_{ds}} + \frac{1}{(\gamma Q_{ds})^2}\right] \tag{7-120}$$

when the generator resistance is

$$R_G = R_d \sqrt{1 + (\gamma Q_{ds})^2} \tag{7-121}$$

Example. Find the noise temperature and gain (with $R_G = R_L$) of an up-converter amplifier when the signal frequency is 2 Gc, the cutoff frequency of the varactor is 30 Gc, the output frequency is 14 Gc, γ is 0.25, and $T_d = 78°$K.

Solution. When $Q_{ds} = 15$, T_{min} from (7-120) $= 53°$K. This is obtained with $R_G = 3.9R_d$. From (7-117) the gain $G = 1.25$.

$$\omega_s C = \frac{1}{R_d Q_{ds}} \qquad \omega_+ C = \frac{\omega_+}{\omega_s} \frac{1}{R_d Q_{ds}}$$

An *electron-beam parametric amplifier* (EBPA) is a special type of degenerate parametric amplifier in which a fast cyclotron wave of an electron beam is used to produce the amplification (Adler, Hrbek, and Wade, 1959). The EBPA type is useful in the frequency range from 200 to 4,000 Mc and has a noise temperature of about 150°K. In calculating the sensitivity of the EBPA receiver, T_R must be multiplied by $\sqrt{2}$, as in the case of a diode degenerate amplifier.

7-3d. Maser Amplifiers. In a maser† transitions between the quantum-energy states of atomic particles are used to achieve amplification (Gordon, Zeiger, and Townes, 1954; Siegman, 1964). To explain the maser consider a system of particles in a static magnetic flux density B_0. Suppose that the particles can occupy either of two energy states, or levels, 1 and 2 with the energy at level 2 greater than that at level 1 and with the energy difference between levels equal to $\Delta\mathcal{E}$. In thermal equilibrium the ratio of the *population densities* n_1 and n_2 of the two states is given by Boltzmann statistics as

$$\frac{n_2}{n_1} = e^{-\Delta\mathcal{E}/kT} \tag{7-122}$$

If electromagnetic radiation of frequency ν and photon energy $\Delta\mathcal{E} = h\nu$ (h = Planck's constant) is incident on the system, transitions from level 1 to level 2 can occur if photons are absorbed. Transitions can also occur from level 2 to level 1 with emission of photons. Such transitions may occur spontaneously, or they may be stimulated by an incident wave with photons of energy $\Delta\mathcal{E}$, which triggers the emission of photons from the particles. Normally n_1 is greater than n_2, and net absorption occurs. But if n_2 can be raised above n_1, emission with amplification of the incident wave can take place.

Masers have been constructed using a ruby crystal with an impurity such as chromium in which the chromium ions act as the active particles

† Acronym for "Microwave Amplification by Stimulated Emission of Radiation."

(Bloembergen, 1956). Three energy levels are employed (Fig. 7-41a). If a strong wave of frequency ν_{13} is absorbed, it can saturate the ν_{13} transition, making the n_1 and n_3 populations equal (Fig. 7-41b). This action of the wave frequency ν_{13} is called *pumping* and ν_{13} the *pump frequency*. The result is that n_1 may become less than n_2, permitting amplification of a wave of frequency ν_{12} at the *signal frequency*.

The lower the ambient temperature of the crystal, the larger the population difference $n_2 - n_1$ becomes. In addition to the transitions stimulated by external signals there are also thermal transitions corresponding to the energy exchanges with the crystal lattice. In order to be able to saturate the pump transition the amount of this exchange must be kept small, which also requires a low operating temperature.

In a cavity maser the crystal is placed in a resonant cavity in a d-c magnetic field. The cavity must be resonant on both pump and signal frequencies in order to achieve strong enough magnetic fields at these frequencies in the crystal. The required low operating temperature is obtained by placing the cavity in liquid helium (4.2°K). With a large enough pump signal the input admittance of the cavity has a negative real part at the signal frequency and so can be used like a negative-resistance amplifier.

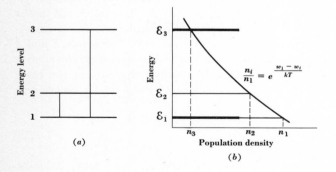

Fig. 7-41. Energy levels in a three-level maser.

Because of the gain mechanism and the low operating temperature the noise temperature of the negative resistance is small, only a few degrees Kelvin. The resonant cavity makes the bandwidth of the amplifier narrow (of the order of 1 percent) which may be undesirable for many radio-astronomy applications.

In a *traveling-wave maser* (TWM) (DeGrasse, Schulz-DuBois, and Scovill, 1959; Walling and Smith, 1965) a slow-wave structure is used to achieve an electromagnetic-field concentration in the maser material. In the slow-wave structure the incident power W, the group velocity of the wave v_g, and the stored energy per unit length $\mathcal{E}$ are related by the following equation

$$W = v_g \mathcal{E} \tag{7-123}$$

Thus, large RF fields are obtained by making v_g small (in practice v_g can be of the order of 1 percent of the free-space velocity of light). An element of length dz of a TWM is shown in Fig. 7-42. Power emitted by the material per unit length is W_e. Since the increments of the signal field add in phase in the forward direction only, virtually all the energy emitted by the element

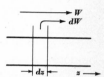

Fig. 7-42. An element of a traveling-wave maser.

travels in the forward direction. Power absorption per unit length W_a also takes place in the element because of ohmic and dielectric losses. The change dW in power level W in the element is thus

$$dW = (W_e - W_a)\, dz \tag{7-124}$$

W_e can be related to a magnetic quality factor Q_m of the maser material

$$Q_m = \frac{\omega_s \mathcal{E}_s}{W_e} \tag{7-125}$$

where ω_s is the angular signal frequency and $\mathcal{E}_s$ stored energy per unit length at the signal frequency. W_a can be expressed by the intrinsic quality factor Q_0 of the propagating structure

$$Q_0 = \frac{\omega_s \mathcal{E}_s}{W_a} \tag{7-126}$$

From (7-123) to (7-125)

$$dW = \frac{W}{v_g} \omega_s \left(\frac{1}{Q_m} - \frac{1}{Q_0} \right) \tag{7-127}$$

The net power gain of the TWM of length l is equal to

$$G = \exp\left[\frac{\omega_s l}{v_g} \left(\frac{1}{Q_m} - \frac{1}{Q_0} \right) \right] \tag{7-128}$$

or, in decibels,

$$G_{\mathrm{dB}} = 4.34 \frac{\omega_s l}{v_g} \left(\frac{1}{Q_m} - \frac{1}{Q_0} \right)$$

The decibel gain is thus directly proportional to the length l of the TWM. The electronic gain G_e of the TWM is the gain with $Q_0 = \infty$.

The instantaneous bandwidth of a TWM is determined by the paramagnetic resonance line width $1/\pi t$. The half-power bandwidth is (see DeGrasse, Schulz-DuBois, and Scovill, 1959)

$$\Delta \nu = \frac{1}{\pi t} \sqrt{\frac{3}{G_e - 3}} \tag{7-129}$$

where G_e is the electronic gain (in decibels) of the maser. A typical value for $\Delta \nu$ is 20 Mc with 30 dB gain. The bandwidth of the TWM can be increased at the expense of the gain by making the d-c magnetic field slightly nonuniform. Also a gain-equalization technique can be used (Tabor and Sibilia, 1963). Here the peak gain at the center frequency of the uniform field TWM is compensated by a suitable staggered tuning in the IF amplifier. The instantaneous signal frequency band of the TWM can be tuned some 10 percent by changing the d-c magnetic field.

The noise temperature of the negative resistance of the maser proper can be shown to be (Walling and Smith, 1965; Ditchfield, 1962)

$$T_- = \left| \frac{\Delta n_0}{\Delta n} \right| T_{MP} \tag{7-130}$$

where Δn_0 is the population-density difference between energy states 2 and 1 in thermal equilibrium and Δn that with pump connected. T_{MP} is the physical temperature of the maser. $|\Delta n_0/\Delta n|$ can be 0.4, and $T_{MP} = 4.2°K$. Hence, T_- is very small, and in practice the losses of the input feed line largely determine the noise temperature.

7-4   Calibration of Receivers

7-4a. Noise Generators. In radio-astronomy receivers signals usually consist of noise. Hence noise generators which produce a known amount of noise power are useful in receiver testing, and they can also be used as a comparison load in switched radiometers.

A standard-noise source consists of a carefully matched load held at an accurately known physical temperature T. It contributes a noise power $W = kT \Delta \nu$. In a cold load the terminal resistance is placed, for example, in liquid nitrogen, in which case its noise temperature is close to 78°K, or in liquid helium, giving a noise temperature of about 4°K. In a hot load the termination is placed in an oven, where the temperature is controlled accurately. The precise value of the noise temperature is dependent on the actual physical temperature of the termination and the feed line (Stelzried, 1961; Penzias, 1965).

In the frequency range of 1 to 1,000 Mc a noise diode or a vacuum-tube diode operated at a temperature-saturation current is a suitable secondary standard or noise generator (Fig. 7-43). R is a shunt resistance for matching the noise generator to the transmission line. The mean-square current corresponding to the diode noise is

$$\overline{g^2} = 2eI_0\,\Delta\nu \tag{7-131}$$

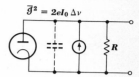

Fig. 7-43. Diode noise generator.

where I_0 is the diode direct current. The maximum frequency is limited by electron transit-time effects and by the anode-cathode stray capacitance. The exchangeable power from the noise generator is

$$W = \frac{eI_0R\,\Delta\nu}{2} + kT_0\,\Delta\nu \tag{7-132}$$

or, expressing it by the equivalent noise temperature T_G of the resistance R, we have

$$T_G = \frac{W}{k\,\Delta\nu} = \frac{eI_0R}{2k} + T_0 = (20I_0R + 1)\,T_0 \tag{7-133}$$

With a 50-ohm resistance

$$T_G = (1,000I_0 + 1)\,T_0 \tag{7-134}$$

The noise temperature of a diode noise generator can be conveniently controlled by varying the filament current. A typical noise diode is a Type 5722, which has a rated maximum anode current of 35 ma, giving a maximum noise temperature of 10,400°K.

A gas-discharge tube is a suitable secondary standard-noise source at microwave frequencies from 100 Mc upward. The effective electron temperature in the positive column of the discharge is frequency-independent, and it emits thermal noise as if it were a resistance at this elevated temperature. In a waveguide noise source the gas-discharge tube is inserted through the broad faces of the guide and is tilted slightly with respect to the longitudinal axis of the guide, as suggested in Fig. 7-44. The noise tube must offer a good impedance match in both the fired condition and in the cold condition to make accurate measurements possible. The attenuation caused by the fired tube is a measure of the coupling of the noise to the waveguide and must be of the order of 20 dB. The cold tube must cause a small attenuation, because losses having a temperature much less than the discharge temperature decrease the effective noise temperature (White and Greene, 1956).

An argon-filled gas-discharge tube has a noise temperature of about 10,000°K. The noise temperature is almost independent of the operating temperature of the tube from -50 to $+100$°C. Gas pressure and direct current of the tube are not critical (Hughes, 1956). A variable noise source from 290 to 10,000°K is obtained by connecting the gas-discharge noise

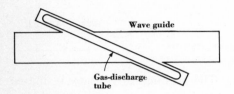

Fig. 7-44. Gas-discharge noise generator.

generator through a precision adjustable attenuator. If the attenuation is L (>1), then assuming an attenuator temperature T_0, the actual output noise temperature T_G' is given by

$$T_G' = \frac{1}{L}\,T_G + \left(1 - \frac{1}{L}\right)T_0 = \frac{1}{L}\,(T_G - T_0) + T_0 \qquad (7\text{-}135)$$

Example. A gas-discharge noise generator has an excess noise ratio $(T_G - T_0)/T_0$ of 15.2 dB. Find the size of an attenuator bringing the output noise temperature of the noise generator down to 1000°K.

Solution. The unattenuated excess noise temperature of the noise generator is $T_G - T_0 = 33.1T_0$. Hence $T_G = 9900$°K, and the required attenuation L is equal to

$$L = \frac{T_G - T_0}{T_G' - T_0} = 13.5 = 11.3 \text{ dB}$$

7-4b. Measurement of Receiver Noise Temperature. The noise temperature of a receiver can be measured by connecting a noise generator to the receiver input and a power-measuring device to the receiver IF output, as in Fig. 7-45. The noise generator must have the internal impedance

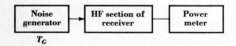

Fig. 7-45. Noise-temperature measurement of a receiver.

of the antenna and its impedance must stay unchanged under operating and nonoperating conditions. A square-law detector can be used as the power-measuring device. Its direct current or voltage is proportional to the noise power in the square-law region. If the output indication with the noise generator unfired ($T_G = T_0$) is I_1 and with noise generator in operation is I_2, then

$$\frac{I_2}{I_1} = \frac{T_G + T_R}{T_0 + T_R} = K_1 \tag{7-136}$$

and the receiver noise temperature T_R is equal to

$$T_R = \frac{I_1 T_G - I_2 T_0}{I_2 - I_1} \tag{7-137}$$

If an adjustable noise generator is used, T_G is adjusted so that $I_2 = 2I_1$, in which case

$$T_R = T_G - 2T_0 \tag{7-138}$$

Adjustable noise generators are often calibrated for noise figure F. The receiver noise temperature is then $T_R = (F - 1)T_0$. For best accuracy in noise-temperature measurements T_G should be of the same order of magnitude as T_R.

Example.   A noise temperature T_R in a low-noise receiver is measured using a room-temperature load $T_G = T_0 = 290°K$ and a cold load $T_G = 78°K$. The corresponding output indications are 1.00 and 0.40. Find T_R.

Solution.   From (7-137)

$$T_R = \frac{0.4 \times 290 - 78}{0.6} = 63°K$$

In accurate measurements of the noise temperature of a low-noise amplifier a cold load and a hot (or room-temperature) load are connected alternately to the receiver input.   Care must be taken to prevent errors due to the impedance differences of the loads.   Impedance variations change the gain and cause a corresponding error in the measurement.

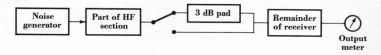

Fig. 7-46.   Measurement of receiver noise temperature using uncalibrated output indicator.

A measurement procedure using an uncalibrated indication of receiver output power is shown in Fig. 7-46.   An adjustable attenuator (adjustable from L to $L + 3$ dB) is connected into the IF amplifier or in some other suitable location in the HF section of the receiver.   An unfired noise generator gives an output indication I when the attenuation is L.   Then 3 dB more attenuation is switched in, and T_G is adjusted so that the same indication is obtained.   The receiver noise temperature is then given by (7-138).

When a two-channel receiver is measured, the noise generator will feed noise power in both channels equally. If output indications are I_2 when the noise generator is operating and I_1 when the noise generator is unfired, then

$$\frac{I_2}{I_1} = \frac{2T_G + T_R}{2T_0 + T_R} \tag{7-139}$$

where T_R is the one-channel noise temperature of the receiver. If T_G is adjusted so that $I_2 = 2I_1$, then

$$T_R = 2(T_G - 2T_0) \tag{7-140}$$

The two-channel noise temperature is

$$T_{R_2} = T_G - 2T_0 \tag{7-141}$$

If the noise generator is calibrated to give the noise figure of the one-channel receiver, it will show an apparent value

$$F_{R_2} = 1 + \frac{T_G - 2T_0}{T_0} = \frac{T_G}{T_0} - 1 \tag{7-142}$$

Solving for T_G and substituting in (7-140) gives

$$T_R = 2T_0(F_{R_2} + 1) - 4T_0 \tag{7-143}$$

From this the dependence of F_{R_2} on F_R is

$$F_{R_2} = \tfrac{1}{2}(F_R + 1) \tag{7-144}$$

Example. The apparent noise figure measured for a two-channel receiver is 3 dB. Find the noise figure of the receiver.
Solution. From (7-144) when $F_{R2} = 2$, $F_R = 3$, or 4.8 dB.

With a square-wave or a pulse-modulated noise generator (noise temperature variation from T_G to T_0) and a measurement of the IF output signal the receiver noise temperature (7-138) can be determined. When T_G is known, the output meter indication can be calibrated to read the noise figure directly. Figure 7-47 shows a suitable arrangement for measuring a low-noise receiver using a direct-reading noise-figure meter. The modulated noise generator is connected to the receiver through a directional coupler, and a cold load is connected to the receiver input. The maximum noise-temperature input to the receiver is

$$\frac{1}{L_D} T_{GN} + \left(1 - \frac{1}{L_D}\right) T_{GC}$$

where L_D = coupling loss of directional coupler
        T_{GN} = operating noise temperature of noise generator
        T_{GC} = temperature of cold load

The minimum noise-temperature input is correspondingly $(1/L_D)T_0 + (1 - 1/L_D)T_{GC}$. Hence K_1 in (7-136) is equal to

$$K_1 = \frac{(1/L_D)T_{GN} + (1 - 1/L_D)T_{GC} + T_R}{(1/L_D)T_0 + (1 - 1/L_D)T_{GC} + T_R} \tag{7-145}$$

In calibrating the meter it is assumed that the input varies between T_G and T_0, giving an apparent noise figure F', from which the actual noise figure or the receiver noise temperature can be obtained.

Example. A direct-reading noise-figure meter uses a noise generator with $T_{GN} = 9900°K$, a 20-dB directional coupler ($L_D = 100$), and a cold load with $T_{GC} = 78°K$. In measuring a low-noise receiver the noise-figure indication is 17 dB. Find the receiver noise temperature.

Solution. When $F'_{dB} = 17$ dB, $F' = 50$, $T'_R = 14,200°K$, and K_1 from (7-136) is 1.66. Substituting K_1 into (7-145) and solving for T_R gives $T_R = 65°K$.

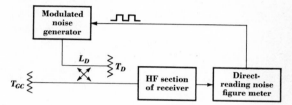

Fig. 7-47. Direct reading of a noise-figure measurement of a low-noise receiver.

7-4c. Calibration Methods. Calibration of radio-astronomy receivers is necessary to provide an absolute scale of antenna temperature. Calibration should be checked frequently because of possible receiver gain and noise-temperature variations. It is common practice to check the calibration just before and after an observation or to use a calibration signal switched on at regular time intervals during observations.

Standard-noise sources and noise generators are used to obtain an accurately known amount of noise power for receiver calibration. For continuum receivers a suitable arrangement is to connect a noise source at the antenna terminals through a directional coupler which has an accurately known coupling (Fig. 7-48a). Thus the calibration can be made without disturbing the receiver. The calibration signal, or the excess noise temperature T obtained when the noise generator is in operation, is, from (7-135),

$$T = \frac{T_G - T_0}{LL_D} \tag{7-146}$$

where L is the attenuation of the attenuator and L_D is the coupling loss of the directional coupler.

Example. An argon gas-discharge noise generator with $T_G = 9900°K$, an attenuator, and a 20-dB directional coupler are used to obtain a calibration signal of $1 \pm 0.05°K$. Find the required attenuation.

Solution. From (7-146) when $T = 1°K$, $T_G = 9900°K$, and $L_D = 100$, L must be 96.1 or 19.8 $\pm$ 0.2 dB.

The arrangement shown in Fig. 7-48b is suitable for an overall calibration of a radio telescope. An auxiliary antenna is used to feed a calibration signal to the radio telescope. The attenuation between the auxiliary antenna and the telescope antenna must be known accurately for absolute calibration.

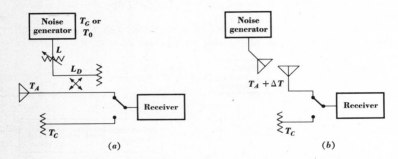

Fig. 7-48.   Calibration circuits.

A radio source with an accurately known flux can also be used to calibrate a radio telescope. The effective aperture of the antenna must also be known accurately. The calibration temperature is given by

$$\Delta T = \frac{A_e S}{2k} \tag{7-147}$$

where A_e = antenna effective aperture
S = source flux density (point source)
k = Boltzmann's constant

References

ADLER, R., G. HRBEK, and G. WADE: The Quadrupole Amplifier: A Low-noise Parametric Device, *Proc. IRE*, vol. 47, pp. 1713–1723, October, 1959.

BELL, D. A.: "Electrical Noise," D. Van Nostrand Company, Inc., Princeton, N.J., 1960.

BENNETT, W. R.: "Electrical Noise," McGraw-Hill Book Company, New York, 1960.

BLACKWELL, L. A., and K. L. KOTZEBUE: "Semiconductor Diode Parametric Amplifiers," Prentice-Hall, Inc., Englewood Cliffs, N.J., 1961.

BLOEMBERGEN, N.: Proposal for a New Type of Solid State Maser, *Phys. Rev.*, vol. 104, pp. 324–327, Oct. 15, 1956.

BLUM, E. J.: Sensibilité des radiotelescopes et récepteurs à correlation, *Ann. Astrophys.*, vol. 22, pp. 140–163, 1959.

CCIR: Documents of the Xth Plenary Assembly, Geneva, 1963, *Rept.* 322, ITU, Geneva, 1964.

CHOW, W. F., and E. CASSIGNAL: "Linear Signal-flow Graphs and Applications," John Wiley & Sons, Inc., New York, 1962.

CHOW, W. F.: "Principles of Tunnel Diode Circuits," John Wiley & Sons, Inc., New York, 1964.

COLVIN, R. S.: A Study of Radio Astronomy Receivers, *Stanford Electron. Lab., Radio Sci. Lab., Sci. Rept.* 18, Stanford, Calif., 1961.

COOPER, B. F. C.: Use of a Y-type Circular Switch with a 21 cm Maser Radiometer, *Rev. Sci. Instr.*, vol. 32, p. 202, February, 1961.

CROOM, D. L.: Naturally Occurring Thermal Radiation in the Range 1–10 Gc/s, *Proc. Inst. Elec. Engrs.*, London, vol. 111, pp. 967–980, May, 1964.

DE GRASSE, R. W., E. O. SCHULZ-DUBOIS, and H. E. D. SCOVILL: The 3-level Solid State Traveling Wave Maser, *Bell System Tech. J.*, vol. 38, pp. 305–334, March, 1959.

DE JAGER, J. T., and B. J. ROBINSON: Sensitivity of the Degenerate Parametric Amplifier, *Proc. IRE*, vol. 49, pp. 1205–1206, July, 1961.

DE JAGER, J. T.: Maximum Bandwidth Performance of a Nondegenerate Parametric Amplifier with Single-tuned Idler Circuit, *IEEE Trans. Microwave Theory Tech.*, vol. MTT-12, pp. 459–467, July, 1964.

DICKE, R. H.: The Measurement of Thermal Radiation at Microwave Frequencies, *Rev. Sci. Instr.*, vol. 17, pp. 268–275, July, 1946.

DICKE, R. H., P. J. E. PEEBLES, P. G. ROLL, and D. T. WILKINSON: Cosmic Black-body Radiation, *Astrophys. J.*, vol. 142, pp. 414–419, 1965.

DITCHFIELD, C. R.: Noise Limits of a Maser Amplifier, *Solid-State Electron.*, vol. 4, p. 171, 1962.

FAY, C. E., and R. L. COMSTOCK: Operation of the Ferrite Junction Circulators, *IEEE Trans. Microwave Theory Tech.*, vol. MTT-13, pp. 15–27, January, 1965.

FEAD, P. M. G.: Versatile Amplifier for Precise Radiometry, *Rev. Sci. Instr.*, vol. 35, pp. 821–828, July, 1964.

FRATER, R. H.: Accurate Wideband Multiplier-square-law Detector, *Rev. Sci. Instr.*, vol. 35, pp. 810–813, July, 1964.

FRIIS, H. T.: Noise Figures of Radio Receivers, *Proc. IRE*, vol. 32, pp. 419–422, July, 1944.

FUJIMOTO, K.: On the Correlation Radiometer Technique, *IEEE Trans. Microwave Theory Tech.*, vol. MTT-12, p. 203, March, 1964.

GORDON, J. P., H. J. ZEIGER, and C. H. TOWNES: Molecular Microwave Oscillator and New Hyperfine Structure in the Microwave Spectrum of NH_3, *Phys. Rev.*, vol. 95, pp. 282–284, July 1, 1954.

GRAHAM, M. H.: Radiometer Circuits, *Proc. IRE*, vol. 46, p. 1966, December, 1958.

HAUS, H. A., and R. B. ADLER: "Circuit Theory of Linear Noisy Networks," The Technology Press of Massachusetts Institute of Technology, Cambridge, Mass., and John Wiley & Sons, Inc., New York, 1959.

HEINLEIN, W., and P. G. MEZGER: Theorie des parametrischen Reflexionsverstärkers, *Frequenz*, vol. 16, nos. 9–11, 1962.

HUGHES, V. A.: Absolute Calibration of a Standard Temperature Noise Source for Use with S-band Radiometers, *Proc. Inst. Elec. Engrs. London, pt. B*, vol. 103, pp. 669–672, 1956.

IRE: Standards on Receivers, Definition of Terms, *Proc. IRE*, vol. 40, pp. 1681–1685, December, 1952.

IRE: Standards of Measuring Noise in Linear 2-ports, *Proc. IRE*, vol. 48, pp. 60–68, January, 1960.

KAC, M., and A. J. F. SIEGERT: On the Theory of Noise in Radio Receivers with Square Law Detectors, *J. Appl. Phys.*, vol. 18. p. 383, 1947.

KELLY, E. J., D. H. LYONS, and W. L. ROOT: The Sensitivity of Radiometric Measurements, *J. Soc. Ind. Appl. Math.*, vol 11., pp. 235–257, June, 1963.

Ko, H. C.: The Distribution of Cosmic Radio Background Radiation, *Proc. IRE*, vol. 46, pp. 208–215, January, 1958.

KRAUS, J. D., and H. C. Ko: Celestial Radio Radiation, *Ohio State Univ. Radio Obs. Rept. 7*, May, 1957.

KUHN, N.: Simplified Signal Flow Graph Analysis, *Microwave J.*, vol. 10, p. 59, November, 1963.

LAX, B., and K. J. BUTTON: "Microwave Ferrites and Ferrimagnetics," McGraw-Hill Book Company, New York, 1962.

MACHIN, K. E., M. RYLE, and D. D. VONBERG: The Design of an Equipment for Measuring Small Radio-frequency Noise Powers, *Proc. Inst. Elec. Engrs. London*, vol. 99, pp. 127–134, May, 1952.

MACKEY, M. B.: Crystal Mixer Receivers Installed on the CSIRO 210-foot Radio Telescope, *Proc. IRE Australia*, vol. 25, pp. 515–520, August, 1964.

MANLEY, J. M., and H. E. ROWE: Some General Properties of Non-linear Elements, Part I: General Energy Relations, *Proc. IRE*, vol. 44, pp. 904–913, July, 1956.

MATTHAEI, G. L.: A Study of the Optimum Design of Wide-band Parametric Amplifiers and Up-converters, *IRE Trans. Microwave Theory Tech.*, vol. MTT-9, pp. 23–38, January, 1961.

MESSENGER, G. C., and C. T. McCOY: Theory and Operation of Crystal Diodes as Mixers, *Proc. IRE*, vol. 45, pp. 1269–1283, September, 1957.

O'DONNELL, S. R.: A Comparison of Radiometers, M.Sc. thesis, The Ohio State University, Columbus, Ohio, 1963.

ORHAUG, T., and W. WALTMAN: A Switched Load Radiometer, *Publ. Natl. Radio Astron. Obs.*, vol. 1, pp. 179–204, 1962.

PENFIELD, P., and R. P. RAFUSE: "Varactor Applications," The M.I.T. Press, Cambridge, Mass., 1962.

PENZIAS, A. A.: Helium-cooled Reference Noise Source in a 4 KMc Waveguide, *Rev. Sci. Instr.*, vol. 36, p. 68, 1965.

PENZIAS, A. A., and R. W. WILSON: A Measurement of Excess Antenna Temperature at 4080 Mc/s, *Astrophys. J.*, vol. 142, pp. 419–421, 1965.

PIERCE, J. R.: "Traveling-wave Tubes," D. Van Nostrand Company, Inc., Princeton, N.J., 1950.

RHEINFELDER, W.: "Design of Low-noise Transistor Input Circuits," Iliffe Books, Ltd., London, 1964.

RICE, S. O.: Mathematical Analysis of Random Noise, *Bell System Tech. J.*, vol. 23, pp. 283–332, July, 1944; vol. 24, pp. 46–156, January, 1945.

RYLE, M.: A New Radio Interferometer and Its Application to the Observation of Weak Radio Stars, *Proc. Roy. Soc. London Ser. A.*, vol. 211, pp. 351–375, 1952.

SELING, T. V.: Some Results on the Use of a Tunnel Diode Amplifier on a Radio Astronomy Receiver, *Proc. IEEE*, vol. 52, pp. 423–424, 1964.

SIEGMAN, A. E.: "Microwave Solid-state Masers," McGraw-Hill Book Company, New York, 1964.

STEINBERG, J. L.: "Les Recepteurs de bruits radioélectriques," *Onde Elec.*, vol. 32, pp. 519–526, 1952.

STELZRIED, C. T.: A Liquid-helium-cooled Coaxial Termination, *Proc. IRE*, vol. 161, p. 1224, July, 1961.

STRUM, P. D.: Some Aspects of Mixer Crystal Performance, *Proc. IRE*, vol. 41, pp. 875–889, July, 1953.

TABOR, W. J., and J. T. SIBILIA: Masers for the Telstar Satellite Communications Experiment, *Bell System Tech. J.*, vol. 42, pp. 1863–1886, 1963.

TIURI, M. E.: Radio Astronomy Receivers, *IEEE Trans. Antennas Propagation*, vol. AP-12, pp. 930–938, December, 1964.

UENOHARA, M., and J. P. ELWARD, JR.: Parametric Amplifiers for High Sensitivity Receivers, *IEEE Trans. Antennas Propagation*, vol. AP-12, pp. 939–947, December, 1964.

YAROSHENKO, V.: Influence of the Fluctuating Factor of Amplification on the Measurement of Weak Noiselike Signals, *Radiotechnica*, vol. 7, pp. 749–751, 1964.

WALLING, J. C., and F. W. SMITH: Solid State Masers and Their Use in Satellite Communication Systems, *Philips Tech. Rev.*, vol. 25, no. 11–12, January, 1965.

WHITE, W. D., and J. G. GREENE: On the Effective Noise Temperature of Gas Discharge Noise Generators, *Proc. IRE*, vol. 44, p. 939, July, 1956.

VAN DER ZIEL, A.: "Noise," Prentice-Hall, Inc., Englewood Cliffs, N.J., 1954.

Problems

7-1. (a) Find the system noise temperature referred to the antenna terminals for a radio-telescope receiver with a 50°K antenna temperature, a receiver with a 1-dB noise figure, and a transmission line between antenna and receiver with 0.5 dB attenuation. The physical temperature of the transmission line is 290°K. (b) How much additional temperature is introduced by the transmission-line loss?

Ans. (a) 169°K; (b) 44°K.

7-2. What is the system noise temperature if a preamplifier of 0.5 dB noise figure is installed (a) ahead of the receiver of Prob. 7-1 (between transmission line and receiver) or (b) at the antenna terminals (ahead of transmission line)? The preamplifier has 15 dB gain. *Ans.* (a) 127°K; (b) 89°K.

7-3. (a) A total-power receiver with a 100°K system noise temperature has an equivalent predetection bandwidth of 5 Mc and a postdetection equivalent integration time of 10 sec. If the average predetection gain is 20 dB and the effective value of the receiver gain fluctuation is 0.5 dB, calculate the minimum detectable temperature of the receiver. (b) Compare this value with the minimum detectable temperature if the gain fluctuation is zero. *Ans.* (a) 1.1°K; (b) 0.014°K.

7-4. A two-channel receiver has a system noise temperature of 500°K with a bandwidth of 10 Mc per channel. Find the system noise temperature required with a single-channel receiver of 8 Mc bandwidth in order to achieve the same sensitivity.

7-5. A 2-Gc liquid-nitrogen-cooled parametric-amplifier receiver has a noise temperature of 50°K and a bandwidth of 10 Mc. The noise temperature of a TWT amplifier receiver is 450°K. The antenna noise temperature is 25°K. Neglecting other sources of noise, what should the bandwidth of the TWT receiver be to give the same sensitivity as that of the parametric-amplifier receiver when used with the same antenna?

7-6. Find the noise temperature of the reflection amplifier of Fig. 7-30, using the exchangeable-power concept.

7-7. A reflection amplifier using a three-port circulator has 15 dB gain. The normalized input conductance of the postamplifier is 1.5. Find the standing-wave ratio of the antenna feed line causing oscillation. *Ans.* VSWR ≥ 17.2

7-8. A liquid-nitrogen-cooled parametric amplifier uses a varactor diode with a cutoff frequency of 50 Gc. The signal frequency is 2.4 Gc. Find the theoretical noise temperature with $\gamma = 0.25$ for pump frequencies of 12 and 24 Gc.

8
Radio Sources

8-1 Introduction In this chapter a summary is given of some of the principal observational results of radio astronomy. Following brief descriptions of the radio sky and source spectra, some fundamental source mechanisms are discussed. Brief treatments are then presented of the optical and radio sun, of the moon and planets, and of galactic radio sources. The latter are divided into thermal sources, nonthermal sources, line-emission clouds, flare stars, and background radiation. The chapter concludes with a discussion of extragalactic radio sources and cosmological considerations. The treatments are not extensive but will, it is hoped, give the reader a background for pursuing the topics further. The observational status of radio astronomy is advancing so rapidly that for the latest information the reader should refer to the current scientific journals. A few of the most important ones in the field of radio astronomy are *The Astrophysical Journal, The Astronomical Journal, Monthly Notices of the Royal Astronomical Society, Nature, Science,* and the *Australian Journal of Physics.* Numerous references are given in this chapter to these and many other journals.

8-2 The Radio Sky The appearance of the radio sky at 250 Mc is shown by the map in Fig. 8-1 (Ko and Kraus, 1957). The contours indicate the equivalent brightness temperature of the sky background, and the small circles show some of the stronger discrete radio sources. The contours are in units of 6°K above the coldest parts of the sky (about 80°K near the galactic poles). Thus, contour 7 is at a temperature of 122°K, i.e., $7 \times 6° + 80°$. The discrete sources are on a relative magnitude scale (4 dB per magnitude) with smaller magnitudes for stronger sources, analogous to the magnitude scales used for optical stars.

When the contour map of Fig. 8-1 is converted to shades of black and white, the result is Fig. 8-2, which shows how the sky would appear if our eyes were sensitive to radio waves instead of to light. The white areas indicate strong radio emission and the dark areas weak emission. The bright points are the discrete radio sources. These are sometimes called *radio stars,* although none shown on this map corresponds to any optical star. The broad band of emission arching across the picture corresponds

294

to the plane of our galaxy with the nucleus, or center, of our galaxy at the lower right. The presentation in Fig. 8-2, as in Fig. 8-1, is a Mercator projection. One may visualize the true (spherical) projection by imagining the right and left edges of Fig. 8-2 to be joined behind one's head while the upper edge is pulled over the top. From the center of a sphere whose inside surface is covered in this manner by Fig. 8-2, the bright band or plane of our galaxy would appear to be a strip running around the sphere like the rim of a wheel as seen from the hub.

A further insight into Fig. 8-2 may be obtained by reference to the simplified sketch of our galaxy shown in Fig. 8-3. The galaxy consists of an aggregation of billions of stars (about 10^{11} stars) arranged like a great flat wheel turning slowly in space. There is a complex central region with a number of spiral arms extending outward. The overall diameter of the system is about 100,000 light-years, and the wheel makes one revolution in about 300 million years. The earth and solar system are situated in one of the spiral arms at a distance of nearly 30,000 light-years from the center. From our position in the galaxy the strongest radio background emission comes from the center of the galaxy. There is also a strong band of radiation everywhere in the plane of the galaxy. In the Cygnus direction we are looking along a spiral arm and observe a peak of radiation (near 20 hr RA and $+40°$ dec in Fig. 8-1). In the direction of the Crab Nebula (M 1) (5.5 hr RA and $+21°$ dec) we are looking almost directly away from the galactic nucleus, or toward the anticenter direction. Directions perpendicular to the plane of the galaxy correspond to the galactic poles, or the coldest regions of the sky. At frequencies higher than 250 Mc the galactic plane tends to stand out more sharply with a higher ratio of temperature between the equator and poles. On the other hand, at lower frequencies the sky tends to be more uniformly bright and because of absorption by regions of ionized hydrogen the galactic plane tends to appear as a trough instead of a ridge.

Discrete radio sources numbering in the thousands have been located. As larger telescopes go into operation, this number continues to increase. With higher-resolution antennas and more sensitive receivers it is now possible to map the sky in much more detail than suggested in Fig. 8-1. Examples of maps with greater detail are presented in Fig. 8-4 (Kraus, 1964; Dixon, Meng, and Kraus, 1965). These maps, made at 600 and 1,415 Mc with the Ohio State University 260-ft radio telescope in 1964, cover about 700 deg^2 of sky with a resolution of over 6,000 beam areas at the higher frequency (about nine beam areas per square degree). The Andromeda and Perseus regions covered by these maps lie between right ascensions of 23^h40^m and 06^h00^m and declinations of 36 to $46°$ (north). Over 125 sources appear on these maps, of which 16 have 3C numbers and 110 have OA (Ohio list A) numbers. Detailed structure also appears

Fig. 8-1. Radio sky at 250 Mc as observed with the Ohio State University 96-helix radio telescope. (*After Ko and Kraus,* 1957.)

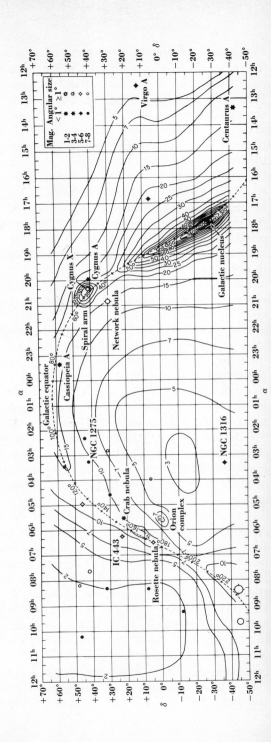

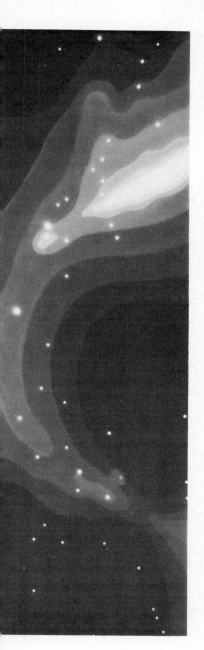

Fig. 8-2. Radio-sky panorama obtained by converting the contour map of Fig. 8-1 to shades of black and white. The figure gives an impression of how the sky would appear if our eyes were sensitive to radio waves instead of to light.

around a number of extended sources, such as the Andromeda nebula M 31 and the California nebula NGC 1499. There is also considerable detail evident in the region of the galactic plane. This detail is in contrast to the contours of Fig. 8-1, made with the Ohio State University 96-helix radio telescope at 250 Mc some nine years earlier (in 1955) at a resolution of 0.1 beam area per square degree. This earlier map shows only a very gradual background variation in the region of the galactic plane and only three discrete sources in the areas of the newer maps of Fig. 8-4.

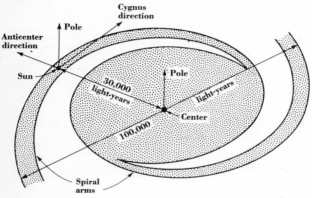

Fig. 8-3. Simplified, idealized sketch of our galaxy.

The Andromeda region in still more detail is revealed by the high-sensitivity profiles and contour map of Fig. 8-5, made with the Ohio State University 260-ft radio telescope at 1,415 Mc (Kraus and Dixon, 1965; Kraus, Dixon, and Fisher, 1966). By repeated scans an rms noise temperature of 0.01°K was achieved in this survey. At this sensitivity the results are close to the resolution or confusion limit of the telescope at 1,415 Mc. Digital output was used, with the data completely processed by the university's IBM 7094 computer. The profiles in Fig. 8-5a are as plotted by the computer, while the contours of Fig. 8-5b were hand-drawn with the aid of digits printed by the computer on a map grid.

The maps of Fig. 8-4 have over 125 listed sources appearing in about 700 deg² of sky. At 1,415 Mc there are over 6,000 beam areas in this sky region, so that there are about 50 beam areas per source, or about 5 deg per source. A resolution of 50 beam areas per source should be adequate to avoid confusion effects except where sources are complex or closely grouped. If the sky area covered by the maps of Fig. 8-4 is considered to be typical as to number of sources per unit area, we should expect to find about 8,000 (41,253/5) sources in the entire sky at this level of sensitivity and certainty

The revised third Cambridge catalog (Bennett, 1962) lists 328 sources. With present-day instruments, lists totaling at least ten times this number would appear to be within reach, and as larger and more sensitive radio telescopes go into operation, the number of radio sources that may be tabulated with confidence should rise into the tens or even hundreds of thousands. The task of doing this, however, is formidable, since as the telescope resolution increases (beam area decreases), the length of time required to survey a unit area tends to increase.

8-3 Source Spectra Ground-based radio astronomical observations can be made over a nominal wavelength range from a few millimeters to some tens of meters. The short wavelength limit is caused by molecular absorption in the atmosphere and the long wavelength limit by ionospheric reflection. This relatively transparent region of the earth's ionosphere is sometimes called the *radio window*. The relation of this window to the one at optical wavelengths is discussed in Chap. 1.

By measuring the flux density of a radio source over as much of a wavelength range as possible, a spectrum may be determined for the source. Measured spectra typical of a variety of radio sources are shown in Fig. 8-6a. Sources like Cassiopeia A and Cygnus A have spectra in which the flux density falls off with frequency, but for objects such as the moon and Mars the flux density increases with frequency.

According to the Rayleigh-Jeans law (Chap. 3) the flux density S of an object at a wavelength λ in the radio spectrum is given by

$$S = \frac{2k}{\lambda^2} \iint T \, d\Omega \tag{8-1}$$

where k = Boltzmann's constant (= 1.38×10^{-23} joule °K^{-1})
   T = equivalent blackbody temperature, °K
   $d\Omega$ = element of solid angle
The integration is carried out over the solid angle subtended by the object. If the temperature is uniform over the source, (8-1) reduces to

$$S = \frac{2k}{\lambda^2} T\Omega_s \tag{8-2}$$

where Ω_s = source solid angle
Let the variation of the flux density S with wavelength be expressed by the proportionality†

$$S \propto \lambda^n \tag{8-3}$$

† This definition, adopted by Conway, Kellermann, and Long (1963), results in a positive spectral index for nonthermal sources, which constitute the majority of all radio sources. The inverse definition ($S \propto \lambda^{-n} \propto \nu^n$) is also in use. It gives a negative index to nonthermal sources.

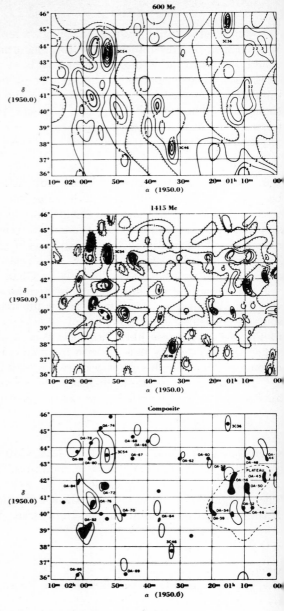

Fig. 8-4(a). Maps of the Andromeda region made with the Ohio State University 260-ft radio telescope at 600 and 1,415 Mc. The contour intervals are 0.03 and 0.05°K at 600 and 1,415 Mc, respectively. The composite map shows the features of principal significance on the other two maps by solid areas for 1,415 Mc and open contours for 600 Mc. (*After Kraus*, 1964.)

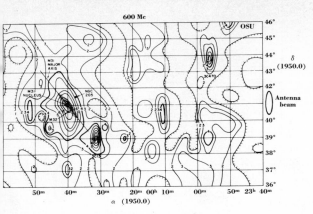

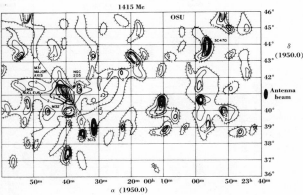

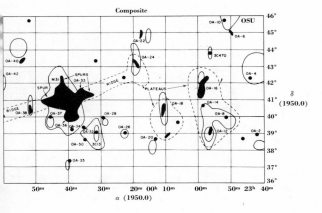

301

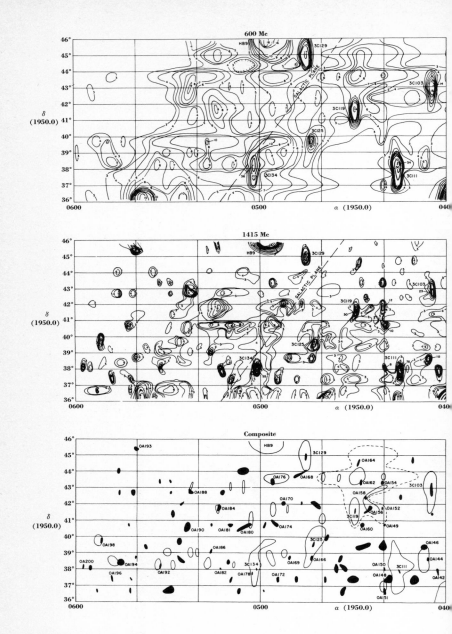

Fig. 8-4(b). Maps of the Perseus region made with the Ohio State University 260-ft radio telescope at 600 and 1,415 Mc. The contour intervals are 0.25 and 0.06°K at 600 and 1,415 Mc, respectively. The composite map shows the features of principal significance of the other two maps by solid areas for 1,415 Mc and open contours for 600 Mc. *(After Dixon, Meng, and Kraus, 1965.)*

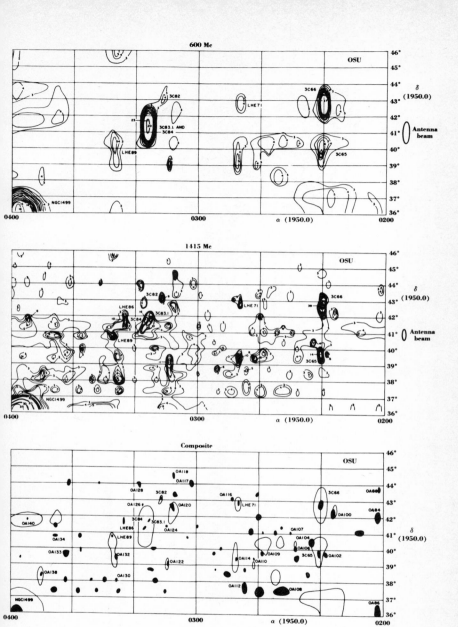

303

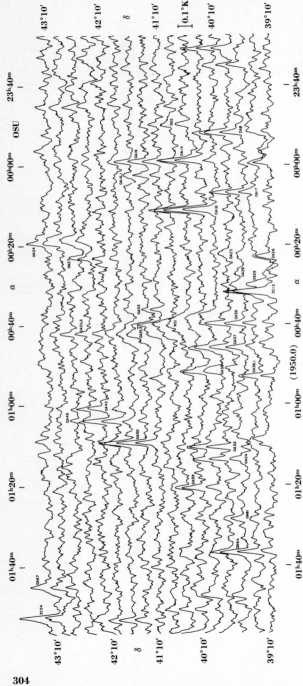

Fig. 8-5(a) High-sensitivity profiles of the Andromeda (M 31) region made with the Ohio State University 260-ft radio telescope at 1,415 Mc. M 31 is near the center, with other sources in the vicinity. (*After Kraus and Dixon, 1965, and Kraus, Dixon, and Fisher, 1966.*)

where n = spectral index, dimensionless

Suppose that the temperature of a source were constant with wavelength. Then, from (8-2), S would vary inversely with the square of the wavelength, and for this case the spectral index n would be equal to -2. This variation is characteristic of the *thermal radiation* from a blackbody. If the slope of the $n = -2$ line in Fig. 8-6b is compared with the slopes of the spectra for the moon and Mars in Fig. 8-6a, we may conclude that the radiation of these two solar-system objects results from *thermal* emission.

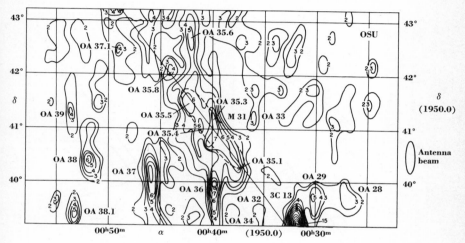

Fig. 8-5(b). Contour map of the Andromeda galaxy (M 31) made with the Ohio State University 260-ft radio telescope at 1,415 Mc. The contour interval is 0.025°K. (*After Kraus and Dixon, 1965.*)

Since the oppositely sloping spectrum of Cassiopeia A and Cygnus A suggests an entirely different mechanism, these sources with a positive index are referred to as *nonthermal* types. The spectrum of Cassiopeia A has an index of about $+0.8$. Although the spectrum of Cygnus A is curved, its index at the lower frequencies is nearly the same as for Cassiopeia A. The spectral index for most nonthermal sources lies between about $+0.3$ and $+1.3$, with an average value near $+0.8$. Calculated spectra with indices of $+0.5$, $+0.7$, and $+1.0$ are illustrated in Fig. 8-6b. The synchrotron process is believed to be dominant in nonthermal sources. This mechanism is discussed in the next section.

The spectrum for the Orion nebula in Fig. 8-6b appears to be of the thermal blackbody type $(n = -2)$ at the lower frequencies, but it becomes flat at the higher frequencies. This nebula is a large hydrogen cloud about 2 light-years in diameter at a distance of about 1,500 light-years. The hydrogen is ionized by the ultraviolet radiation from hot stellar sources at the center of the cloud. Radio radiation is by a thermal mechanism in which free electrons passing near protons are accelerated and caused to emit. This so-called *free-free* mechanism is discussed further in the next section. The effective temperature is given by

$$T = T_c(1 - e^{-\tau}) \tag{8-4}$$

where T_c = cloud temperature
$\qquad \tau$ = optical depth

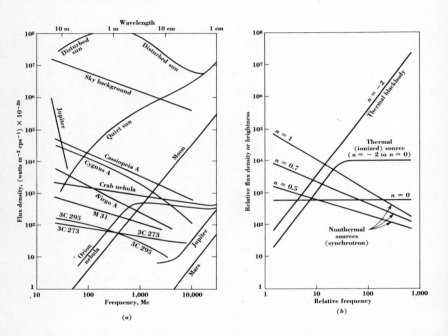

Fig. 8-6. (*a*) Spectra of typical radio sources; (*b*) calculated spectra for various values of spectral index *n*.

If the cloud is dense ($\tau \gg 1$), T is a constant ($= T_c$) and from (8-2) and (8-3) the spectral index $n = -2$, as for a true blackbody radiator. This situation holds for the Orion nebula at the lower frequencies. However, at the higher frequencies τ, being proportional to the square of the wavelength, becomes small ($\tau \ll 1$), and T becomes a function of the wavelength. For this condition we have

$$T = T_c\tau \tag{8-5}$$

or

$$T \propto \lambda^2 \tag{8-6}$$

Introducing (8-6) into (8-2) yields a spectral index $n = 0$, so that the spectrum is flat. A calculated spectrum with $n = -2$ at the lower frequencies and $n = 0$ at the higher frequencies is illustrated in Fig. 8-6*b* for comparison with the measured Orion nebula spectrum in Fig. 8-6a. Such bent spectra characterize the radio emission from ionized-hydrogen clouds. These clouds are referred to as emission nebulosities and are common close to the plane of our galaxy.

A typical radio-source spectrum (synchrotron type) is illustrated in Fig. 8-7, together with the spectrum of a thermal blackbody at 6000°K. This graph embracing both the radio and optical parts of the electromagnetic spectrum helps to explain why the radio sky appears so different from the optical sky. A hot blackbody (or star) is a strong emitter in the optical spectrum but weak at radio wavelengths. On the other hand, radiation from relativistic electrons moving in weak interstellar magnetic

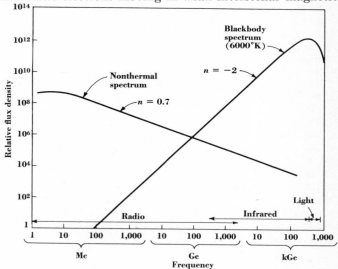

Fig. 8-7. Spectrum of nonthermal radio source compared with spectrum of a thermal blackbody at 6000°K.

fields (typical of nonthermal radio sources) is strong in the radio spectrum but weak at optical wavelengths. Hence, the optical picture is largely one of stellar objects while the radio picture is more one of clouds of relativistic electrons. In general, the distribution of the two is quite different.

The spectra of a number of nonthermal sources are illustrated in Fig. 8-8. Sources with straight spectra are presented in (a); those with curved spectra are shown in (b). The flux densities of these sources at 400 Mc are listed in Table 8-1. The spectral index, or range of index for curved spectra, is also given. The data in Fig. 8-8 and Table 8-1 are largely from Conway, Kellermann, and Long (1963). Supernova remnants, like Cassiopeia A, appear to have straight spectra (constant indices) to very short wavelengths. A number of sources, like 3C 48, 147, 295, and Cygnus A, have spectra which are considerably flatter and straighter at frequencies below 300 Mc than they are above. These sources are believed to be at great distances and to have very large power output. It is suggested that these are younger sources, where the rates of injection of the relativistic electrons and of radiation loss have not yet reached equilibrium in the synchrotron process (Kellermann, 1964). The spectra with the most

Table 8-1
*Flux density and spectral-index data
on some nonthermal radio sources*

Source	Flux density at 400 Mc†	Spectral index
Cassiopeia A	6,100	0.77
Cygnus A	4,500	0.7–1.2
Hydra A	133	0.87
Taurus A	1,230	0.27
Virgo A	580	0.83
3C 28	66	1.10
3C 48	36	0.2–0.7
3C 98	25	0.70
3C 147	52	0–0.7
3C 273	59	0.33
3C 286	23	0.1–0.7
3C 295	52	0.4–0.8
3C 298	24	0.3–1.0
3C 310	25	0.94
3C 452	29	0.78
CTA 21	9	−0.2−+0.9
CTA 102	6	−0.3−+0.5

† In flux units (10^{-26} watt m^{-2} cps^{-1}).

curvature are those for CTA 21 and 102 (California Institute of Technology list A). These have a flux-density spectrum which falls off both above and below a frequency around 1,000 Mc. Such a spectrum may indicate a source with a small energy spread of the relativistic electrons. Fig. 8-9 shows the spectra of two OA (Ohio list A) sources and three sources with reverse curvature (concave upward). Of the latter, 3C 273 shows only a slight amount of curvature,† but Perseus A and 3C 279 show a much more pronounced curvature, with a tendency for the spectrum to become flat at the high-frequency end (Dent and Haddock, 1965). As radio spectra are extended to very long and very short wavelengths, many tend to show radical changes in slope, suggesting that a number of different mechanisms are dominant in different parts of the spectrum.

In a study of over 200 nonthermal sources Kellermann (1964) found that the distribution of sources as a function of spectral index was much flatter for sources near the galactic plane ($b < 10°$) than for sources at

† 3C 273 has been resolved into two components with very different spectral indices, the spectrum in Fig. 8-9 being a composite (see page 389).

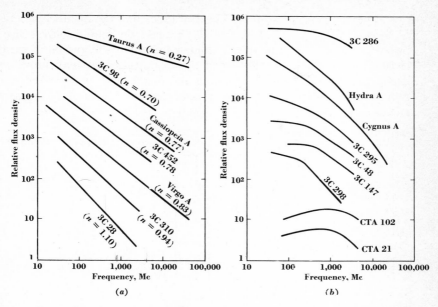

Fig. 8-8. Spectra of some strong nonthermal radio sources. (*After Conway, Keller-mann, and Long,* 1963.)

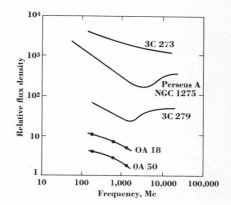

Fig. 8-9. Spectra of three sources (3C 273, Perseus A, and 3C 279) having reversed curvature and of two OA sources.

higher galactic latitude ($b > 10°$). This effect is illustrated in Fig. 8-10. The average index is about 0.75, and all indices are less than 1.3.

In comparing the power output of radio sources with their spectral index there appears to be a trend for the weaker sources to have a lower index (0.3 to 0.5) and the stronger sources a higher index (0.7 to 1.0) (Heeschen, 1960; Kellermann, 1964). This comparison is, of course, restricted to the identified extragalactic sources, since the distance must be known to determine the power output and, to date, distance determinations are dependent on optical red-shift measurements.

A comparison between the spectral index and the distance separation of the two components of identified extragalactic double sources also appears to indicate a trend toward higher index with increasing separation (Kellermann, 1964).   If this trend is substantiated, it may prove to be useful as a method for determining the distance of unidentified double radio sources entirely by means of radio measurements of the separation angles and spectra.

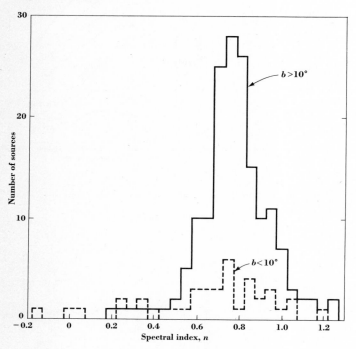

Fig. 8-10.   Number of nonthermal sources as a function of spectral index for galactic latitudes above 10° (solid curve) and below 10° (dashed curve).   (*After Kellermann*, 1964.)

8-4   Mechanisms   Three important mechanisms of continuum emission from radio sources are (1) thermal blackbody radiation, (2) thermal emission from ionized gas, and (3) synchrotron radiation.   Thermal blackbody emission has already been discussed in Chap. 3 in connection with the Planck and Rayleigh-Jeans radiation laws.   The other two mechanisms have been mentioned frequently, and highly simplified discussions of them are presented in this section.   Line emission is treated in Section 8-9*c*. Mechanisms involving plasma and gyro oscillations are considered in Chap. 5.   These mechanisms, and possibly others, such as Cerenkov radiation, are important in radiation from the sun and probably also from flare stars.

8-4a. *Thermal Emission from Ionized Hydrogen.* Line emission from neutral hydrogen and other atoms or molecules is associated with transitions between energy levels in the atoms or molecules, and an analysis involves quantum-mechanical considerations. Thermal radio emission from ionized gas, on the other hand, comes from electrons detached from the atoms.† These free electrons possess no definite energy levels, and their radiation is in the form of a continuous spectrum. Radiation occurs from an electron during acceleration, as when it is deflected in passing near a proton. The electron is only temporarily influenced by the proton. The interaction is termed a *free-free transition* since the electron is free, or unbound, before and after the interaction. At radio wavelengths, where $h\nu/kT \ll 1$, classical theory is adequate. For diffuse ionized hydrogen the absorption coefficient‡ is (Shklovsky, 1960)

$$K' = 9.8 \times 10^{-13} N^2 T_e^{-3/2} \nu^{-2} [19.8 + \ln (T_e^{3/2} \nu^{-1})] \tag{8-7}$$

where N = free-electron density, number m^{-3}

T_e = electron (kinetic) temperature, °K

ν = frequency, cps

This expression may be simplified by evaluating the slowly varying logarithmic term on the assumption that $T_e = 10,000°$K and that the frequency is in the microwave region (of the order of 1,000 Mc), giving the approximation

$$K' = 1.3 \times 10^{-11} N^2 T_e^{-3/2} \nu^{-2} \quad \text{neper m}^{-1} \tag{8-8}$$

The optical depth is given by the integral of K' over the absorbing path, or

$$\tau = \int_0^l K' \, ds \quad \text{nepers}$$

$$= 1.3 \times 10^{-11} T_e^{-3/2} \nu^{-2} \int_0^l N^2 \, ds \tag{8-9}$$

The integral in (8-9) is called the *emission measure* (EM) of the ionized region. It is customarily expressed in electrons per cubic centimeter for N and parsecs for the path length. Making this change and also expressing ν in megacycles per second (instead of cps) and setting $T_e = 10,000°$K, the optical depth becomes (Greenstein and Minkowski, 1953)

$$\tau = 0.4\nu^{-2} \int_0^l N^2 \, ds = 0.4\nu^{-2} \text{EM} \tag{8-10}$$

† Line emission may occur from ionized hydrogen regions during a transient recombination of an ion and electron. (See footnote to Table 8-6.)

‡ $K' = K\rho$, Sect. 3-13.

where ν = frequency, Mc
$\quad N$ = free-electron density, number cm^{-3}
$\quad l$ = path length, pc
$\quad$ EM = emission measure
Assuming that N is constant over the path, we have

$$\tau = 0.4\nu^{-2}N^2l \tag{8-11}$$

The brightness temperature T_b of the ionized-hydrogen cloud is

$$T_b = T_e(1 - e^{-\tau}) \tag{8-12}$$

At a wavelength of 10 cm the optical depth is sufficiently small so that (8-12) reduces to

$$T_b = T_e\tau \tag{8-13}$$

Assuming uniform brightness over a source of extent Ω_s, its flux density S is given by

$$S = \frac{2k}{\lambda^2} T_b\Omega_s \tag{8-14}$$

where k = Boltzmann's constant ($= 1.38 \times 10^{-23}$ joule $°K^{-1}$)
Noting that

$$S \propto \frac{T_b}{\lambda^2} \propto \frac{\tau}{\lambda^2} \propto \frac{1}{\nu^2\lambda^2} \tag{8-15}$$

it follows that the flux density should be constant as a function of frequency (spectral index $n = 0$). This is found to be the case in the decimeter range, as shown, for example, by the spectrum for the Orion nebula in Fig. 8-11 (Menon, 1964).

However, at longer wavelengths the optical depth is not small compared to unity, and (8-13), and therefore (8-15), no longer hold. At sufficiently long wavelengths, where the optical depth is large compared to unity, (8-12) reduces to

$$T_b = T_e \tag{8-16}$$

and hence,

$$S \propto \lambda^{-2} \tag{8-17}$$

so that at longer wavelengths the flux density varies inversely as the square of the wavelength (spectral index $n = -2$), as in the case of a thermal blackbody radiator. Hence, the term *thermal emission*.

Thus, a typical region of ionized hydrogen gas is optically thick at longer wavelengths with the spectral index ($n = -2$) of a thermal blackbody but is optically thin at shorter wavelengths with a spectral index

$n = 0$. An example is the spectrum for the Orion nebula shown in Fig. 8-11 (also see Fig. 8-6). The break between the optically thick and optically thin condition occurs at a wavelength of about 30 cm ($\tau \sim 1$). Further discussion is given in Sec. 8-9b.

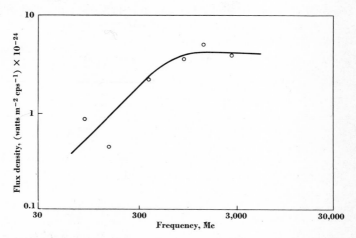

Fig. 8-11. Spectrum of the Orion nebula, a typical ionized-hydrogen radio source. The circles are observed points. (*After Menon*, 1964.)

8-4b. Synchrotron Mechanism. This mechanism is believed to be responsible for the emission of most of the nonthermal radio sources. Such emission was observed from the General Electric synchrotron soon after World War II (Elder, Langmuir, and Pollack, 1948). In the synchrotron, electrons were being accelerated to very high energies in a strong magnetic field, and the radiation was observed in the optical spectrum. The mechanism is the same as in the radio case. Hence, even at radio wavelengths the mechanism is usually referred to as the *synchrotron mechanism* and the radiation called *synchrotron emission*.

In a radio source the magnetic fields are very much weaker than in the synchrotron, a fact which tends to shift the emission to lower frequencies or into the radio spectrum. Although radiation from relativistic (high-energy) particles had been analyzed by Schott (1912), the laboratory production of radiation in the General Electric synchrotron revived interest in the problem (Schwinger, 1949). Then Alfvén and Herlofson (1950) proposed that radiation from radio sources might originate from relativistic electrons moving in the magnetic field of a star. Almost at the same time Kiepenheuer (1950) suggested that the electrons are in fact the electron components of cosmic rays and that the radiation occurs during their interaction with interstellar magnetic fields.   His interpretation gained wide acceptance, and with subsequent refinements provides the theoretical basis for the nonthermal emission (Shklovsky, 1960).   Details of the theory are complex, but the basic principles are straightforward.   The following discussion is a highly simplified treatment of some of these principles.

A particle moving perpendicular to a magnetic field will describe a circle of radius

$$R = \frac{mv}{eB} \qquad (8\text{-}18)$$

where R = radius, m
$\quad m$ = mass of particle, kg
$\quad v$ = velocity of particle, m sec^{-1}
$\quad e$ = charge of particle, coul
$\quad B$ = magnetic flux density, webers m^{-2}, (1 weber m^{-2} = 10^4gauss)
The frequency (revolutions per second) is

$$\nu = \frac{v}{2\pi R} = \frac{1}{2\pi}\frac{e}{m}B \qquad \text{cps} \qquad (8\text{-}19)$$

Radiation or absorption by the particle is at the frequency ν, which is called the *gyro* or *cyclotron frequency* (see Chap. 5).   For example, consider an electron in the earth's ionosphere.   For an electron
$\quad e = 1.6 \times 10^{-19}$ coul
$\quad m = 9.1 \times 10^{-31}$ kg
Taking $B = 5 \times 10^{-5}$ weber m^{-2} yields a frequency $\nu = 1.4$ Mc.   The magnetic flux density in the interstellar medium is much weaker, 10^{-9} to 10^{-10} weber m^{-2}, so that an electron in this medium would have a gyro frequency of only 140 to 14 cps.

The above applies only to low-energy electrons, i.e., ones whose velocity is small compared to the velocity of light ($v \ll c$).   For relativistic electrons, i.e., ones with velocities approaching that of light ($v \rightarrow c$), the situation is different.   Here the particle energy is greater than that given by the product of its rest mass and the square of the velocity of light.   The energy expression is

$$\mathcal{E} = \frac{m_0 c^2}{\sqrt{1-(v/c)^2}} = mc^2 \qquad (8\text{-}20)$$

where $\mathcal{E}$ = energy of particle, joules
$\quad m_0$ = rest mass of particle, kg
$\quad c$ = velocity of light ($= 3 \times 10^8$ m sec^{-1})
$\quad v$ = velocity of particle, m sec^{-1}
$\quad m$ = relativistic mass, kg
Expressing the energy in electron volts, (8-20) becomes

$$\mathcal{E}_v = 6 \times 10^{18} \frac{m_0 c^2}{\sqrt{1 - (v/c)^2}} \tag{8-21}$$

where $\mathcal{E}_v$ = energy of particle, ev

The radiation from the relativistic particle is effectively concentrated in a cone of angle θ centered on the direction of the instantaneous velocity (see Fig. 8-12) as given by

$$\theta = 2\sqrt{1 - (v/c)^2} = 1.2 \times 10^{19} \frac{m_0 c^2}{\mathcal{E}_v} \quad \text{rad} \tag{8-22}$$

For an electron with $\mathcal{E}_v = 10^9$ ev, i.e., a velocity within about 1 part in a million of the velocity of light, the cone angle θ is only 10^{-3} rad, or 3.4 min of arc.

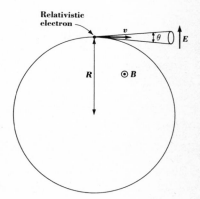

Fig. 8-12. Orbital plane of a relativistic electron showing cone of angle θ, in which radiation is concentrated. The radiated electric field E is polarized parallel to the orbit plane.

An observer in the plane of the orbit will receive pulses from the electron of length

$$\Delta t \simeq \frac{R\theta}{c} \left[1 - \left(\frac{v}{c}\right)^2 \right] \quad \text{sec} \tag{8-23}$$

Now

$$R \simeq \frac{c}{2\pi\nu} \tag{8-24}$$

and

$$\nu = \frac{1}{2\pi} \frac{e}{m_0} B \sqrt{1 - \left(\frac{v}{c}\right)^2} \tag{8-25}$$

At very high energy $(v \rightarrow c)$ the frequency ν tends to zero as the radius R approaches infinity. Substituting (8-21), (8-22), (8-24), and (8-25) into (8-23) yields

$$\Delta t \simeq \frac{2m_0}{eB} \left[1 - \left(\frac{v}{c}\right)^2 \right] \tag{8-26}$$

The maximum radiation from the particle (power per unit bandwidth) occurs at the frequency ν_{max}, given by

$$\nu_{max} \simeq \frac{1}{2\pi\Delta t} = \frac{1}{4\pi}\frac{eB}{m_0}\frac{1}{1-(v/c)^2} \qquad \text{cps} \tag{8-27}$$

where e = charge of particle, coul
    B = magnetic flux density, webers m^{-2}
    m_0 = rest mass of particle, kg
    v = velocity of particle, m sec^{-1}
    c = velocity of light $(= 3 \times 10^8 \text{ m sec}^{-1})$

Evaluating (8-27) for an electron and expressing its energy in electron volts

$$\nu_{max} = 0.06 B \mathcal{E}_v^2 \tag{8-28}$$

where ν_{max} = frequency of maximum radiation from electrons, cps
    B = magnetic flux density, webers m^{-2}
    $\mathcal{E}_v$ = energy of electron, ev

If the energy is expressed in billion electron volts (gigavolts), B in gauss, and ν_{max} in megacycles, we have

$$\nu_{max} = 6 \times 10^6 B_g \mathcal{E}_{GV}^2 \tag{8-29}$$

where ν_{max} = frequency of maximum radiation of electrons, Mc
    B_g = magnetic flux density, gauss
    $\mathcal{E}_{GV}$ = energy of electron, Gev

Evaluating (8-29) for a 1-Gev electron in an interstellar magnetic field of 10^{-5} gauss, we obtain a frequency of maximum radiation of 60 Mc.

The spectrum of a relativistic electron is effectively a continuous one, with the form shown by Fig. 8-13. The spectrum of the radiation from an assemblage of such relativistic electrons will be a function of the energy distribution (or energy spectrum) of the electrons. Following the proposal by Kiepenheuer (1950) that cosmic rays are the source of the nonthermal galactic emission, an energy distribution characteristic of primary cosmic rays would appear to be appropriate for the assemblage. Such an energy spectrum is of the form

$$N(\mathcal{E}) = (\text{const})\mathcal{E}^{-\alpha} \tag{8-30}$$

where $N(\mathcal{E})$ = number of electrons as a function of the energy
    $\mathcal{E}$ = energy of electron
    α = energy-spectrum index

The total power radiated by a distribution of relativistic electrons will then be

$$W = (\text{const}) \int_0^{\mathcal{E}} W(\mathcal{E})N(\mathcal{E}) \, d\mathcal{E} \qquad \text{watts} \tag{8-31}$$

where $W(\mathcal{E})$ = power radiated per electron
    $N(\mathcal{E})$ = energy spectrum, number per energy interval

The power radiated by a single electron is a function of the square of its energy. Assuming that all this radiation is at the frequency of maximum emission and a cosmic-ray type of energy distribution is involved, (8-31) becomes

$$W = (\text{const}) \int_0^{\mathcal{E}} \mathcal{E}^2 \, \mathcal{E}^{-\alpha} \, d\mathcal{E} = (\text{const}) \quad \mathcal{E}^{3-\alpha} \tag{8-32}$$

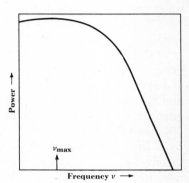

Fig. 8-13. Spectrum of relativistic electron. The ordinate is the radiated power and the abscissa the frequency, both to logarithmic scales.

But as given by (8-27), (8-28), or (8-29), the frequency of maximum radiation of an electron is proportional to the square of its energy, or the energy is proportional to the square root of the frequency. Thus, (8-32) can be written

$$W = (\text{const}) \, \nu^{(3-\alpha)/2} \tag{8-33}$$

where ν = frequency

The variation of the total power radiated by an assemblage of relativistic electrons is then

$$\frac{dW}{d\nu} = (\text{const}) \, \nu^{(1-\alpha)/2} = (\text{const}) \, \lambda^{(\alpha-1)/2} \tag{8-34}$$

where λ = wavelength

The brightness and the flux density of a radio source due to such an assemblage of relativistic electrons is therefore

$$S \propto B \propto \lambda^{(\alpha-1)/2} \tag{8-35}$$

where S = flux density
B = brightness
λ = wavelength
α = energy spectral index

A more detailed development yields

$$S \propto \lambda^{(\alpha-1)/2} \, B^{(\alpha+1)/2} \tag{8-36}$$

where B = magnetic flux density

For cosmic rays an appropriate value of α is 2.4. Accordingly, a radio source with such a cosmic-ray distribution should have the flux-density variation with wavelength given by

$$S \propto \lambda^{(2.4-1)/2} = \lambda^{0.7} \qquad (8\text{-}37)$$

or a (flux-density) spectral index $n = 0.7$. Observations show that this index value is typical of nonthermal radio sources. Curved or bent radio spectra imply a more complex type of energy distribution (Kellermann, 1964).

Twiss (1954) has shown that below a critical frequency, which depends on a number of characteristics of the emitting medium, the spectral index of a synchrotron source will reverse sign because of a *self-absorption* process and that under such (optically thick) conditions the index at lower frequencies will be -2.5. This frequency of reversal (at which the flux density is a maximum) is not the same as the frequency of maximum power radiated by a single electron as given by (8-28) and (8-29). Williams (1963) has listed a number of sources of small angular extent and high surface brightness which exhibit a sharp decrease in flux density below about 100Mc. This suggests the presence of *synchrotron self-absorption* in these sources at the lower frequencies.

Synchrotron radiation is linearly polarized with the electric-field E direction parallel to the orbital plane, as suggested in Fig. 8-12. Polarization observations of nonthermal radio sources do show linear polarization but only of small degree, indicating that the magnetic fields in the sources may be oriented in an irregular manner, or that there has been depolarization of the waves in passing through intervening magneto-ionic media, or both.

8-5 The Sun: General Features   The sun is the central body of the solar system. It has a diameter of about 864 thousand miles and is at a mean distance from the earth of about 93 million miles (1.50×10^8 km). Its angular diameter as seen from the earth is about $\frac{1}{2}°$. Actually the angular diameter varies slightly during the year because of the small eccentricity of the earth's orbit and attendant variation in the earth's distance. For example, during 1959 the sun's diameter ranged from $32'35''.7$ on January 3, to $31'31''.3$ on July 5, the earth being a few percent closer to the sun on January 3 than on July 5. The sun's mass is 329,400 times that of earth and about 750 times the total mass of all the planets of the solar system.

The sun is of great interest to astronomers in their study of the universe because it is the only star close enough to be studied in any great detail. It is a rather average star of the yellow-dwarf class (spectral type G2V). In the Hertzsprung-Russell spectral-magnitude classification of stars it is situated toward the cooler, fainter end of the principal star group called the *main sequence*.

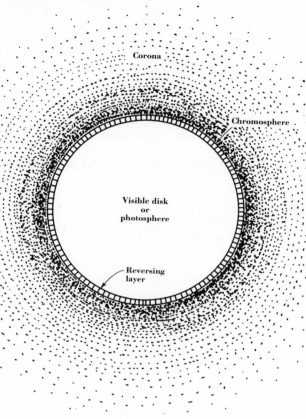

Fig. 8-14. Sketch of the sun showing its visible disk, or photosphere, and layers of the atmosphere.

  The angular diameters in the first paragraph refer to the visible disk of the sun, which is called the *photosphere*. Above the photosphere, as shown in Fig. 8-14, lies the *solar atmosphere* which is divided into two parts: (1) the lower atmosphere, or *chromosphere*, extending up to a height of several thousand kilometers, and (2) the upper atmosphere, or *corona*, which extends to very great heights. During a total eclipse the corona is observed visually to extend at least 1 million kilometers above the photosphere. At radio frequencies there is evidence that the corona extends much farther than this, and it is convenient in such discussions to speak of its extent in terms of solar radii, using the photospheric radius as the unit of distance. To set a definite limit to the extent of the solar corona would be quite arbitrary. It is a gaseous envelope that becomes more tenuous with distance from the sun, and the corona is of appreciable density even at the distance of the earth. At the base of the solar atmosphere, between the chromosphere and photosphere, is a thin region a few hundred kilometers in thickness which is called the *reversing layer*. The gas in this layer is cooler than the photosphere and absorption of the continuous photospheric light emission occurs at the spectral wavelengths of the gases in the reversing layer. The reversed, or solar absorption, spectrum, first observed by Fraunhofer, was once thought to be produced in this layer. Other layers in the atmosphere are now believed also to be responsible.

The principal visual surface features of the sun may be grouped into the following categories:

(1)   Sunspots
(2)   Faculae and flocculi
(3)   Granules and granulation
(4)   Flares
(5)   Spicules
(6)   Prominences and filaments

Sunspots appear as dark areas on the disk of the sun because they are regions of somewhat lower temperature than the surrounding photosphere. They appear to be disturbances which break through the photosphere from lower levels. They are known to possess intense localized magnetic fields (of the order of 4,000 gauss). Figure 8-15 is a photograph of the sun on April 7, 1947, showing a large sunspot group below the center of the disk. Figure 8-16 is an enlargement of this sunspot group. Figure 8-17 is a series of daily photographs showing the progress of the same sunspot group across the disk for more than one full rotation of the sun. The north pole is up, with rotation from left to right as seen from the northern hemisphere of the earth with a terrestrial (noninverting) telescope. The spot group is first seen near the eastern edge, or limb, of the sun on March 5, crossing the central meridian of the sun (meridian facing the earth) on March 9, and disappearing around the western limb on March 16. On April 2 it has reappeared on the eastern limb. It crosses the central meridian on April 6 and disappears again around the western limb on April 13. It is about 28 days between the central meridian passages of the group on March 9 and April 6. This apparent or *synodic rotation period* of the sun is about two days longer than its sidereal period, or rotation period with respect to the stars. This difference is due to the fact that during a sidereal rotation of the sun the earth advances about one-twelfth of its way around its orbit, so that the sun must turn more than 360° to bring the spot back to the central meridian (meridian facing the earth).

By observations of the motion of sunspots, it has been found that the sidereal period of the sun is about 25 days for spots at the equator, about 26.5 days for ones 30° north or south of the equator and about 27.5 days for spots 45° from the equator. It will be noted that the spot group of March and April, 1947, was about 20° south of the equator and had a sidereal period of about 26 days. The difference in rotation period of spots as a function of latitude is taken as evidence that the solar surface, or photosphere, is gaseous. Some of the planets, such as Jupiter, also have a rotation period which varies as a function of latitude.

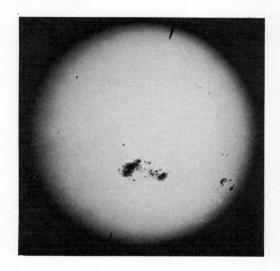

Fig. 8-15. The sun on April 7, 1947, showing a large sunspot group below the center of the disk. (*Mount Wilson and Palomar Observatories photograph.*)

Occasionally sunspots may be so large and prominent as to be visible with the unaided eye. With the magnification of a telescope, a sunspot is seen to possess considerable detail, as shown in Fig. 8-16. The dark central region of the spot is called the *umbra*, and the lighter border zone between the umbra and the bright photosphere is called the *penumbra*.

A given sunspot may last from one day to several months. The number of spots which appear on the sun's surface at any time varies greatly and follows a well-established 11-year cycle discovered by Schwabe in 1843. During a period of minimum spottedness the solar disk may be practically free from spots for months, while at the maximum of the spot cycle it may never be without spots or groups of them. Sunspot cycles are discussed in more detail in Sec. 8-6, and the relation of these cycles to solar radio emission is treated in Sec. 8-7.

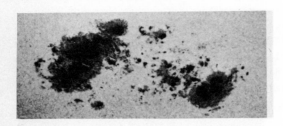

Fig. 8-16. Enlargement of the large sunspot group of April 7, 1947. The overall extent of this group (left edge of large left-hand spot to right edge of large right-hand spot) is about 170,000 miles. (*Mount Wilson and Palomar Observatories photograph.*)

Faculae (or *plages*) are bright mottled patches near sunspots. They commonly are discernible on the sun's disk at locations where sunspots later appear and may also persist at places where spots have disappeared. The bright elements of faculae are called *flocculi*. When the photosphere is observed under high magnification and steady seeing conditions, it appears to be covered with many small *granules*.

Flares are sudden, localized increases in the light intensity in the neighborhood of sunspots. They are best observed in monochromatic radiation of Hα, Ca II, and other lines. These sudden brightenings may cover from 50 to 1,000 millionths of the area of the solar hemisphere for time intervals of the order of tens of minutes to several hours (Dodson, 1958). Flares are very common, and in periods of great solar activity one may occur on the average every 2 hr. Flares are of particular interest in radio astronomy because of their close association with the occurrence of enhanced radio emission from the sun. There is also evidence that terrestrial ionospheric disturbances occur in close time association with many flares. Ejection of matter from the flare region generally accompanies the flare. The velocity of the particles ejected from the sun may range from a few hundred kilometers per second to the order of 150,000 km sec^{-1}, or one half the velocity of light.

Spicules, or jets, are fine bright spikes which emerge from the chromosphere. They occur in large numbers and may be responsible for ejecting large numbers of particles into the corona (Roberts, 1955).

Prominences are bright clouds which may rise far above the photosphere. Flare brightenings are commonly accompanied by prominence activity. Rapidly rising or *eruptive prominences* are sometimes observed near the onset of flare brightenings (Dodson, 1958). Some eruptive prominences appear to be ejected from the sun at high velocities, of the order of 1,000 km sec⁻¹, and are associated with sudden large increases in solar radio emission.

Filaments are thin threadlike markings on the sun's disk. They are believed to be prominences as seen from above instead of in profile.

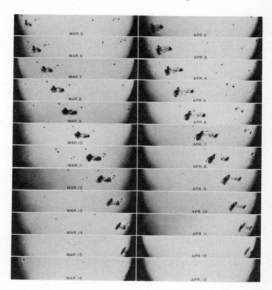

Fig. 8-17. Progress of sunspot group of Figs. 8-15 and 8-16 across the solar disk as shown by daily observations over a 5-week period. (*Mount Wilson and Palomar Observatories photograph.*)

8-6 Sunspot Cycles In 1843 Schwabe (Kiepenheuer, 1953), discovered a periodicity in the numbers of sunspots which he set at about 10 years. Following extensive observations, R. Wolf, of Zurich, in 1852 set the period at 11.1 years. Using older data, he also extended the sunspot-number data back to 1749 and the maximum and minimum dates of the sunspot cycle back to 1610. Wolf also devised a method of arriving at a sunspot number, which is in wide use and known commonly as the *Zurich number*. It is defined as follows

$$N = k(10g + f) \tag{8-38}$$

where N = Zurich sunspot number

 f = number of individual spots which show umbrae

 g = number of groups in which these spots are arranged

 k = a factor involving the instrument used, method of observations, and characteristics (fatigue effects, etc.) of the observer

Although the "number" N arrived at as in (8-38) is somewhat arbitrary, it is nevertheless extremely useful as a basis for comparing solar activity over long periods of time.

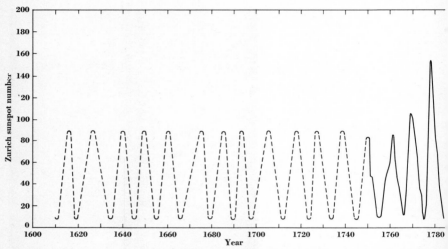

Fig. 8-18. Observed sunspot numbers from 1749 to 1964 (solid curve). The dashed, smoothed curve from 1610 to 1749 is significant only as regards the dates of spot maxima and minima for this period. (Prior to 1749 no quantitative spot-number data are available, and the maximum and minimum values shown are entirely arbitrary.)

  In Fig. 8-18 the annual mean sunspot numbers in the Zurich (or Wolf) scale are plotted from 1749 to 1964. In addition, a smoothed, dashed curve is sketched between 1610 and 1749 which indicates the dates of maximum and minimum spots in this period. Prior to 1749 no quantitative records are available of the maximum and minimum spot numbers, and the values shown are entirely arbitrary. The spot data prior to 1952 are from Waldmeier (Kiepenheuer, 1953) and those after 1952 from the Zurich-number data as published in *Sky and Telescope*. The Zurich sunspot numbers are also given with other solar data in the *Quarterly Bulletin on Solar Activity* (International Astronomical Union), published by the Eidgenössische Sternwarte Zurich. Between the minima of 1755.2 and 1954.2 there are 18 complete sunspot cycles, yielding an average of 11.05 years per cycle. It is of particular interest that, as shown in Fig. 8-18, the last spot maximum (1957) is the highest recorded to date.

Sunspots show a magnetic-polarity reversal between hemispheres with each 11-year cycle, an effect discovered by Hale (Kiepenheuer, 1953). A full magnetic spot cycle therefore covers two ordinary (11-year) cycles and lasts about 22 years. This polarity reversal effect is illustrated in Fig. 8-19, in which the annual mean Zurich sunspot numbers are plotted with alternate 11-year cycles reversed in sign. The designation of a particular cycle as positive and the adjacent ones as negative is entirely arbitrary. Where the spot-number curve crosses zero on this graph the total

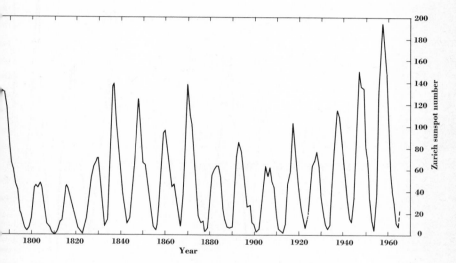

number of spots (irrespective of sign) may not be zero, but the implication is that the algebraic sum is zero. Actually, near spot minimum the spot number is a mixture of spots of opposite polarity belonging to two overlapping cycles. Bracewell (1953) has pointed out that the analysis of spot numbers may be improved if the numbers are plotted on the basis of a 22-year cycle, as in Fig. 8-19.

Although the 11-year cycle is the most prominent periodicity in Fig. 8-18 (and Fig. 8-19), a longer period is also evident. Wolf in 1862 placed it at 78 years. However, the trends in later cycles would seem to indicate an average for this period more like 88 years based on long-period maxima in the years 1781, 1855, and 1953, and minima in the years 1814 and 1904. Actually these dates suggest that the long period may be lengthening from about 75 years in the early 1800's to about 95 years in the latest cycle.

In addition to the 11- (or 22-) year cycle and the 88-year cycle, other shorter periodicities may be apparent. For example, a period of about 3 months has been noted (see Fig. 8-23). Still shorter periodicities are sometimes evident. For example, a periodicity of about 27 days may occur, but this would most likely be related to the rotation period for a large active spot group.

Another interesting aspect of the 11-year cycle is shown by Fig. 8-20. Here smoothed spot-number curves based on the yearly Zurich mean values are compared in two groups according to polarity for all cycles from 1745 to the present. A strong correlation is seen to exist between the shape of a curve for one cycle and its maximum value. Thus, the cycles with higher maxima are more asymmetrical, with a more abrupt

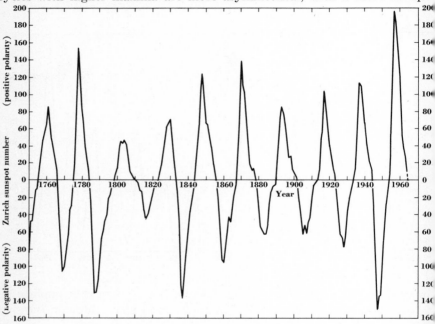

Fig. 8-19. Observed spot numbers from 1749 to 1964 plotted with alternate 11-year cycles reversed in sign to illustrate the 22-year cycle.

rise in the early part of the cycle followed by a more gradual decline. This effect, noted by Waldmeier in 1935 (Kiepenheuer, 1953), is of value in predicting trends within any particular cycle. Comparing the curves of Fig. 8-20a with those of Fig. 8-20b, it is to be noted that the positive cycles (a) commonly appear to have larger and more sharply peaked maxima than the negative cycles. Also the negative cycles (b) have more of a tendency for double peaks. The present cycle is seen to have a considerably higher maximum than any previous cycle, positive or negative.

By measurements of solar photographs the total sunspot area may be determined. At spot minimum the total area covered is of the order of

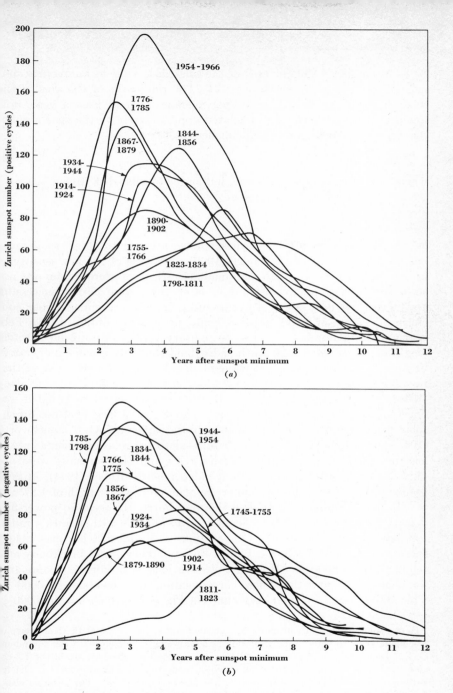

Fig. 8-20. Observed sunspot numbers for 11-year cycles from 1745 to 1966 arranged in two groups (a) and (b) according to polarity. The initial minimum of each cycle is adjusted to zero on the time scale.

10 to 100 millionths of the area of the solar disk while at an average spot maximum the area is of the order of 2,000 millionths of the area of the solar disk. When dealing with spot averages over at least a year, it is found that the area covered is related approximately to the spot number by the empirical equation (Kiepenheuer, 1953)

$$A = 17N \tag{8-39}$$

where A = area covered by spots, millionths of solar-disk area
      N = Zurich sunspot number

8-7 The Radio Sun The earliest known attempt to detect solar radio emission was by Sir Oliver Lodge (1900). Because of the low sensitivity of his equipment and severe man-made electrical interference, he was not successful. Attempts by others† in later years were also unsuccessful and it was not until about 40 years after Lodge's experiments that conclusive evidence of radio emission from the sun was obtained. In February, 1942, a number of GL-type military radars working on a wavelength of about 5 m in the south of England suddenly became inoperative because of a very strong noise type of signal. At first it was feared that the Germans had developed a new radar-jamming system, but J. S. Hey (1946), on investigating reports of the interference, concluded that it was due to strong solar radio emission associated with a large sunspot on the solar disk at the time. Independently, in the same year, G. C. Southworth (1945) at the Bell Telephone Laboratories detected solar emission at wavelengths in the 10-cm region. Southworth's and Hey's results were not published until 1945 and 1946, and independently Reber (1944), in September, 1943, detected strong solar emission on a wavelength of 1.87 m.

If the sun radiated only as a thermal source, the received power flux density should vary with wavelength, in accordance with Planck's radiation law, as given by (3-49). Observations with optical telescopes yield a spectrum which does follow a Planck radiation curve for a blackbody at a temperature of 6000°K. However, observations with radio telescopes give flux densities greater than those corresponding to a blackbody at 6000°K. This is illustrated by the heavy curve in Fig. 8-21, which fits the 6000°K

---

† In one such attempt, about 1934, Arthur Adel and J. D. Kraus, at the University of Michigan, tried to detect solar radiation at a wavelength of 1 cm with the equipment C. E. Cleeton and N. H. Williams (1934) used for the first observations of the 1.3-cm ammonia line. (Theirs was the first microwave-spectroscopy measurement.) The results were negative because the receiver was far too insensitive. The antenna was a 3-ft diameter parabolic searchlight reflector.

‡ A still later publication (Schott, 1947) indicates that noise signals identified as of solar origin were detected on German 1.7-m wavelength radar equipment operating in Denmark as early as 1940.

Planck curve at wavelengths less than 1 cm but branches into two curves, designated *quiet sun* and *disturbed sun*, at wavelengths larger than 1 cm. The quiet-sun curve indicates the minimum background radiation from the sun. At times of sunspot activity the radiation may be greatly enhanced, as indicated by the disturbed-sun curve. For the sun to radiate

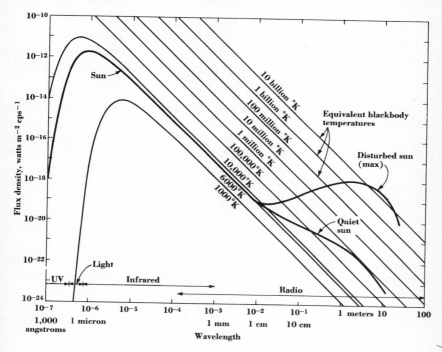

Fig. 8-21. The solar spectrum from the ultraviolet to radio wavelengths (heavy curve). The solar flux density is equal to that from a blackbody radiator at 6000°K at wavelengths less than 1 cm but is greater than this at longer wavelengths. At radio wavelengths the radiation is between that of the quiet- and disturbed-sun curves. The other (lighter) curves give the flux density corresponding to blackbody radiators at various temperatures in degrees Kelvin.

such large fluxes as a blackbody the sun's temperature would need to be millions of degrees. This is suggested in Fig. 8-21 by the curves corresponding to equivalent blackbody temperatures up to 10^{10}°K. These curves are calculated on the basis that the sun's angular extent is equal to that of the visible disk (0.224 deg²) and that the sun's temperature is constant over the disk. Actually, the distribution is not uniform and also varies with wavelength, so that the curves are only approximations to the equivalent blackbody temperatures.

It is convenient to divide the solar radio emission into two categories related to the two conditions just discussed, or

(1)   Emission from the *quiet or undisturbed sun* at times of little or no sunspot activity

(2)   Emission from the *disturbed sun* at times of sunspot activity

Radiation from the *disturbed sun* may, in turn, be divided into

(1)   A *slowly varying component*, evident at wavelengths in the 3- to 60-cm range, which varies over periods of days, weeks, or months

(2)   A *rapidly varying component*, characterized by bursts of radiation varying over intervals of seconds, minutes, or hours

8-7a. The Quiet Sun.   At wavelengths of 1 cm or less the radio sun corresponds in size to the optical disk (photosphere), and the brightness appears to be uniform. At wavelengths of the order of 10 cm the radio sun is slightly larger than the optical disk, and there is a peaking in the radio brightness near the limb of the sun, called *limb brightening*. At wavelengths in the meter range the radio sun is much larger than the optical disk, and the brightness tends to peak at the center. It appears that at the shortest wavelengths the radiation originates close to the photosphere, while at meter wavelengths it is generated high in the corona. Thus, the situation seems to be that at a frequency ν the radiation comes mostly from a layer located just above the *critical layer* for that frequency. The frequency of the critical layer is a function of the electron density (see Chap. 5) and is given by

$$\nu = \frac{e}{2\pi} \sqrt{\frac{N}{\epsilon_0 m}} \tag{8-40}$$

where N = electron density, electrons m^{-3}

ν = critical frequency, cps

ϵ_0 = permittivity of vacuum (= 8.85×10^{-12} farad m^{-1})

m = mass of electron (= 9.1×10^{-31} kg)

e = charge of electron (= 1.6×10^{-19} coul)

The electron density of the solar atmosphere decreases with height, so that the critical layer is close to the photosphere at the highest frequencies but is high in the corona at the lower frequencies. According to the Allen-Baumbach formula (Allen, 1947), the electron density N in terms of the distance from the center of the sun is given by

$$N = (1.55r^{-6} + 2.99r^{-16})10^{14} \tag{8-41}$$

where N = electron density, m^{-3}

r = distance from center of sun in solar radii, (= 1 at photosphere)

Theoretical brightness-temperature distributions (Smerd, 1950) are shown in Fig. 8-22 for wavelengths between 1 cm and 5 m.

At the longer (meter) wavelengths solar emission can be detected at heights of several solar radii. An even greater extent for the corona has been deduced from measurements involving absorption, refraction, and/or diffraction when the solar corona occults a radio source. Observations by Machin and Smith (1952), Hewish (1957), Denisse (1957), and Vitkevitch (1957) of the occultation of the Crab nebula showed effects at distances of 15 to 20 solar radii. Slee (1961) has found evidence of scattering effects by the solar corona on discrete radio sources out to 100 solar radii, with the average scattering varying inversely as the 2.3 power of the distance. A distance of 100 solar radii is nearly ½AU (earth-sun distance).

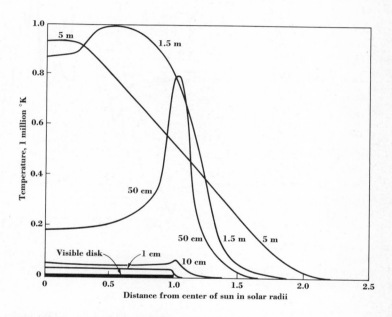

Fig. 8-22. Apparent blackbody temperature across the solar disk in solar radii for several values of wavelength. (*After Smerd,* 1950.)

8-7b. *The Slowly Varying Component.*   The slowly varying component of solar radio emission is prominent in the 3- to 60-cm wavelength range.   The variation characteristic of this component is illustrated by the upper curve in Fig. 8-23, observed at 10.7 cm by Covington (1957). The curve gives the monthly averages of the 10.7-cm solar radio intensity from 1952 to 1957 to an arbitrary vertical scale. Below the 10.7-cm curve the Zurich sunspot numbers from 1952 to 1958 are shown for comparison. The sunspots reach a minimum in 1954 and a maximum in 1957.   The 10.7-cm curve follows a similar trend, even to some of the minor fluctuations in the sunspot curve.   This apparent correlation suggests that the 10.7-cm emission is associated with sunspot activity.

Observations by Christiansen and his associates in Australia have demonstrated that the emission originates in condensations, or *radio plages*, situated at distances up to 100,000 km above active plage areas.   Using a crossed-grating interferometer, Christiansen, Mathewson, and Pawsey (1957) (Christiansen and Mathewson, 1958) obtained the radio picture of

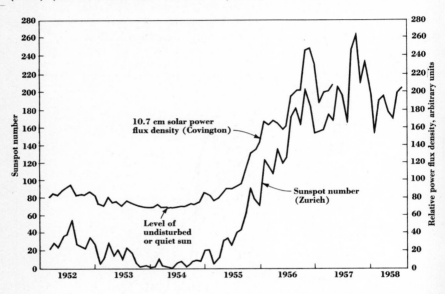

Fig. 8-23.   The slowly varying component of radio radiation from the disturbed sun, as observed by Covington (1957) at 10.7 cm, compared with the Zurich sunspot numbers from 1952 to 1958.

the sun at 21-cm wavelength presented in Fig. 8-24.   The emission is predominantly the slowly varying component.† The close association of this component with spot areas is well illustrated by comparing the radio picture (above) with the visual appearance of the solar disk (below) showing the sunspot areas.   All the radio emitting regions are associated with spot groups, even in the case of the radio region above the southeast-limb, where no spots are visible.   A few days later, because of the sun's rotation, a spot group appeared around the limb at this position.   Thus, it is evident that much of the radio emission originates in regions situated well above the photosphere.

† Although these observations were made at 21 cm, they are not related to the 21.1-cm line radiation of neutral hydrogen.

In general, the slowly varying component displays strong circularly polarized radiation from condensations of about sunspot size, while more randomly polarized emission comes from a region of larger extent, approximating that of the plage area on the sun associated with the sunspot.

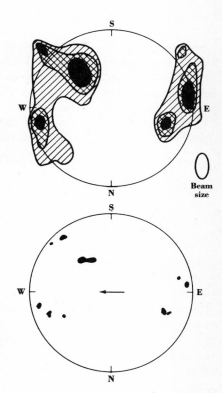

Fig. 8-24. A 21-cm radio picture of the sun for June 27, 1957 (above); the visual appearance on the same date (below). (*After Christiansen and Mathewson, 1958.*)

8-7c. The Rapidly Varying Component. This component consists of bursts of radiation varying over intervals of seconds to hours. Intense and complex groups of bursts often follow the appearance of a flare in the solar chromosphere. These flares, which are readily observed photographically in the Hα line of hydrogen or H or K lines of calcium, show up as brightenings in the plage area near a sunspot. They appear suddenly and fade away gradually, with the complete event lasting some minutes to an hour. A flare may release as much as 10^{25} joules of energy, which is equal to the energy released in the explosion of 2,400 million megatons of TNT (2.4×10^{15} tons). Much of this energy is apparently derived from the solar magnetic field. Following a flare, existing arches or prominences on the sun may erupt.

The radio emission following a flare is highly variable; on meter wavelengths there may be bursts of seconds or minutes duration with the later appearance of storms lasting for hours or days. On centimeter wavelengths there is less activity, but bursts of 1 min to 1 hr may occur. Figure 8-25 is an example of the rapidly varying component of the solar radiation observed on a wavelength of 11 m (27 Mc) at the Ohio State University on August 31, 1956. There is an outburst at 0737.5 EST coinciding with the occurrence of a large (importance 3) solar flare. On the higher frequencies of 167 and 460 Mc, the outburst was observed at the National Bureau of Standards, Boulder, Colorado, to begin at 0737.1 EST, or 0.4min before the start on 27 Mc as observed at Ohio State. The observation of solar radio emission first on higher frequencies and a short interval later on lower frequencies is typical.

From studies of the rapidly varying component over a range of frequencies Wild (1963) has classified these emissions into five principal types:

(1)   Noise-storm bursts (type I)
(2)   Slow-drift bursts (type II)
(3)   Fast-drift bursts (type III)
(4)   Broad-band continuum emission (type IV)
(5)   Continuum emission at meter wavelengths (type V)

The characteristics and interrelationships of these types are well illustrated by considering the sequence of events following a solar flare, as presented in the idealized dynamic spectrum of Fig. 8-26. Wild divides the events into two phases. In *phase* 1 short, strong bursts begin right after the visible flare and move rapidly from around 500 Mc to lower frequencies. These are fast-drift, type III bursts. Their instantaneous emission is in a narrow band a few megacycles wide. The emission is believed to originate in plasma oscillations associated with the ejection of an electron jet at velocities of 100,000 km sec^{-1} or more as a result of the chromospheric explosion of a flare, as suggested in Fig. 8-27a. Often the type III bursts in phase 1 are accompanied by a continuum emission confined to the meter wavelengths and designed as type V. The continuum emission is believed to be produced by synchrotron emission from the fast-rising electron jet.

In small flares only the above (phase 1) sequence may be observed. In large flares, however, a longer-duration sequence, or *phase* 2 sequence, follows phase 1. The phase 2 begins with slow-drift bursts classified as type II. These bursts have drift rates of the order of 20 Mc (sec^{-1}) min^{-1}. This contrasts with the much faster drift rates of the type III bursts, which are of the order of 20 Mc (sec^{-1}) sec^{-1}. The type II bursts typically have a well-defined second harmonic. The emission is believed to be produced by plasma oscillations in a shock front rising ahead of an expanding gas cloud rising at the flare location, as suggested in Fig. 8-27b. Rise velocities of

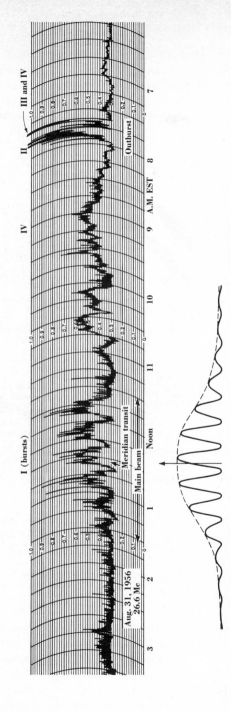

Fig. 8-25. Single-frequency (26.6 Mc) record of a solar outburst as observed at the Ohio State University on Aug. 31, 1956. The radio outburst began at 7:35.5 A.M. and coincided with the occurrence of a large (importance 3) solar flare. The lobe pattern of the interferometer antenna with envelope (dashed) is shown below.

the order of 1,000 km sec⁻¹ are indicated, which are very much less than the velocities of the electron jet which occurs right after the flare onset. Haddock (1958) has reported that close inspection of the type II burst spectra often reveals a fine structure which appears to consist of a large number of short bursts of the type III variety.

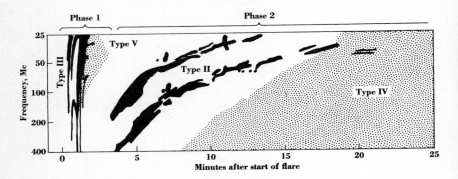

Fig. 8-26.   Dynamic radio-spectrum record of a large solar outburst.   (*After Wild,* 1963.)

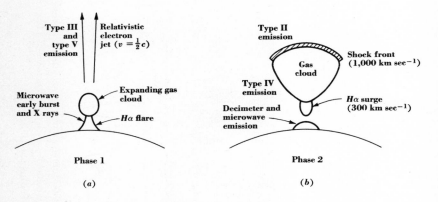

Fig. 8-27.   Idealized diagrams of phase 1 and phase 2 of a solar outburst. (*After Wild,* 1963.)

Sometimes the type II bursts of phase 2 are followed by a very broad-band, stable, continuum emission that may last for hours or even days. This emission is classified as type IV.   Presumably the emission is of the synchrotron type and comes from the gas cloud (Fig. 8-27b) rising above the flare location.   The type IV emission source rises at velocities of the order of 1,500 km sec⁻¹ for the first half hour or so, the source reaching heights of several solar radii (Boischot, 1959). The source then tends to stop, and may remain relatively stationary for hours or days. At this stage the emission becomes strongly circularly polarized. Although the above discussion is idealized and oversimplified, it does serve to illustrate many of the events which may follow a solar flare. A comprehensive summary of solar radio astronomy is given by Kundu (1964).

Referring once more to the single-frequency (27 Mc) record of a solar radio outburst in Fig. 8-25, the absence of a dynamic spectrum makes any classification uncertain, but it is likely that the initial burst lasting several minutes consisted of type III and type V bursts, characteristic of phase 1 following a flare. The second peak, beginning about 7 min later, probably marked the start of phase 2, as caused by a slow-drift type II burst. Then about 8:02, or about 25 min after the start of the entire sequence, emission of the type IV variety began, which persisted for many hours. The sharp, short-duration spikes on this type IV background might be classified as type I. Six lobes of the interferometer pattern of the antenna can be seen between 11:30 and 13:30 as the sun passed through the main beam. Meridian transit was at 12:32. Since the antenna response is many decibels down 5 hr from the meridian, the intensity of the outburst was actually of the order of 50 times the maximum noon intensity, instead of twice as large, as indicated by the deflection of the record of Fig. 8-25. Between 8:30 and 11:00 it would appear that solar activity was at a relatively high level (considerably above the noon level) but well below the initial peaks of the outburst.

Observations of the sun at a wavelength of 3.2 mm by Simon (1965) indicate that at this wavelength emission correlates with active Hα regions and regions of strong magnetic fields. Particularly high 3.2-mm enhancements were detected in regions that later flared.

The various types of solar radio emission are summarized in Table 8-2.

8-8   Radio Emission from the Moon and the Planets

8-8a. Introduction. At optical wavelengths the moon and the planets are seen mainly in the sun's reflected light. Very little light is radiated from these objects acting as blackbody emitters. At radio wavelengths the situation is reversed, and the sun's reflected radiation is extremely small compared to the thermal blackbody emission.[†] The temperature of the moon and planets is of the order of 100°K (actual range

---

[†] Reflection from the moon and planets is important, however, in radar astronomy (Evans, 1961; Goldstein, 1964; James, 1964).

Table 8-2
Types of solar radio emission

Type	Duration	Band-width	Drift rate	Polari-zation	Mecha-nism	Temper-ature, °K
Quiet sun	Constant (or 11-year period)	Con-tinuum		Random	Thermal	10^6
Slowly varying component	Days or months	Con-tinuum		Random (CP† at cm)	Thermal	$<2 \times 10^6$
Rapidly varying component (after flares): Phase 1:						
Type III	Seconds	5 Mc	20 Mc sec^{-2}	Random	Plasma	$>10^{11}$
Type V	Minutes	Con-tinuum		Random (usually)	Synchro-tron	10^{11}
Phase 2: Type II	Minutes	50 Mc	20 Mc sec^{-1} min^{-1}	Random	Plasma	$<10^{11}$
Type IV	Hours	Con-tinuum		Random to CP	Synchro-tron	10^{11}
Type I	Hours	Con-tinuum		Random to CP	?	10^9

† Circularly polarized.

50 to 700°K). It is evident from the 100°K curve in Fig. 3-13 that at this temperature a blackbody radiates most strongly in the infrared and short-radio-wavelength region and very weakly in the visible region. In the case of Jupiter, mechanisms other than thermal blackbody radiation are involved.

8-8*b*. *The Moon*.‡ The moon's thermal radio emission was first observed by Dicke and Beringer (1946) using a wavelength of 1.25 cm. Later, Piddington and Minnett (1949) conducted measurements at the same wavelength over the entire lunar cycle and found that the temperature varied with the lunar phase. However, the temperature variation was smaller than had been observed by Pettit and Nicholson (1930) and Pettit (1935) in the infrared region (8 to 14 microns), and, furthermore, it

‡ An excellent summary of thermal radio emission from the moon and planets is given by Mayer (1964).

showed a retardation in phase of about 45°. This is illustrated by Fig. 8-28, which compares Piddington and Minnett's temperatures at 1.25-cm wavelength (dashed curve) with Pettit and Nicholson's infrared temperatures (solid curve). The dotted and dash-dot curves give, respectively, the temperature measured at 0.86 cm by Gibson (1958) and at 3.15 cm by Mayer, McCullough, and Sloanaker (1961).

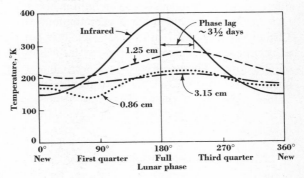

Fig. 8-28. Lunar temperature in degrees Kelvin as a function of lunar phase, showing the temperature variation at infrared wavelengths and at wavelengths of 0.86, 1.25, and 3.15 cm. The temperatures are those of an equivalent blackbody radiator.

The infrared temperature is symmetrical with respect to full moon, reaching a maximum at full moon and a minimum at new moon. The exact shape and extreme values shown in Fig. 8-28 are only approximate, but these are secondary in importance to the symmetry. The microwave temperatures, on the other hand, exhibit not only a phase lag of about 3.5 days, but a definite asymmetry with respect to the maximum temperature, as shown by Gibson's 0.86-cm data. At longer wavelengths the moon's temperature variation with phase appears very small, if present at all. At 75 cm Seeger, Westerhout, and Conway (1957) report no variation exceeding 10 percent.

The smaller temperature range of the radio temperatures as compared to the infrared values is taken to indicate that the microwave radiation originates at some depth below the surface of the moon, whereas the infrared radiation comes from a thin surface layer. Further support for this explanation is found in the absence of significant changes in microwave temperature during an eclipse of the moon. The sun's radiation is cut off for only an hour or so during an eclipse, while during the lunar night it is cut off for 2 weeks. Gibson (1958) reports that during two such eclipses the temperature at 0.86 cm showed no important changes. However,

Pettit (1940) reported that the infrared temperature at the center of the disk dropped from 371°K before an eclipse to 175°K during totality.

High-resolution observations by Coates (1961) at 4.3 mm and by Salomonovich and Losovskii (1963) at 8 mm indicate that at these wavelengths the radio emission from the moon varies over its surface and, furthermore, that the darker areas, or maria, show larger temperature fluctuations with phase than the lighter areas.

Although the moon has a very tenuous atmosphere, measurements by Elsmore (1957) at 3.7 m indicate that it is not negligible, having a density at the lunar surface equal to the earth's atmosphere at heights of the order of 500 km. These observations were made by comparing the time of obscuration of the Crab nebula during a lunar occultation with computed values based on a refracting lunar atmosphere. A lunar surface-atmospheric density about 2×10^{-13} that of the earth's sea-level-atmospheric density is indicated.

8-8c. Mercury. Radio emission at 3 cm wavelength has been detected from Mercury by Howard, Barrett, and Haddock (1962). A blackbody temperature of 400°K was deduced. Observations of Mercury are difficult because it is so close to the direction of the sun. Even at maximum elongation the angular separation is only 28°, and the power received from the sun through the minor lobes of the telescope antenna pattern may be comparable to the power received from Mercury in the main lobe.

8-8d. Venus. The first detection of Venus at radio wavelengths was at 3 and 9 cm by Mayer, McCullough, and Sloanaker (1958a) near the time of the 1956 inferior conjunction. A blackbody temperature of about 600°K was indicated. Subsequently Gibson and McEwan (1959) observed Venus at 8.6 mm and found a blackbody temperature of 400°K. These and other measurements of Venus near inferior conjunction indicate blackbody temperatures of around 600°K for wavelengths of 2 to 20 cm but a decreasing temperature below 2 cm reaching about 350°K at 5 mm (Mayer, 1964). Barrett (1961) has accounted for this spectrum by assuming a surface temperature of 580°K and a dense absorbing-emitting atmosphere at a lower temperature. Calculations for a carbon dioxide–nitrogen atmosphere with a surface pressure of the order of 20 atm agree quite well with the measurements.

At inferior conjunction Venus is closest to the earth, with its dark side facing the earth. Higher temperatures might be expected for the side facing the sun, and this was borne out by measurements at 3 cm by Mayer, McCullough, and Sloanaker (1963), shown in Fig. 8-29, which were over about half the phase variation of Venus before and after inferior conjunction. Although no measurements were made of the fully illuminated side of Venus (at superior conjunction), the measurements extended over a long enough period to show a significant trend in temperature. The best-fitting (least-square) sine curve to the experimental points, shown by the solid curve in Fig. 8-29, indicates blackbody temperatures of about 730 and 550°K for the bright and dark sides of the planet. A phase shift in the minimum temperature of about 12°K after inferior conjunction is also apparent, although the spread in the experimental data makes for considerable uncertainty. However, if this phase shift is admitted, it can be

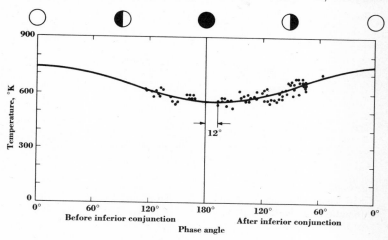

Fig. 8-29. Variation of the apparent blackbody temperature of Venus at 3.15 cm as a function of the phase of solar illumination. (*After Mayer, McCullough, and Sloanaker,* 1963.)

interpreted as indicating that Venus rotates in the retrograde sense, i.e., turns on its axis in the opposite sense to the earth's rotation. Radar observations by Goldstein and Carpenter (1963) yield a retrograde rotation, with a 250-day period, which is consistent with the phase shift in Fig. 8-29.

Measurements similar to those in Fig. 8-29 but at 10 cm have been made by Drake (1962). His results are shown in Fig. 8-30. A somewhat smaller temperature variation with phase than at 3 cm is indicated with bright- and dark-side temperatures of 660 and 580°K. The average over the phase cycle is 620°K. The apparent phase lag of the minimum after inferior conjunction is 17 deg. The smaller temperature variation with phase may be interpreted to mean that the 10-cm radiation is emitted from layers further below the planetary surface than the 3-cm radiation.

It would be desirable to extend the phase measurements closer to superior conjunction than shown in Figs. 8-29 and 8-30, but this is difficult because Venus is over 6 times as far away at superior conjunction as at inferior conjunction, so that the flux density will be reduced by a factor of about 6^2 ($= 36$) because of the increased distance.

If the surface emissivity of Venus is less than unity, the actual temperatures of the surface materials will be higher than the blackbody values quoted above.† For example, if the emissivity is taken as 0.9, as indicated by radar observations, the true temperatures will be 11 percent higher than the blackbody values.

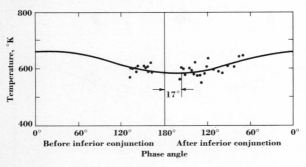

Fig. 8-30. Variation of the apparent blackbody temperature of Venus at 10 cm as a function of the phase of solar illumination. (*After Drake*, 1962.)

8-8e. The Earth. It is of interest to consider how the earth might appear as a radio source to a distant observer, as, for example, on a space vehicle. From direct measurements we have much detailed information concerning the earth's surface temperature. These values must be smoothed and averaged to yield quantities such as might be measured by a distant observer. The average surface temperature is about 290°K. Some bright-to-dark-side or phase variation in the blackbody temperature is to be expected, especially at the shorter radio wavelengths. Mechanisms other than thermal blackbody radiation may also play a significant role in the earth's radio emission. For example, the integrated effect of electrical-discharge radiation from lightning in thunderstorms may be significant at decameter wavelengths if appreciable amounts can propagate out through the ionosphere. At hectometer wavelengths, i.e., at wavelengths of hundreds of meters, the earth's ionosphere will probably be so opaque that the earth's temperature observed from space will be that of the ionospheric electrons rather than that of the earth's surface. Radiation by nonthermal mechanisms from electrons moving in the earth's magnetosphere may also be significant. Thus, one may expect the earth to have a

† Apparent temperature = actual temperature × emissivity

complex radio spectrum which tends to a thermal-blackbody type at centimeter wavelengths. Only natural mechanisms have been considered. If man-generated radio radiation is included, the spectrum will be even more complex.

8-8f. Mars. Even at nearest approach, or opposition, Mars is about twice as far as Venus at inferior conjunction. The diameter of Mars is about half that of Venus, and its temperature is less, even for the sunlit or bright side which we observe at opposition. These factors combine to make the maximum flux density of Mars an order of magnitude less than the maximum for Venus, and the measurements are correspondingly more difficult. The first measurements were made at a close approach in 1956 by Mayer, McCullough, and Sloanaker (1958b) at 3 cm and yielded a blackbody temperature of 218 ± 76°K. More accurate measurements by Giordmaine, Alsop, Townes, and Mayer (1959) in 1958 at 3 cm gave a blackbody temperature of 211 ± 28°K. This value is less than the infrared blackbody temperature of the sunlit side (250°K), implying that the radio emission originates further below the surface than the infrared radiation. Kuiper (1952) estimates an average infrared temperature of 217°K over both the sunlit and dark hemispheres. The subsurface temperature measured at 3 cm should be close to this average value, which is borne out by the radio observations.

Some of the difficulty inherent in the planetary-temperature measurements is illustrated by noting that the antenna temperature T_A of Mars measured by Mayer, McCullough, and Sloanaker (1958b) was only 0.24°K. When the planet solid angle is small compared to the antenna beam area, as here, the blackbody temperature T of the planet is given by

$$T = \frac{\Omega_A}{\Omega_s} T_A \tag{8-42}$$

where Ω_A = antenna beam area

Ω_s = source disk area

In the Mars measurements the ratio Ω_A/Ω_s was 910, yielding the planet temperature of 218°K (= 910 × 0.24). The large Ω_A/Ω_s ratio means that a 0.1°K probable error in the measurements results in a 91°K probable error in the planetary temperature.

8-8g. Jupiter. During the early part of 1955 Burke and Franklin (1955) observed a strong, fluctuating noise at a declination of +22° during a sky survey at 22 Mc. The noise was noted on 10 out of 31 nighttime records. Nine of the events occurred at about the same sidereal time and lasted no longer than required for a sidereal source to pass through the antenna beam.

The source position agreed with that of Jupiter. Further observations confirmed that Jupiter is a sporadic, fluctuating, and intense source of decametric radiation. The equivalent blackbody temperature at decameter wavelengths is many millions of degrees, which is much too large to admit a thermal mechanism. Measurements over a range of frequencies show that the decametric spectrum is decidedly nonthermal, with a spectral index n which may be as much as $+5$ or more (Smith, 1963; Carr, et al., 1964).

The first detection of thermal radiation from Jupiter was achieved in 1956 by Mayer, McCullough, and Sloanaker (1958b). They measured a blackbody disk temperature $140 \pm 56°K$ at a wavelength of 3 cm. Further observations of improved accuracy in 1957, when Jupiter was near opposition, gave a blackbody disk temperature of $145 \pm 26°K$. This temperature agrees closely with the infrared value, which is presumably related to the thermal temperature of the cloud surface surrounding the planet.

Later measurements at longer centimeter wavelengths indicated a marked increase in blackbody disk temperature with wavelength. Sloanaker (1959) found a temperature at 10 cm of 600°K. Subsequent observations by Roberts and Stanley (1959), Epstein (1959), Drake and Hvatum (1959), McClain (1959), and McClain, Nichols, and Waak (1962) indicated equivalent blackbody disk temperatures of about 2000°K at 21 cm and 5000°K at 31 cm. These temperatures appeared too high to be accounted for by a thermal type of mechanism, and the inference was that the centimeter radiation from Jupiter was a mixture of thermal radiation, dominant at shorter wavelengths, such as 3 cm, and nonthermal radiation, dominant at longer wavelengths, such as 30 cm. Interferometer measurements by Radhakrishnan and Roberts (1961) at 31 cm revealed that at this wavelength the radio emission from Jupiter comes from a region which is 3 times the visible diameter of the planet in the plane of the equator of Jupiter and about the same dimension as the visible diameter in the polar direction. Furthermore, the radiation was observed to have a degree of linear polarization d_l of about 0.3. These results may be interpreted to indicate that at 31 cm the radiation is largely synchrotron type from relativistic electrons trapped near the plane of the equator in the Jovian magnetic field.

Radiation from Jupiter between 40 and 400 Mc is very weak and has not as yet been detected. This leaves a gap in Jupiter's spectrum (Fig. 8-6) between the intense, sporadic, and fluctuating decameter radiation and the relatively steady centimeter radiation.

Let us return to a further discussion of the decametric radiation from Jupiter. Following Burke and Franklin's announcement, Shain (1956) confirmed their findings by searching through some 18.3-Mc sky-survey records made in 1950 and 1951. These records showed a series of bursts, previously assumed to be terrestrial interference, which could be ascribed to Jupiter.   By plotting periods of occurrence of radiation, which were almost always 2 hr or less in duration, vs. longitude of the central meridian of Jupiter at the time of observation, Shain demonstrated that a minimum longitude drift with date occurred for a Jovian rotation period of $09^h55^m13^s$. Another important deduction was that the radiation probably originated from one or more localized regions and not from the planet as a whole. Shain's period was nearly the same as the *system* II *rotation period* of $09^h55^m40^s\!.6$ for visual markings on the temperate zone.   Visible markings in the equatorial regions rotate with a shorter period, and a *system* I *period* of $09^h50^m30^s$ has been adopted for them.   At first it was thought that the radio source or sources might correspond to some visible markings in the temperate region, such as the great red spot, the white oval spots, or the so-called south tropical disturbance.   However, observations over several years showed that the sources of the radio emission on Jupiter drifted with respect to all the surface markings, and it is now thought that the radio period is related to the solid body of the planet and, perhaps, more particularly to its magnetic-field configuration.   The visual features of the surface are believed to be atmospheric-cloud markings which drift continually with respect to the solid body of the planet.

A *system* III *period for the radio emission* of $09^h55^m29^s\!.37$ was adopted in 1962 by the IAU Commission 40 (1962).   The system III epoch is 1957, January 1, at 0000 UT (JD 2,435,839.5).   System III coincides with system II at epoch, and the central-meridian longitude at epoch is $108°02$. The system III period was based on observations between 1950 and 1960. Douglas and Smith (1963b) have found evidence for a $0^s\!.8$ lengthening of the period sometime after about 1961–1962.   This does not necessarily imply a change in the rotation period.   A movement or drift of the main source in longitude could give the same effect.

The multiple character of the source of Jovian radio emission is well illustrated by the occurrence-probability histogram of Fig. 8-31, made by Douglas and Smith (1963a) from their 22-Mc observations during 1961. There are three principal regions of activity: region 1, centered around 120° longitude, region 2, centered around 225°, and region 3, centered around 300° (180° from region 1).   This polar-probability-occurrence histogram is reminiscent of an antenna radiation pattern with a main lobe, side lobes, and high front-to-back ratio, but this does not necessarily imply a single source on Jupiter with this pattern.   Actually what is plotted is occurrence probability or relative frequency of storms with flux greater than 10^{-21} jan and not the intensity of the emissions.   Thus, activity in regions 1 and 3 may be as intense as in region 2 but less frequent.   The pattern does not necessarily imply a single source or even three distinct sources.   There are many possible interpretations (Douglas and Smith, 1963a).

Slee and Higgins (1963) observed Jupiter on 19.7 Mc with an interferometer and found that the fringe visibility was not reduced with a
32-km-base-line separation. This puts an upper limit of one-third of the
diameter of the planet on the source size.

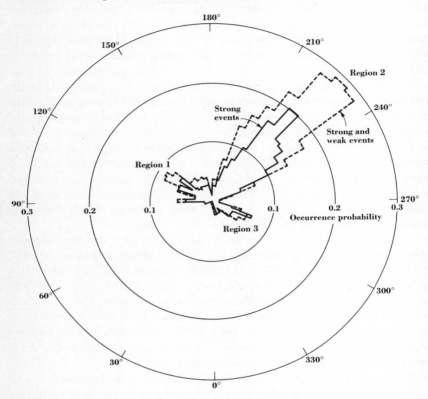

Fig. 8-31. Occurrence probability or relative frequency of Jupiter decametric
radio emission as a function of longitude at 22 Mc during 1961. The solid line is
for strong events only; the dashed line includes strong and weak events. (*After
Douglas and Smith*, 1963a.)

The time structure of the decametric Jupiter radiation is very complex. The individual noise-storm durations, such as used in the histogram
of Fig. 8-31, are rarely more than 2 hr in length, and typically are measured
in tens of minutes. A period of activity on Jupiter may last for several
days or more, but the combined effects of rotation of Jupiter and the earth
mean that the observed activity amounts to a periodic sampling. Thus,
for example, when Jupiter activity is observed, there is an increased
probability of observing activity at the same longitude one rotation later.

A typical record of Jupiter activity is shown in Fig. 8-32, as recorded on on the night of February 13–14, 1956, at the Ohio State University on 26.6 Mc (Kraus, 1958). Between 11:30 and 11:45 P.M. (EST) there was strong activity, followed by about 15 min of inactivity, after which an intense active period began that lasted almost without interruption for about 1 hr and 10 min. Some of the breaks in the activity on the record occurred when Jupiter drifted through nulls of the interferometer antenna pattern. The calculated null positions are shown in Fig. 8-32 by the small solid

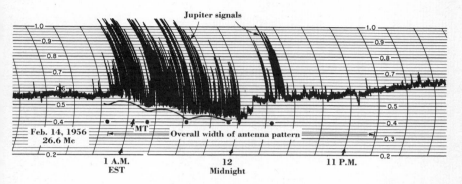

Fig. 8-32. Jupiter radiation recorded in 26.6 Mc at the Ohio State University on the night of Feb. 13–14, 1956. (*Kraus, 1958.*)

circles. The null at 12:28 A.M. is indistinct, possibly because the Jupiter emission was exceptionally strong at the time. The recorder time constant was about $\frac{1}{2}$ sec, and it is to be noted that even with so short a time constant the background or base level was noticeably raised for intervals of nearly 25 min while Jupiter was between the nulls of the antenna pattern, the rise and fall of the base level corresponding closely with the interferometer pattern. A wavy line following this is sketched below the pattern in Fig. 8-32.

Finer structure in the Jupiter activity is illustrated in Fig. 8-33. Here 4 min of record made at the Ohio State University on January 25, 1956, at 26.6 Mc are presented (Kraus, 1958). The radiation comes in burst groups containing individual pulses of the order of 1 sec in length. This is most apparent in Fig. 8-33 for the group around 3:41. Higher-speed recordings sometimes reveal even finer structure, with durations of the order of 10 msec (Kraus, 1956, 1958; Gallet, 1961; Riihimaa, 1964). Thus, the Jupiter noise storms may be subdivided into structures with three widely separated periods of (1) minutes, (2) seconds, and (3) milliseconds. Gallet (1961) refers to the pulses of the order of 1 sec duration as L pulses and the pulses of the order of 10 msec duration as S pulses.

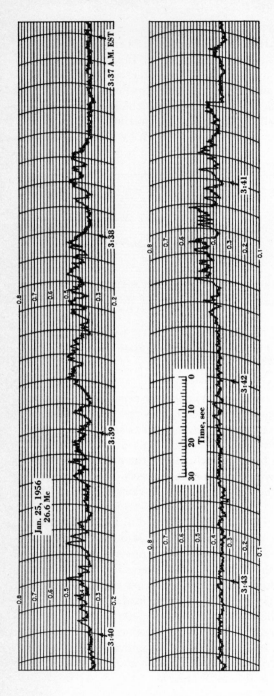

Fig. 8-33. Jupiter radiation recorded on 26.6 Mc at the Ohio State University on Jan. 25, 1956, showing the fine structure of about 4 min of activity. (*Kraus*, 1958.)

While the shape of the occurrence-probability histogram (Fig. 8-31) tends to persist for years, there is a change in its amplitude, which has a long-period variation of the order of 10 years (Douglas, 1964). This variation shows an inverse correlation with the 11-year period of mean solar activity. Jupiter's period of revolution (11.9 years) is of the same order and, if a factor, would complicate the interpretation. Another possible factor is that the radio source on Jupiter is sufficiently directional so that from time to time its beam may be shifted out of the plane containing the earth. A tendency of this kind could make the frequency with which Jupiter storms are observed dependent on the earth's declination viewed from Jupiter.

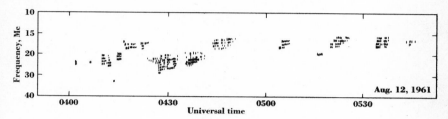

Fig. 8-34. Dynamic spectrum of Jupiter radiation as recorded at the High Altitude Observatory illustrating a drift of emission from high to low frequencies. (*After Warwick*, 1963.)

From swept-frequency observations with a phase-switching interferometer operating in the 8- to 41-Mc range Warwick (1963) has found that the decametric Jupiter radiation possesses a more or less permanent *dynamic spectrum*. A sample record of one of Warwick's records made at the High Altitude Observatory (Boulder, Colorado) is shown in Fig. 8-34. The frequency scale (ordinate) is from 10 to 40 Mc, with time as abscissa. The Jupiter emission shows up as dark areas comprised of nearly horizontal bands, whose bandwidth at any time appears to be less than 10 Mc. A tendency for the activity shown in Fig. 8-34 to drift to lower frequencies is also to be noted. This activity occurred when region 2 was facing the earth. A downward frequency drift is characteristic of this region. On the other hand, region 1 is found to have an upward frequency drift. Warwick has found that activity areas of characteristic shapes, which he calls *landmarks*, tend to reoccur on the records. These suggest a permanent dynamic spectrum.

Plotting the system III central-meridian longitude of Jupiter as the time of landmark occurrence vs. days from opposition, Warwick obtained the results shown in Fig. 8-35. The occurrences are divided into two groups, one associated with region 1 (lower) and the other with region 2 (upper).   The spread in occurrences is mostly less than ±10°, which suggests a narrow beam of radiation.   Region 2 shows no significant drift in longitude with time, but there appears to be some tendency to drift by region 1.

At 4.8 Mc, Ellis (1962) obtained a histogram of occurrence probability vs. longitude that has no pronounced peaks.   However, a mean-power histogram has two main peaks at about 160 and 330°.   Neither coincides with any of the three regions observed at higher frequencies.

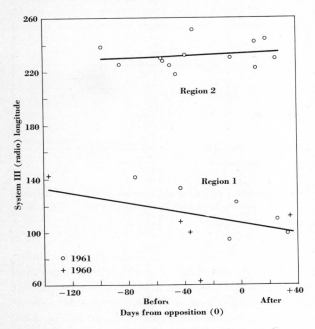

Fig. 8-35. Central-meridian longitude of Jupiter for landmark occurrences as a function of the days from opposition. (*After Warwick*, 1963.)

In 1956 and 1957 Kraus (1958) noted intense Jupiter decameter activity 3 to 5 days after very large solar flares and commented as follows: "Although these sequences may be entirely coincidental, they do suggest the possibility that the mechanism producing the Jupiter radiation may be initiated or triggered by particles emitted from the sun."   Subsequent studies by Warwick (1960) and by Carr, Smith, Bollhagen, Six, and Chatterton (1961) indicate that there indeed may be a correlation between periods of Jupiter decameter events and prior solar activity, with delay times ranging from a few days to a week or so.   The variability in delay can be accounted for by differences in velocity of the solar plasma, or particle, clouds. However, more data are required definitely to establish such a short-period effect. Over long periods (years) there is an inverse correlation with sunspot number.

An interesting characteristic of the Jupiter decametric radiation is that it may have a high degree of circular polarization. Near 20 Mc the sense is predominantly right-handed, with left-handed polarization observed only occasionally. Sometimes the polarization shifts from right to left and back in a period of a few minutes.

Interpretation of the decametric Jupiter emission in terms of what is inherent in the radiation from Jupiter and what has been imposed by propagation through the interplanetary medium or the earth's ionosphere is a difficult problem. Present thinking is that the emission is generated in the atmosphere and/or magnetosphere of the planet and that the radiation is intrinsically polarized and directional, with some effects induced over the Jupiter-earth propagation path. Thus, the S and L pulses and rapid polarization shifts may be propagation effects.

Morris and Berge (1961) found that the plane of polarization of the Jupiter radiation at 1,390 Mc rocks through an angle of about 18° with rotation of the planet, from which it may be inferred that the Jovian magnetic dipole is tilted half this amount, or about 9°, from the axis of rotation. To explain the decametric emission properties Warwick (1963) has postulated that the magnetic dipole is displaced from the center of the planet and, to account for the decimeter rocking effect, is also tilted.

Bigg (1964) has demonstrated a correlation between the strength and duration of the decameter radiation and the position of Jupiter's satellite Io. It has been suggested by Tiuri and Kraus (1965) that the satellite-ionization phenomenon (Kraus, Higgy, and Crone, 1960; Kraus and Tiuri, 1962; Tiuri and Kraus, 1963) involving the interaction of a satellite with an electron shell in the magnetosphere of a planet may be involved.

An excellent review article on the Jupiter decametric radiation has been written by Douglas (1964).

8-8h. Saturn and the Other Planets.

The detection of 4-cm thermal emission from Saturn in 1957 was reported by Drake and Ewen (1958). Measurements at 3 cm by Cook, Cross, Bair, and Arnold (1960) yielded a blackbody disk temperature of 106 ± 21°K, or about 20°K less than the infrared disk temperature. Measurements by Rose, Bologna, and Sloanaker (1963) at 9 cm using two different linear polarizations indicated a blackbody disk temperature of 213°K when the plane of polarization (electric vector) was aligned with Saturn's polar axis but only 140°K when aligned with the equator. As yet no conclusive observations have been reported of decametric radiation from Saturn. And to date there have been no reports of radiation at any radio wavelength from the other planets.

Data on the radio and infrared temperatures of the moon and the planets are summarized in Table 8-3.

Table 8-3
Radio and infrared temperatures
of the moon and the planets

Object	Diameter, miles	Apparent diameter from earth, sec of arc	Blackbody temperature, °K		Wavelength, cm
			Infrared	Radio	
Moon	2,160	1,870 (avg)	150–375	145–280	~1
Mercury	3,000	5–13	610 ss†	400	3
Venus	7,600	11–67	370 ss	550–700	3–10
			250 ds†	500	2
Earth	7,900		287 avg		
Mars	4,200	4–24	250 ss	211	3
			217 avg		
Jupiter	89,000	32–50	140	145	3
				600	10
				>600	>10
Saturn	75,000	14–20	125	106	3
				140–213	9
Uranus	31,000	3.8	90		
Neptune	33,000	2.5	70		
Pluto	3,500(?)		?		

† ss, sunlit side; ds, dark side.

8-9　Galactic Radio Sources　Radio emission from within our galaxy but outside the solar system may come from several types of discrete sources: (1) *supernova remnants*, such as Cassiopeia A and the Crab nebula, with nonthermal spectra and prominent at longer wavelengths, (2) *ionized hydrogen* (H II) *clouds*, such as the Orion nebula, with thermal spectra and prominent at shorter wavelengths, (3) *line emission* from neutral hydrogen (H I) clouds at 21 cm, from hydroxyl radical (OH) clouds at 18 cm, or from clouds emitting other lines, and (4) *flare stars*. In addition to the discrete sources there is a background of emission, which may consist of unresolved discrete sources or a more or less continuous distribution of radiation, due, for example, to a blend of thermal emission from ionized hydrogen and nonthermal radiation from relativistic particles moving in the galactic magnetic field.

8-9a. Nonthermal Sources. The most intense discrete source in in the sky (other than the sun) is Cassiopeia A. It has a nonthermal spectrum with index $n = +0.8$. It was identified by Baade and Minkowski (1954) with wisps of faint optical-emission nebulosities believed to be the remnants of a supernova explosion which probably occurred about A.D. 1700. A photograph with the Mount Palomar 200-in. telescope is shown in Fig. 8-36a. The source is at a distance of a few thousand parsecs and subtends an angle of about 4 min of arc. The wisps of nebulosity show a large radial velocity of over 5,000 km sec^{-1}, indicating a rapid expansion. Shklovsky (1960) believes the emission is of the synchrotron type, and he predicted that the flux density should decrease by an amount $2(1 + 2n)/$ (age in years), where n is the spectral index. For $n = 0.8$ and an age of 265 years this yields about a 2 percent per year decrease. Measurements of the flux density indicate a decrease of 1 to 2 percent per year (Högbom and Shakeshaft, 1961; Findlay, Hvatum, and Waltman, 1965).[†] The measurements were made by comparing the flux density of Cassiopeia A with that of Cygnus A and noting the change in ratio, assuming that the flux density of Cygnus A remains constant. Absolute radio flux-density measurements to a 1 percent accuracy are very difficult with presently available techniques.

Recently, using the new Cambridge University radio telescope, Ryle, Elsmore, and Neville (1965) have made observations of Cassiopeia A in the continuum at 21 cm wavelength with the high resolution of about 25 sec of arc. Their observations reveal a radio source having a ringlike structure, with localized peaks, and a sharp outer boundary. Their results are illustrated by the contour map of Fig. 8-36b. The sharp outer boundary appears to be a characteristic feature of a supernova remnant. Most of the radio emission can be accounted for by a spherical shell of uniform and isotropic emissivity with a shell thickness ¼ its radius. The positions of the optical wisps of nebulosity (Fig. 8-36a) are also indicated in Fig. 8-36b. It is evident that none of these correspond in position with the peaks of the radio emission. It may be that the radio and optical emissions come from different parts of the object. On the other hand, much of the optical nebulosity may be obscured from view by intervening gas and dust.

The new Cambridge telescope consists of three 60-ft-diameter steerable dish antennas, two fixed and one on rails, operating in an interferometer aperture-synthesis arrangement. The antennas track the source region for an extended period, while the earth's rotation changes the interferometer base-line orientation and effective length (Ryle, 1962).

A number of other nonthermal galactic radio sources, such as the Cygnus loop and IC 443, appear to be supernova remnants similar to

† Findlay et al. give 2,370 flux units at 1,440 Mc (1963.0) with 1.75 percent decrease per year.

Cassiopeia A. The Cygnus loop has a low expansion velocity, less than 100 km sec^{-1}, and is believed to be about 50,000 years old.

The first identification of a radio source with an optical object (other than the sun) was made by Bolton (1948), who identified the strong nonthermal source Taurus A with the Crab nebula (M 1). This source has a spectral index $n = 0.27$. The photograph of Fig. 8-37, taken by Dr. Walter Baade in the Hα line of hydrogen, shows a remarkable filamentary structure. The Crab nebula is the remnant of a supernova explosion that

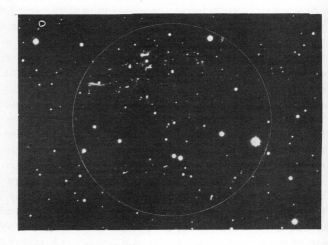

Fig. 8-36. (*a*) (above) A Mount Palomar photograph of the Cassiopeia A nebulosities. The circle indicates the approximate extent of the radio source. (*b*) (facing page) High-resolution radio map of Cassiopeia A made with the Cambridge University radio telescope. The contour interval is 3000°K. The crosses show the positions of some of the brighter stars. (*After Ryle, Elsmore, and Neville, 1965.*)

occurred in A.D. 1054. According to Chinese chronicles, the explosion was witnessed on July 4, 1054, when a very bright "guest star" appeared that was so bright it was visible even in broad daylight. It gradually decreased in brightness and is now a relatively faint optical object of ninth magnitude. The Crab nebula is at a distance of 1,000 pc or more and subtends an angle of about 4 min of arc. The Crab-nebula explosion is both older than the one in Cassiopeia A (911 years as compared to 265) and has a lower spectral index (0.3 as compared to 0.8). It is also thought that the supernova explosion involved in the Crab nebula is fundamentally different from the one in Cassiopeia A. The Cassiopeia A explosion apparently involved a very massive star with approximately one solar mass being ejected (Minkowski, 1959). The Crab nebula probably involved, however, only about one-tenth this mass. The Crab nebula is classified as a *supernova of type* I

and Cassiopeia A as a *supernova of type* II.   Type I supernovas involve
the ejection of a small fraction of a solar mass, while type II supernovas
may throw off several solar masses.   The two types of supernovas also
appear to have different kinds of light curves (luminosity versus time
graph), but the observational data on which to base such a distinction are

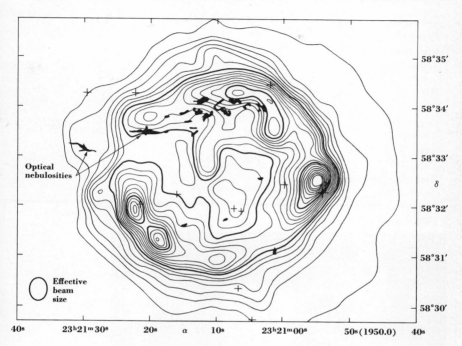

inadequate by modern standards for the Crab nebula and entirely missing
for Cassiopeia A.   The Cassiopeia A explosion is believed to have occurred
about A.D. 1700, but there is no record that the event was observed,
probably because it is in a highly obscured part of the sky.   The radio
emission from the Crab nebula, like that from Cassiopeia A, is believed to
be due to the synchrotron process.   At 11 cm the degree of linear polarization d_l of the Crab nebula is only 0.03, but at 3 cm it is 0.07 (Mayer,
McCullough, and Sloanaker, 1957, 1964) (see Fig. 5-4).   A comparable
degree of linear polarization of the optical emission has also been measured
(Oort and Walraven, 1956), tending to confirm the belief that the synchrotron mechanism is responsible.   The fact that very energetic particles are
required to produce the synchrotron emission suggests that the Crab
nebula may also be a source of cosmic rays.

In addition to Cassiopeia A and the Crab nebula, a number of other
nonthermal galactic radio sources have been identified with what appear
to be supernova remnants.   Data on some of these sources are summarized
in Table 8-4.

Table 8-4
Galactic nonthermal sources

Object	Flux density, flux units†	Frequency, Mc	Distance, pc	Type SN remnant	Age, years (1965)	Spectral index
Cassiopeia A	11,000	178	3,400	II	~265	0.77
Puppis A	7,000	100		II		
Crab nebula (M 1), supernova of A.D. 1054	1,000	1,000	1,100	I	911	0.27
HB 21	500	100		II		
IC 443	210	178	2,000	II		0.28
Cygnus loop (network nebula)	300	100	770	II	~50,000	0.4–0.5
Tycho's supernova (A.D. 1572)	134	178	~360	I	393	0.6
Kepler's supernova (A.D. 1604)	80	100	1,000	I	361	0.7

† 10^{-26} watt m^{-2} cps^{-1}.

Fig. 8-37. Photograph of the Crab Nebula (M 1) in the Hα line of hydrogen. (*Mount Wilson and Palomar Observatories photograph.*)

8-9b. Thermal Sources. The interstellar hydrogen in our galaxy tends to be distributed in clouds. In the absence of any exciting sources the hydrogen remains in its neutral state, and the clouds emit only at 21 cm

When a hot stellar source is in or near such a cloud, however, its ultra-violet radiation tends to ionize the cloud, causing it to emit continuum radiation characteristic of a thermal source. Hot O and B stars are often the exciting sources, and cloud temperatures of 10,000°K are typical. The effective range or radius of excitation of the exciting source, which depends on the type of star, determines the size of a sphere called the *Strömgren sphere* (Strömgren, 1948). From the size of the Strömgren sphere and a measurement of the radio flux density S it is possible to determine the average electron density. In the relation

$$S = \frac{2k}{\lambda^2} T_b \Omega_s \tag{8-43}$$

where S = flux density, jan
$\quad k$ = Boltzmann's constant ($= 1.38 \times 10^{-23}$ joule °K^{-1})
$\quad \lambda$ = wavelength, m
$\quad T_b$ = average brightness temperature over the sphere, °K
$\quad \Omega_s$ = solid angle subtended by Strömgren sphere, rad^2
all quantities except T_b are known or can be measured, so that T_b can be determined. Assuming a value for the electron temperature T_e, the optical depth τ can then be deduced from

$$T_b = T_e(1 - e^{-\tau}) \tag{8-44}$$

or, for the optically thin case ($\tau \ll 1$), from

$$T_b = T_e \tau \tag{8-45}$$

In the Rosette nebula a brightness temperature of 200°K was measured at 242 Mc (Ko and Kraus, 1955). Assuming an electron temperature of 10,000°K, this leads to an optical depth of 0.02 ($= 200/10,000$). From (8-10) we obtain an emission measure

$$\begin{aligned} \text{EM} &= 2.5\tau\nu^2 \\ &= 2.5 \times 0.02 \times 242^2 = 2{,}930 \end{aligned} \tag{8-46}$$

Taking 37 pc as the diameter of the Strömgren sphere yields an average free-electron density

$$N = \sqrt{\frac{\text{EM}}{l}} = \sqrt{\frac{2{,}930}{37}} \simeq 9 \text{ cm}^{-3}$$

Two examples of thermal hydrogen-cloud (H II) radio sources are the Orion nebula and the Rosette nebula, shown in Figs. 8-38 and 8-39. At longer wavelengths an optically thick spectrum ($n = -2$) would be expected. This is observed in the case of the Orion nebula at the lower frequencies (Fig. 8-40). The break between the optically thin ($n = 0$) and optically thick condition occurs at a wavelength of about 30 cm. By way of contrast, the flux-density spectrum of the Rosette nebula, shown in Fig. 8-40 (Menon, 1962), is substantially flat ($n = 0$) to a wavelength of at least 3 m.   Assuming $\tau \sim 1$ at the break of the spectrum, we have, from (8-46), that EM $= 2.5\nu^2$, where ν is the frequency of the break.   Thus, the lower break frequency (< 100 Mc) for the Rosette nebula as compared to the break frequency ($\sim 1{,}000$ Mc) for the Orion nebula indicates a much lower emission measure for the Rosette nebula.

Fig. 8-38.   Photograph of the Orion nebula. (*Official U.S. Navy photograph, courtesy of Dr. K. Aa. Strand.*)

Whereas the Orion nebula has an electron density of several thousand electrons per cubic centimeter at its center decreasing to about 10 cm^{-3} at the edge, the average density for the Rosette nebula is only about 10 cm^{-3} (Ko and Kraus, 1955) with a minimum density at the center.   Thus, these two ionized-hydrogen clouds differ significantly in form.   The Rosette nebula is also much more massive, containing about 10,000 solar masses, as compared to about 100 for the Orion nebula.   Upper limits for the ages of the nebulas have been placed at 10,000 years for the Orion nebula and 50,000 years for the Rosette nebula (Menon, 1964).   These ages refer to the length of time during which ultraviolet excitation from the exciting

arly-type) stellar sources has been important.   As a result of the excita-
on, mass motions take place in the nebula which are believed to result in
ne irregular electron-density distributions found, for example, in the
.osette nebula.

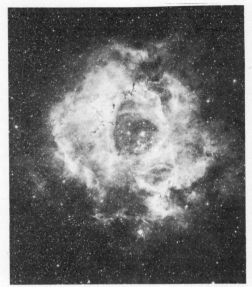

Fig. 8-39.   Photograph of the Rosette nebula.
(*Mount Wilson and Palomar Observatories photograph.*)

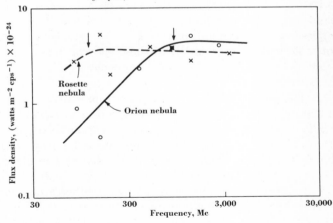

Fig. 8-40.   Spectra of the Orion and Rosette nebulas.   The circles
are measured values for the Orion nebula, and the crosses are
measured values for the Rosette nebula.   The arrows indicate the
break frequencies of the spectra.   (*After Menon*, 1962, 1964.)

Since the first measurements of ionized-hydrogen clouds by Haddock, Mayer, and Sloanaker (1954) a total of several dozen sources have been identified with such clouds. Data on several of these sources, including the Orion and Rosette nebulas, are summarized in Table 8-5.

Table 8-5
Galactic thermal sources (H II regions)

Object	Flux density, flux units†	Distance, pc
Cygnus X	~5,000	~1,000
Omega nebula (M 17)	1,000	1,700
North America nebula	550	900
Orion nebula (M 42)	520	500
Rosette nebula	260	1,400
Lagoon nebula (M 8)	260	1,200

† 10^{-26} watt m^{-2} cps^{-1} at 1,400 Mc.

8-9c. Line-emission Clouds and Galactic Structure. Observations of the continuum radiation of thermal and nonthermal radio sources of the type we have been discussing are usually conducted using receivers of wide bandwidth, with the exact frequency of operation being noncritical. This continuum radiation extends over essentially all frequencies with varying intensity, the emission coming from nonresonant processes. In contrast to this broadband continuum radiation there is also narrow-band emission, or absorption, related to the difference in energy level of atoms or molecules. Observations of these *spectral lines* are made with narrow-band receivers, with the exact frequency of observation being of prime importance.

The possibility of a spectral line from neutral interstellar hydrogen at 1,420 Mc (21 cm) was reported by van de Hulst (1945) of Leiden, Netherlands, in 1945.† The line was detected in emission in 1951 by Ewen and Purcell (1951) at Harvard and a few weeks later by groups in the Netherlands and in Australia. The line was detected in absorption in 1954 by Hagen and McClain (1954) at the U.S. Naval Research Laboratory. Although attempts were made to detect other atomic or molecular lines, such as the 327-Mc line of deuterium and the 1,665- and 1,667-Mc lines of the OH radical, all were unsuccessful until 1963, when Weinreb, Barrett, Meeks, and Henry (1963) detected the close pair of OH ($O^{16}H^1$) lines in absorption. They noted the decrease in intensity of Cassiopeia A at or

† A bit of historical background on this is given in Chapter 1.

close to the OH-line frequencies caused by absorption of OH over the path between the earth and Cassiopeia A. Absorption by neutral hydrogen over the same path had been observed by Hagen and McClain in 1954.

The neutral-hydrogen (H I) transition is associated with a transition between two closely spaced energy levels of the ground state related to the electron spin, or magnetic dipole orientation. When the magnetic dipole moment of the electron is parallel to that of the nucleus, the energy of the hydrogen atom is very slightly greater than when the dipole moments are opposed. The transition between these two substates is called a *hyperfine transition*. The probability of spontaneous emission is so low that an atom in the higher state remains there for 11 million years, on the average, before falling to the lower state, with the emission of a 21-cm photon. This low probability and the lack of knowledge of the density of neutral hydrogen in interstellar space made it quite uncertain whether the line could be detected. However, the suggestion by van de Hulst (1945) that a search be made brought positive results 6 years later.

The OH line involves an electric dipole transition which is much more intense than the magnetic dipole transition of the hydrogen line. The OH transition probability is also higher. As a result, it has been possible to detect the OH line even though the OH abundance appears generally to be very much less than that of neutral hydrogen. Over the Cassiopeia A path the neutral hydrogen atoms are 20 million times more abundant than the OH radicals (Barrett, 1964). Australian observers (Robinson et al., 1964) indicate that in the direction of the galactic nucleus this ratio is much smaller (relatively more OH). They also find curious concentrations of OH near the nucleus that differ significantly in radial-velocity distribution from the neutral-hydrogen concentrations. The Australian group (Gardner, Robinson, Bolton, and van Damme, 1964) has also detected the two OH satellite lines at 1,612 and 1,720 Mc. These lines are not so strong as the close pair at 1,665 and 1,667 Mc.

Barrett (1964) has given an excellent discussion of the problems of detection of the OH and other molecular lines and points out that the possibility of detecting the 1,637- and 1,639-Mc lines of the hydroxyl radical $O^{18}H^1$ is promising and recommends that a search be made. Barrett also discusses the possibilities for detection of lines from SH, SiH, CH, CN, and NH. The frequencies for a number of atomic and molecular lines are listed in Table 8-6. Recently, Hoglund and Mezger (1965) have detected a high-level transition of ionized hydrogen (H II) from quantum energy level 110 to 109 at a frequency of 5,008.932 Mc in the Omega and Orion nebulas. This transition and other high-level transitions at 1,424.7, 5,736, and 8,872 Mc were predicted by Kardashev (1960).

Observations of the 21-cm hydrogen line have enabled radio astronomers to deduce for the first time a picture of the spiral structure of our galaxy. · From profiles of intensity vs. frequency (or radial velocity) it is found that in some directions the neutral hydrogen is moving toward us and in other directions away from us, with velocities up to several hundred kilometers per second. From such observations around the galactic plane and a suitable velocity-distance model, it is possible to construct an approximate map of the distribution of neutral hydrogen in our galaxy. Such a picture assembled from Dutch and Australian observations is presented in Fig. 8-41 (Kerr and Westerhout, 1964). The positions of the sun and galactic nucleus are shown. Their distance of separation is about 30,000 light-years. In directions near the galactic center the radial velocity due

Table 8-6
Atomic and molecular lines

Type	Frequency, Mc except as noted	Remarks
H I	1,420	Detected
Deuterium	327	
OH:		
$O^{16}H^1$	1,612	Weak, detected
	1,665	Strong, detected
	1,667	Strong, detected
	1,720	Weak, detected
$O^{18}H^1$	1,584	Weak
	1,637	Strong
	1,639	Strong
	1,692	Weak
H II†	1,425	
H II	5,009	Detected
H II	5,736	
H II	8,872	
SH	111	
SiH	~2,400	
CH	~1,000	
CH	~3,180	
CN	~113 Gc	
NH	~950 Gc	

† *Hydrogen recombination line.* The designation here of H II indicates that the emission occurs in regions where most of the hydrogen is ionized (H II regions). Actually, the emission (or absorption) is from neutral hydrogen atoms during a transient recombination of an ion with an electron. The atom is in an excited state (at a high energy level) which is in contrast with the neutral hydrogen (H I) atom in the unexcited or ground state responsible for the 1,420 Mc line. The circumstances also differ from those of thermal continuum emission, where an electron is deflected in passing near a proton (see Sec. 8-4a).

to galactic rotation tends to approach zero and to become confused with radial velocities due to local turbulences. Hence, a sector toward the galactic center is left blank in Fig. 8-41. In spite of this problem, observations of the galactic nucleus (Sagittarius A) at the hydrogen line and also in the continuum indicate the existence of a remarkable structure at the center of our galaxy (Rougoor and Oort, 1960). There appears to be a small source at the center of a flat disk and a ring of neutral hydrogen, with all three components (source, disk, and ring) embedded in a nonthermal

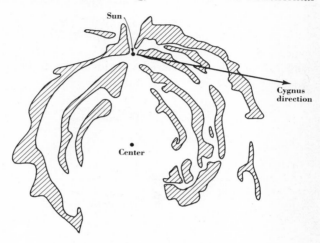

Fig. 8-41. Distribution of neutral hydrogen in our galaxy, showing its complex spiral structure. Shaded areas correspond to the regions of greatest neutral hydrogen concentration as deduced from 21-cm observations. (*After Kerr and Westerhout*, 1964.)

source produced by relativistic electrons moving in a magnetic field. Regions near the center show both high rotational velocity and rapid expansion radially outward. The reasons for this complex structure are not known, but some astronomers suggest that the structure is the result of an explosion in the nucleus in an earlier epoch (Westerhout, 1964).

The profile (intensity vs. frequency) obtained when a stationary region of neutral hydrogen is scanned in frequency with a narrow-band receiver is as suggested in Fig. 8-42a. The profile is centered on 1,420 Mc, and its width is a measure of the receiver bandpass characteristics. It is assumed that the neutral-hydrogen cloud is stationary (with respect to the observer) and has no internal turbulence or thermal broadening. If the cloud has no internal turbulence or thermal broadening but is moving as a whole away from the observer, the profile of Fig. 8-42a would be displaced down in frequency, as suggested in Fig. 8-42b, where the frequency shift $\Delta\nu$ is related to the velocity of recession v by

$$\mp v = c\,\frac{\pm\Delta\nu}{\nu} \qquad (8\text{-}47)$$

where v = velocity of approach $(-)$ or recession $(+)$
    c = velocity of light
    $\Delta\nu$ = frequency shift
    ν = rest frequency ($= 1{,}420$ Mc)

At the hydrogen-line frequency a frequency shift of 5 Kc corresponds to a velocity of 1.056 km sec^{-1}. Equation (8-47) is a simplification of a more general relation and holds only for $v \ll c$. If the receiver had a frequency response approaching zero bandwidth, the profiles in Fig. 8-42a and b would be delta functions, as indicated.

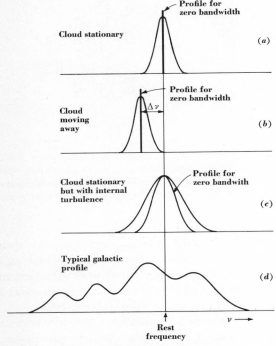

Fig. 8-42. Idealized hydrogen-line profiles.

If the cloud is stationary as a whole with respect to the observer but has thermal broadening and/or turbulence of a symmetrical type, then the profile will be broadened, but the peak will not be displaced, as suggested in Fig. 8-42c. A broadened profile would also be obtained with a zero-bandwidth receiver, but the profile would be less broad. The zero-bandwidth profile is the true profile, which may be approached but not attained in practice as the bandwidth is reduced. The observed profile is the convolution of this true profile and the receiver profile.

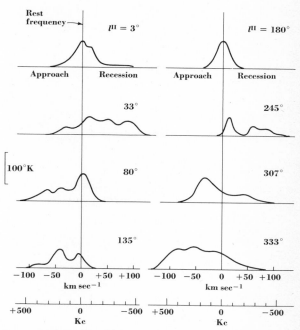

Fig. 8-43. Hydrogen-line profiles at different longitudes in the plane of our galaxy. (*After Kerr and Westerhout, 1964.*)

In general, in measurements of galactic hydrogen the gas may be in relative motion with respect to the observer and also have internal motions. The profile of Fig. 8-42*d* is typical. It suggests that there are four regions, or clouds, in the direction of the antenna beam, with three clouds receding from the observer at different velocities and one cloud approaching. All four peaks are broadened, suggesting also that the clouds have internal motions.

In the measurements of the neutral-hydrogen radiation from our galaxy, profiles were obtained along the equator at different galactic longitudes, of which those in Fig. 8-43 are typical (Kerr and Westerhout, 1964). Referring to Fig. 8-43 and to the sketch of Fig. 8-44, showing the coordinates in the galactic plane, we note that in a direction close to the center of the galaxy ($l^{II} = 3°$) the peak is centered on the rest frequency (1,420

Mc), indicating no relative motion. In the direction $l^{II} = 33°$ the profile is displaced to the right, indicating recession. According to Fig. 8-44, this would be expected if the emission is from regions inside of a circle having the galactic nucleus–sun distance as radius. Assuming the sun to be stationary, motion inside the circle would be clockwise, as indicated, on the assumption that the galaxy is winding up (note hypothetical path of spiral arm through sun shown by dashed line). In the directions $l^{II} = 80$

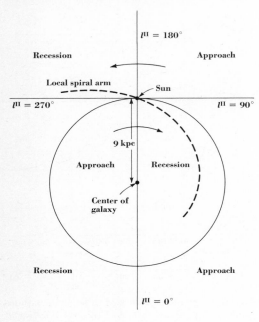

Fig. 8-44. Simplified plan-view sketch of our galaxy showing regions of relative approach and recession with respect to the sun. The local spiral arm is indicated by the dashed line.

and 135° the profiles show approach, while at 180°, in the direction of the galactic anticenter, there is no relative motion. At $l^{II} = 245°$ the profile indicates recession, but the profiles at 307 and 333° show approach.

The peaks in the profiles of Fig. 8-43 can be interpreted as related to different spiral arms. A detailed interpretation, however, requires that a model be assumed for the galactic motions. Such a simple, idealized model is shown in Fig. 8-45. It is assumed that the galactic motions are circular and that the angular velocity ω is constant at any radius R. As shown by van de Hulst, Muller, and Oort (1954), the relative velocity v of the point F with respect to S (the sun) is given by

$$v = \omega R \sin \delta - \omega_0 R_0 \sin \gamma \tag{8-48}$$

where ω_0 = angular velocity at sun's radius
    R_0 = sun's radius from center of galaxy
    γ = angle between antenna beam and galactic nucleus
    δ = angle at point P between antenna beam and galactic nucleus

From the theorem of sines

$$\frac{\sin \gamma}{R} = \frac{\sin (180^\circ - \delta)}{R_c} = \frac{\sin \delta}{R_0} \tag{8-49}$$

or

$$R \sin \delta = R_0 \sin \gamma \tag{8-50}$$

From (8-50) and (8-48) we then obtain

$$v = (\omega - \omega_0) R_0 \sin \gamma \tag{8-51}$$

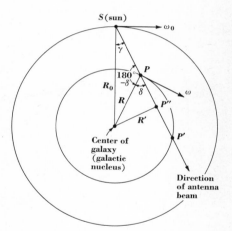

Fig. 8-45. Simplified model of galactic motions with associated geometry.

In (8-51) v can be calculated by (8-47) from the observed hydrogen-line displacement, γ is known from the antenna setting, and ω_0 and R_0 are given by the galactic model. The angular velocity ω of the hydrogen region at the point P can then be calculated from (8-51). From a known or assumed relation between ω and R, the radius R of the point P can be determined and, hence, the distance SP. Knowing the distance SP, the direction γ, and the flux density, a map of the neutral-hydrogen distribution, such as Fig. 8-41, can be constructed.

The above discussion is somewhat oversimplified. Thus, the sun's motion in the solar neighborhood makes it desirable to refer the motions to the local standard of rest (see Chap. 2) instead of to the sun. The motion of the observer on the earth with respect to this local standard of rest also needs to be taken into account. Furthermore, referring to Fig. 8-45, it is evident that there will be an ambiguity in the distance determination, since for a given ω and R the hydrogen-line region could be either at P or P'. However, if the largest radial velocity in the line profile is taken, it should correspond without ambiguity to the point P'' at a radius R'. Outside of the radius R_0 the position ambiguity does not occur. A graph of ωR versus R from Rougoor and Oort (1960) is shown in Fig. 8-46, along with a curve of ω versus R. It is the latter curve that is useful in converting ω to R.

The above discussion has been concerned with line emission. Let us analyze next the problem of *line absorption* (Purcell and Ewen, 1951; van de Hulst, Muller, and Oort, 1954; Hagen, Lilley, and McClain, 1955). Suppose that a homogeneous line cloud filling the antenna beam is located

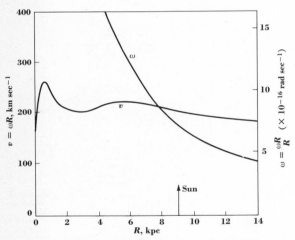

Fig. 8-46. Circular velocity v and angular velocity ω in our galaxy as a function of distance R from the center. (*After Rougoor and Oort,* 1960.)

between the observer and a discrete source, as in Fig. 8-47. If T_s' is the observed antenna temperature due to the source in the continuum immediately adjacent to the line, then in the cloud's line frequency range (signal band) the antenna temperature is given by

$$T_l = T_s'e^{-\tau} + T_c'(1 - e^{-\tau}) \tag{8-52}$$

where T_l = signal band or line temperature

τ = optical depth of cloud in the line ($\tau = f(\nu)$)

T_s' = apparent source temperature, or antenna temperature due to source ($= T_s\Omega_s/\Omega_A$; $\Omega_s \ll \Omega_A$)

$T_c{}'$ = apparent cloud temperature, or antenna temperature due to cloud in the line ($= T_c\Omega_M/\Omega_A$)

T_s = true source temperature

T_c = true cloud temperature in the line

Ω_s = source solid angle

Ω_A = antenna beam solid angle

Ω_M = main beam solid angle

The first term in (8-52) involves line absorption, while the second term involves both line emission and absorption. For frequencies just removed from the line (comparison band) $\tau = 0$, and we have

$$T = T_s{}' \tag{8-53}$$

where T = comparison-band temperature (off of line)

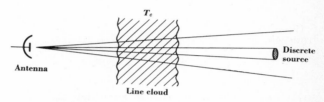

Fig. 8-47. Line cloud (hydrogen line or other atomic or molecular line) and discrete source in antenna beam of radio telescope.

It is assumed that the source temperature is substantially constant across both bands (signal and comparison). By arranging the radiometer to measure the difference of the signal and comparison bands ($T_l - T = \Delta T$) the radiometer (on-source) output is

$$\Delta T = T_s{}'e^{-\tau} + T_c{}'(1 - e^{-\tau}) - T_s{}' \tag{8-54}$$

or

$$\Delta T = (T_c{}' - T_s{}')(1 - e^{-\tau}) \tag{8-55}$$

Three cases are of interest:

(1) If $T_c{}' > T_s{}'$ (apparent cloud line temperature greater than apparent source temperature), the difference ΔT is positive, and the line is seen in *emission.*

(2) If $T_c{}' = T_s{}'$, the output is zero.

(3) If $T_c{}' < T_s{}'$, the difference ΔT is negative, and the line is seen in *absorption.*

If the antenna beam is moved off the discrete source but is still filled by the cloud, we have, assuming a negligible change in cloud effect, that the radiometer (off-source) output is

$$\Delta T' = T_c{}'(1 - e^{-\tau}) \tag{8-56}$$

Equation (8-56) describes the situation for a simple emission measurement, while (8-55) describes the situation for an absorption measurement making use of a discrete source. If the optical depth of the cloud is small ($\tau \ll 1$), we have approximately that

$$\Delta T' = T_c'\tau \qquad \text{off source} \tag{8-57}$$

and

$$\Delta T = (T_c' - T_s')\tau \qquad \text{on source} \tag{8-58}$$

From (8-57) and (8-58) it is clear that if $T_s' \gg T_c'$, the line will be much more readily detected in absorption than in emission. The requirement $T_s' \gg T_c'$ can usually be met by using a sufficiently directive antenna and/or selecting a sufficiently strong discrete source. Subtracting (8-56) from (8-55) gives

$$\Delta T - \Delta T' = -T_s'(1 - e^{-\tau}) \tag{8-59}$$

from which

$$\tau = -\ln\left(\frac{\Delta T - \Delta T'}{T_s'} + 1\right) \tag{8-60}$$

From (8-56) and (8-59) we also have that the cloud temperature

$$T_c = \frac{T_s'\Delta T'\Omega_A}{(\Delta T' - \Delta T)\Omega_M} \tag{8-61}$$

where ΔT = radiometer output on source

$\Delta T'$ = radiometer output off source

T_s' = apparent source temperature

Ω_M = main beam solid angle

Ω_A = beam solid angle

Thus, from measurements of T_s', ΔT, and $\Delta T'$ and a knowledge of the antenna beam angles the optical depth τ and line temperature T_c of the cloud can be deduced. The above relations apply to a homogeneous cloud between the source and observer which fills the antenna (main) beam.

If there are several line clouds in the antenna beam, the above analysis requires revision. Following Hagen, Lilley, and McClain (1955), consider the case of two line clouds i and j in the antenna beam subtending solid angles Ω_i and Ω_j, respectively, as in Fig. 8-48. Let the i-cloud optical depth be τ_i and the j-cloud optical depth be τ_j. Assuming further that the i cloud is in line with the discrete source and that the antenna beam is aligned with both, the signal band gives

$$T_l = T_s \frac{\Omega_s - \Omega_i}{\Omega_A} + T_s \frac{\Omega_i}{\Omega_A} e^{-\tau_i} + T_c(1 - e^{-\tau_i}) \frac{\Omega_i}{\Omega_A}$$
$$+ T_c(1 - e^{-\tau_i}) \frac{\Omega_j}{\Omega_A} P_{nj} \tag{8-62}$$

and the comparison band yields

$$T = T_s \frac{\Omega_s}{\Omega_A} \qquad (8\text{-}63)$$

where T_s = source temperature (assumed uniform)

T_c = cloud line temperature (assumed uniform)

Ω_s = source solid angle

Ω_i = i-cloud solid angle

Ω_j = j-cloud solid angle

Ω_A = beam solid angle

τ_i = optical depth of i cloud

τ_j = optical depth of j cloud

P_{nj} = normalized antenna-power-pattern value in direction of j cloud

The first term in (8-62) gives the temperature contribution due to the part of the source not shielded by the i cloud, it being assumed that $\Omega_i < \Omega_s$. The second term is equal to the contribution from the shielded

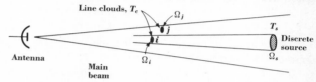

Fig. 8-48. Several line clouds and discrete source in antenna beam of radio telescope.

part of the source as affected by absorption in the i cloud. The third and fourth terms give the temperature contributions due to emission and internal absorption from the i and j clouds, respectively. Subtracting T from T_l yields the radiometer (on-source) output

$$\Delta T = \left(T_c \frac{\Omega_i}{\Omega_A} - T_s' \frac{\Omega_i}{\Omega_s}\right)(1 - e^{-\tau_i}) + T_c(1 - e^{-\tau_j}) \frac{\Omega_j}{\Omega_A} P_{nj} \qquad (8\text{-}64)$$

where T_s' = apparent or antenna temperature due to source $[= T_s\Omega_s/\Omega_A]$

Note that $T_s' \dfrac{\Omega_i}{\Omega_s} = T_s \dfrac{\Omega_s}{\Omega_A} \dfrac{\Omega_i}{\Omega_s} = T_s \dfrac{\Omega_i}{\Omega_A}$

For a number of line clouds (more than the two assumed above) the radiometer (on source) output would be given by summing terms like the first one in (8-64) over all i clouds (in cone subtended by the discrete source) and summing terms like the second one over all j clouds (in antenna beam but not in cone subtended by the discrete source). Rearranging yields

$$\Delta T = -\sum_i T_s'(1 - e^{-\tau_i}) \frac{\Omega_i}{\Omega_s} + \sum_{i,j} T_c(1 - e^{-\tau_{i,j}}) \frac{\Omega_{i,j}}{\Omega_A} P_{ni,j} \qquad (8\text{-}65)$$

In (8-65) the first term is taken over all i clouds and gives the contribution due to the absorption of the source continuum radiation produced by the i clouds. The second term is taken over all clouds (i and j) and represents the contribution due to the line emission from all line clouds in the antenna beam. The latter may be replaced by the mean $\overline{\Delta T'}$ of the line profiles in nearby comparison regions (beam off source) assuming that the comparison regions differ but little from the on-source region. Further, if $\Omega_i \simeq \Omega_s$, (8-65) reduces to

$$\Delta T = -T_s'(1 - e^{-\tau_t}) + \overline{\Delta T'} \tag{8-66}$$

where τ_t = total optical depth of all line clouds between observer and discrete source (within Ω_s)

From (8-66)

$$\tau_t = -\ln\left(\frac{\Delta T - \overline{\Delta T'}}{T_s'} + 1\right) \tag{8-67}$$

where ΔT = radiometer on-source output
   $\overline{\Delta T'}$ = mean radiometer off-source output
   T_s' = apparent or antenna temperature due to discrete source

In neutral-hydrogen measurements in the direction of the discrete source Cassiopeia A, Hagen, Lilley, and McClain (1955) obtained the profiles shown in Fig. 8-49. The on-source (signal-band) output ΔT

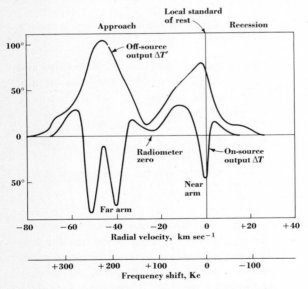

Fig. 8-49. Hydrogen-line emission profiles near Cassiopeia A (upper) and absorption profile in the direction of Cassiopeia A (lower). The ordinate is output temperature in degrees Kelvin. (*After Hagen, Lilley, and McClain, 1955.*)

exhibits three well-defined absorption minima. The pair with the larger frequency displacement is interpreted as being due to two neutral-hydrogen clouds in the far arm of the local galactic structure, while the third absorp-

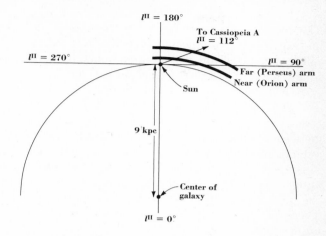

Fig. 8-50. Galactic structure in the direction of Cassiopeia A.

tion minimum is believed due to a cloud in the near arm. Cassiopeia A is situated beyond the second arm, as suggested in Fig. 8-50. From their measurements (Hagen et al., 1955) the data of Table 8-7 were obtained.†

Table 8-7

Object	Radial velocity, km sec⁻¹	$\overline{\Delta T'} - \Delta T$, °K	Optical depth, τ_t (max)
Near-arm cloud (Orion)	−1.0	71.5	0.96
Far-arm cloud I (Perseus)	−38.5	89.4	1.47
Far-arm cloud II (Perseus)	−48.6	100	2.04

A size estimate for the clouds may be deduced from the fact that a cloud in the far arm which just covers Cassiopeia A (angular diameter 6 min of arc) would be about 5 pc in diameter.

The total number of hydrogen atoms N_t in a column of 1-m² cross-section is given by

† More recent data are given by Muller (1957).

$$N_t = 1.835 \times 10^{22} T \int \tau(v) \, dv \tag{8-68}$$

where T = cloud kinetic temperature, °K

$\tau(v)$ = optical depth as a function of velocity

dv = elemental velocity, km sec^{-1}

The integration is carried out over the velocity range of a given hydrogen-line feature to obtain the number of atoms it contains. According to Hagen, Lilley, and McClain, the total number of neutral hydrogen atoms per square meter in the near-arm cloud is about 2×10^{25}. If the cloud is 5 pc thick, the density is about 130 cm^{-3}.†

The lower (absorption) curve in Fig. 8-49 has much sharper features than the upper (emission) curve because the angular diameter of the discrete source (Cassiopeia A) is much less than the beam width of the 50-ft antenna used by Hagen, Lilley, and McClain. The diameter of Cassiopeia A is about 5 min of arc, while the half-power beam width of the 50 ft antenna at 1,420 Mc is about 50 min of arc, or 10 times as great. To achieve the same resolution in emission would require a 500-ft-diameter antenna. The selection of a still smaller discrete source could yield even sharper absorption profiles.

From observations of hydrogen-line absorption in the direction of a number of intense radio sources, Clark (1965) concludes that there are, on the average, about four neutral-hydrogen clouds per kiloparsec in the galactic plane, with perhaps a higher density in the local spiral arm.

8-9d. Flare Stars.

The sun has a maximum flux density of about 10^{-18} watt m^{-2} cps^{-1} at a wavelength of 1 m. It reaches this maximum only sporadically during disturbed conditions usually associated with a solar flare. If the sun were as far away as the nearest stars, or at least 4 light-years away, it would be nearly a million times more distant, and its maximum flux density would be reduced by almost 120 dB, or to about 10^{-30} watt m^{-2} cps^{-1}. This is about 40 dB below the present minimum detectable level of a large radio telescope ($S_{min} \simeq 10^{-26}$ watt m^{-2} cps^{-1}). Hence, a star such as the sun at the distance of the nearest stars would be undetectable with present techniques.

A class of optical objects called *flare stars* has been found, however, which exhibit sudden increases in brightness amounting to 2 or 3 magnitudes (8 to 12 dB) in a period of a few minutes, with rates in some cases of as much as 1 dB sec^{-1}. The prototype of this class of variable stars is UV Ceti, which is the fainter component of the double star Luyten 726-8 at a distance of about 8.5 light-years. UV Ceti is a red dwarf of spectral

---

† By way of comparison, the density of neutral hydrogen is about 1 cm^{-3} in the interstellar medium of the solar neighborhood. On the other hand, the (ionized) hydrogen density in the Orion nebula is $\sim$1,000 cm^{-3}.

type M. An outburst capable of increasing a star's optical output by 8 to
12 dB must be much more violent than a solar flare, which barely affects
the total visual output of the sun. Assuming that a flare star and the sun
have about the same undisturbed radio levels, the more violent outburst
of a nearby flare star might be detected with a radio telescope if it were
40 dB stronger than a solar radio outburst (Struve, 1959).

For several years flare-star watches have been in progress at English
and Australian radio observatories. Flares are sporadic and unpredictable,
and observations of the star, to be significant, must also be made simultaneously by optical observers. Although such cooperative programs are
still in an early phase, evidence has been obtained that the flare stars
UV Ceti (Lovell, Whipple, and Solomon, 1963) and V 371 Orionis (Slee,
Solomon, and Patston, 1963) have detectable radio emission at the time
of optically observed flares, making flare stars the first true radio stars
The rate of occurrence of optical and radio flares for UV Ceti is about 0.7
per day. The radio output of a flare-star flare appears to be 40 to 60 dB
greater than for a solar flare, while the optical output of a flare-star flare
is 10 to 20 dB greater than for a solar flare. A comparison of the observed
starting times of optical and radio flares indicates that the velocity of wave
transmission at optical and radio wavelengths is the same to within 1 part
in 1 million (Lovell, 1964).

8-9e. Background Radiation. In addition to the discrete sources
and line emission of our galaxy there is a background of radiation, which
may consist of unresolved discrete sources but which may also represent a
more or less continuous distribution of radiation from our galaxy. This
continuous spatial distribution is manifested by a narrow ridge extending
all the way around the galactic plane. This shows prominently in the maps
of the radio sky in Figs. 8-1 and 8-2, which are in celestial coordinates
(Mercator projection). In Fig. 8-51 the radio emission is presented in
galactic coordinates. The ridge at the galactic equator is about 5° in half
width at the longer wavelengths, becoming narrower at centimeter wavelengths. At wavelengths longer than about 10 m the ridge becomes double;
i.e., an absorption trough appears where the ridge is located at shorter
wavelengths. An interesting aspect of such absorption features is that if
the absorption by an H II region is essentially complete $(\tau \gg 1)$, any
radiation observed in the direction of the region must be generated between
it and the observer on the earth (Shain, 1957).

Away from the galactic plane radiation falls off more rapidly with
decreasing wavelength than it does in the galactic plane. From more
detailed studies of such trends it may be inferred that the galactic background radiation has two principal components, one concentrated close to
the galactic plane with a thermal spectrum and the other having a maxi-

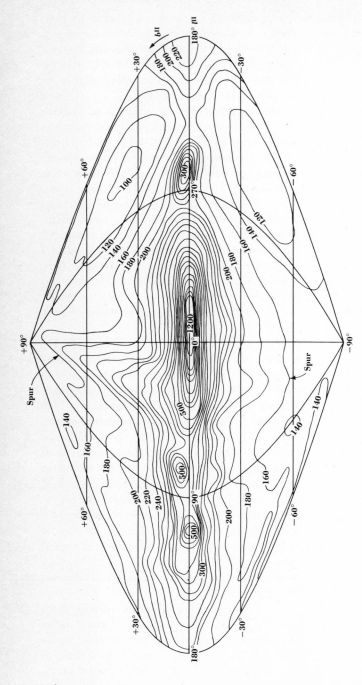

Fig. 8-51. Radio emission from the sky at 200 Mc in new galactic coordinates. Temperatures are indicated in degrees Kelvin. (*After Dröge and Priester*, 1956.)

mum along the galactic plane but of greater extent away from the plane and with a nonthermal spectrum. The thermal component arises from ionized-hydrogen regions and has a spectral index† varying from about $n = +2$ at the shorter wavelengths ($\lambda < 1$ m) to $n = 0$ at the longer wavelengths ($\lambda > 10$ m). The nonthermal component is presumably due to synchrotron radiation from relativistic electrons moving in the galactic magnetic field and has a spectral index† of about $n = +2.5$. The discovery (Westerhout, Seeger, Brouw, and Tinbergen, 1962) of linear polarization in the galactic background radiation provided conclusive evidence that the synchrotron mechanism was responsible. The equivalent temperature of the synchrotron emission depends on both the density of the relativistic electrons and the strength of the magnetic field. If the electron density were constant, the temperature distribution could give, in principle, the magnetic field distribution.

The nonthermal emission from our galaxy is superposed on an isotropic component of extragalactic origin. Although concentrated near the galactic plane in a rather symmetrical fashion, the nonthermal component shows asymmetries and fine structure, which are more apparent on the higher-resolution surveys. One of the most striking features of the background radiation from our galaxy is the spur at $l^{II} = 20°$, which can be clearly seen in Fig. 8-51. It is nearly perpendicular to the galactic plane and extends almost to the north and south galactic poles. In longer-wavelength surveys (15 m) the spur appears to blend into a trough near the south galactic pole (Reber, 1964).

Measurements of the polarization of the background radiation show some systematic trends in the direction angle of the linearly polarized component (Muller et al., 1963; Mathewson and Milne, 1964). Mathewson and Milne find that most of the polarized galactic radio emission at 408 Mc is in a band about 50° wide encircling us. The band passes through the galactic poles and crosses the galactic plane at galactic longitudes of $l^{II} = 160$ and 340°. This polarization distribution is presumed due to synchrotron radiation originating within the local arm of our galaxy in the presence of a relatively uniform magnetic field which extends parallel to the arm (direction $l^{II} = 70$ or 250°) and which pervades its interior. In our vicinity the arm is believed to be approximately cylindrical in form, with our position near the center of the arm.

8-10 Extragalactic Radio Sources Many radio sources have been identified with external galaxies. These are sometimes divided broadly into two groups, *normal galaxies*, such as M 31, with radio power outputs in the range of 10^{30} to 10^{32} watts, and *radio galaxies*, such as Cygnus A,

---

† Spectral index of brightness temperature ($T_b \propto \lambda^n$). See Prob. 8-21.

with radio power outputs in the 10^{35} to 10^{38} watt range. The quasi-stellar radio sources, such as 3C 48, with radio power of 10^{37} to 10^{38} watts, may be regarded as a special class of the radio galaxies. A quasi-stellar radio source is sometimes called a *quasar*, *QSO*, *QSS*, or *QSRS*.

The radio and optical power outputs of astronomical objects are summarized in Table 8-8. In going from the exploded-star, or supernova, remnants, which have the greatest power output of any galactic objects, to the external galaxies the optical and radio power outputs increase by factors of 100 to more than a billion. Since a galaxy may have over a billion solar masses, it is not surprising that the total power output of a galaxy exceeds that of a supernova remnant. But to have powers a billion times that of a supernova remnant means that an entire galaxy must be radiating with the effectiveness of a supernova remnant and that a very significant fraction of the total galactic mass is being converted into radiant energy.

Table 8-8
Power output of astronomical objects

Object	Optical power, watts	Radio power, watts
White dwarf star	10^{23}	?
Sun	4×10^{26}	10^{12}
Supergiant star	10^{31}	?
Flare star	10^{25}	10^{16}
Supernova remnant	10^{29}	10^{28}
Normal galaxy (10^{11} solar masses)	10^{37}	10^{30}–10^{32}
Radio galaxy	10^{37}	10^{35}–10^{38}
Quasi-stellar radio source	10^{39}	10^{37}–10^{38}

In Table 8-9 a number of radio sources identified with external galaxies are listed in order of increasing radio power output (or radio luminosity). This arrangement also results in the sources being in order of increasing distance with but few exceptions. The sources listed are typical of most of the extragalactic radio objects, and some of them are discussed in more detail in the following pages. They are taken up in order of increasing radio power output. The power given assumes isotropic radiation by the source.

The *Magellanic clouds*, with radio power outputs of about 10^{30} watts, are the nearest external systems to our own galaxy. These clouds are dwarf irregular systems typical of many external galaxies. Their closeness makes possible detailed studies of many problems of galactic structure and stellar evolution that would be impossible in more distant systems. For example, there are many individual neutral-hydrogen complexes in the large Magellanic cloud (LMC) that correspond to ionized-hydrogen regions having young stellar associations. On the other hand, in the small

Magellanic cloud (SMC) there are only a few neutral-hydrogen clouds, and these are but loosely associated with ionized-hydrogen regions (Robinson, 1963; Bok, et al., 1964).

Table 8-9
Extragalactic radio sources

Object	Log of radio power,[†] watts	Distance,[‡] Mpc	Relativistic distance,[¶] Mpc	Remarks
SMC	29	0.05		Nearby irregular
LMC	30	0.05		Nearby irregular
M 33	31	0.7		Nearby Sc
M 101	31	3.5		Nearby Sc
M 31	32	0.6		Nearby Sb
M 82	33	3		Irregular exploding galaxy
M 77	33	11		Seyfert galaxy
Perseus A (NGC 1275)	35	55		Seyfert galaxy
Virgo A (M 87)	35	11		Elliptical galaxy with nuclear jet
Centaurus A (NGC 5128)	35	5		Double-double source
Fornax A	35	17		
Hydra A	36	160		
Hercules A	37	475	425	
3C 273	37	475	425	Quasi-stellar radio source[§]
3C 47	37	1,300	1,050	Quasi-stellar radio source[§]
Cygnus A	38	170		Prototype double source
3C 48	38	1,100	900	Quasi-stellar radio source[§]
3C 295	38	1,400	1,100	
3C 147	38	1,600	1,200	Quasi-stellar radio source[§]

† Radio power in watts $= 10^N$, where N = number listed in table.

‡ Distance is luminosity, or Hubble distance ($= cz/H_0$, where $H_0 = 100$km sec^{-1} Mpc^{-1}).

¶ The relativistic distance assumes a uniformly expanding universe (see Sec. 8-11). The two distances differ significantly only for large values.

§ The distances given for the quasi-stellar radio sources depend on the assumption that the red shift results from a cosmological velocity of recession. The power depends on this assumption and also on the premise that the radiation from the object is isotropic. The distances and powers must be revised if these assumptions do not hold (see footnote on page 389).

Mathewson, Healey, and Westerlund (1963) and Westerlund and Mathewson (1965) have identified three emission nebulas in the large Magellanic cloud as supernova remnants of type II. The objects are N 49, N 63A and N 132D in Henize's catalog. They have nonthermal radio spectra with the respective indices 1.0, 0.5, and 0.5. Some data on these objects and two type II supernova remnants in our own galaxy (Cassiopeia A and the Cygnus loop) are compared in Table 8-10. The object N 49 is unique in that 60 solar masses appear to have been ejected. If this object had occurred recently in our own galaxy, it would have dwarfed Cassiopeia A as a radio source.

Table 8-10†
Supernova remnants (*See also Table* 8-4)

Object	Distance, kpc	Diameter, pc	Radio power, watts	Ejected mass, solar masses	Age, years
N 49	55	18	1.2×10^{28}	60	3,000
N 63A	55	7	4×10^{27}	2	3,000
N 132D	55	6	1×10^{28}	3	300
Cassiopeia A	3.4	4	3×10^{28}	1	265
Cygnus loop	0.77	40	5.6×10^{25}	6	50,000

† From Westerlund and Mathewson, 1966.

The *Andromeda galaxy* (M 31) is the nearest external system of a form (Sb) and size similar to our own galaxy. It has a radio power output of about 10^{32} watts. At a distance of 2 million light-years, it was first observed with a radio telescope by Brown and Hazard (1951). Radio contours of M 31 measured with the Ohio State University 260-ft radio telescope in the continuum at 1,415 Mc are shown in Fig. 8-52 (Kraus, 1964).

Profiles and also a contour map of a larger region around M 31, as measured with the Ohio State University 260-ft radio telescope at a sensitivity of 0.01°K, are presented in Fig. 8-5 (Kraus and Dixon, 1965). In addition to M 31, near the center of Fig. 8-5, there are several dozen OA (Ohio list A) sources which appear to be grouped in three principal clusters, one about ½ hr preceding M 31, one about ½ hr following M 31, and one to the south. M 31 has a spectral index of about 0.6. This compares with about 0.5 for our own galaxy.

The neutral- (atomic-) hydrogen distribution in M 31 and its radial velocity, as measured by Argyle (1965), are depicted in Fig. 8-53a and b. The southern part of the galaxy is seen to be approaching (negative velocities) and the northern part receding. The plane of the galaxy is tilted about 15° from our line of sight, so that we observe it not quite edge on. Argyle deduced a mass of 1.9×10^9 solar masses of atomic hydrogen, assuming a distance of 500 kpc to M 31. This atomic-hydrogen mass is less than 1 percent of the total mass of the galaxy. This small value is quite typical

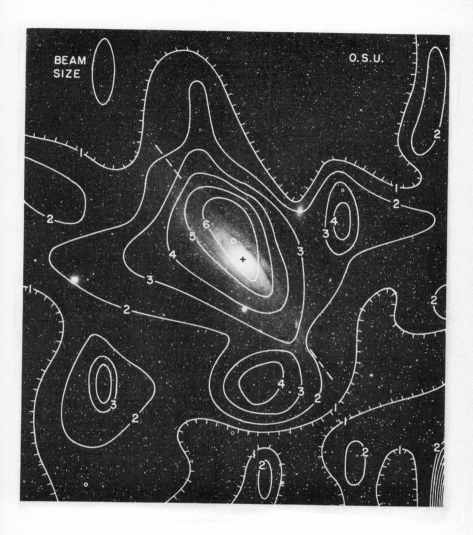

Fig. 8-52. Radio contours of the Andromeda galaxy (M 31), as observed with the Ohio State University 260-ft. radiotelescope at 1415 Mc, superimposed on a Perkins Observatory photograph. The contour interval is 0.05°K antenna temperature. The cross indicates the optical nucleus and the small white circle the approximate radio center.

for spiral galaxies and contrasts with values of 40 percent or more for the Magellanic clouds. The neutral hydrogen of the large and small clouds has been measured as 4×10^8 and 6×10^8 solar masses, respectively (Kerr, Hindman, and Robinson, 1954). These values amount to roughly one-half of the total mass of the respective clouds, indicating that these clouds (or irregular galaxies) are much richer in atomic hydrogen than spiral galaxies

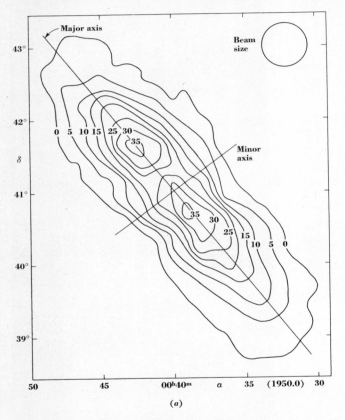

Fig. 8-53a. Distribution of neutral hydrogen in M 31. (*After Argyle,* 1965.)

such as M 31. Details of the neutral-hydrogen distribution near the nucleus of M 31 in 21 km sec⁻¹ velocity bands are shown in Fig. 8-54 (Brundage and Kraus, 1965).

For more distant galaxies, where the size of the galaxy is small compared to the antenna beam size, the mass of atomic hydrogen $M_\odot$, expressed in solar masses, is given by Roberts (1962)

$$M_\odot = 2.36 \times 10^{31} d^2 \int S \, dv \qquad (8\text{-}69)$$

where d = distance of galaxy, Mpc

S = flux density of emission line, watts m^{-2} cps^{-1}

dv = elemental velocity, km sec^{-1}

The integration is equivalent to the area under the emission-line velocity profile for all velocities (limits of integration $-\infty$ to $+\infty$). A small optical depth is assumed. For objects larger than the antenna beam area, as in the

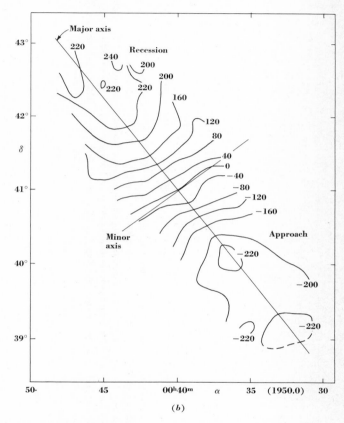

Fig. 8-53b. The mean radial velocity of neutral hydrogen in M 31 (contours in km sec^{-1}). (*After Argyle*, 1965.)

case of Argyle's M 31 observations, the integral in (8-69) is replaced by a summation involving the profiles for the individual elements of an observational grid covering the object.

A comparison of the radio and optical flux densities (radio and optical magnitudes) of a number of objects is presented in Fig. 8-55. In this graph the difference of the optical and radio magnitudes $(m_p - m_r)$ is plotted as a function of the radio magnitude m_r (Brown and Hazard, 1952), where

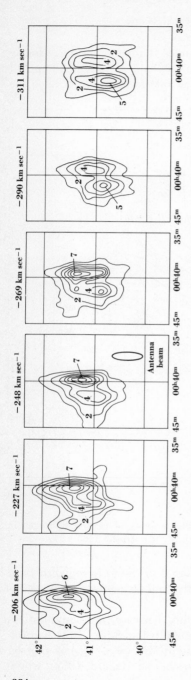

Fig. 8-54. Details of neutral-hydrogen distribution near the nucleus of M 31 (00ʰ40ᵐ; 41°) as revealed by the Ohio State University 260-ft radio telescope. (*Brundage and Kraus*, 1965.) Each map gives the neutral-hydrogen distribution in a 21-km-sec⁻¹ velocity band (100-Kc frequency band) centered on the velocities indicated at the top of each map. The contour interval is 0.5°K antenna temperature. The positions of the peaks are consistent with a ring concentration of neutral hydrogen about 8 kpc in radius lying in the plane of the galaxy with a deficiency of neutral hydrogen inside the ring. A distance of 600 kpc to M 31 is assumed. The coordinates are epoch 1950.0.

$$m_r = -54.4 - 2.5 \log S \qquad (8\text{-}69a)$$

where S = flux density of galaxy at 158 Mc
The Magellanic clouds, M 31, M 33, and other normal galaxies have small, mostly negative magnitude differences. The more powerful radio galaxies, however, have large optical-radio magnitude differences.

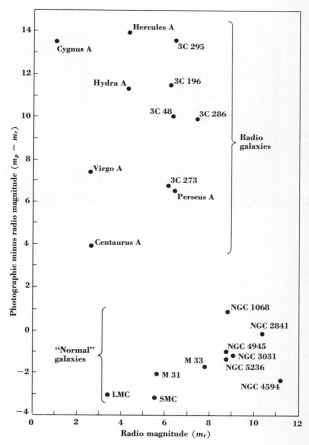

Fig. 8-55. Difference of photographic and radio magnitude versus radio magnitude of galaxies.

The *peculiar galaxy* M 82 has a radio power output of about 10^{33} watts and is at a distance of about 10 million light-years. The galaxy, shown in the photograph of Fig. 8-56, has massive extensions above and below the principal plane. These extensions appear to be expanding from the center with velocities up to 1,000 km sec^{-1}. Evidence points to an explosion in the center of M 82 approximately 1 million years before the present epoch with an expulsion of matter and magnetic field (Lynds and Sandage, 1963). A high degree of polarization of the light radiated in the continuum from the extensions indicates synchrotron emission. The radio power output, or radio luminosity, of M 82 (10^{33} watts) is intermediate between the so-called normal galaxies (10^{31} to 10^{32} watts), such as M 31, and the more powerful radio galaxies (10^{35} to 10^{38} watts).

Fig. 8-56.   Exploding galaxy M 82. (*Mount Wilson and Palomar Observatories photograph.*)

The radio source *Perseus A* (NGC 1275), with a radio power output of 10^{35} watts, is an example of a radio galaxy. Perseus A is located in the Perseus cluster of galaxies, of which NGC 1275 is the brightest member. It is at a distance of about 180 million light-years. The radio structure surrounding Perseus A is complex, and some emission may come from other members of the cluster (Leslie and Elsmore, 1961). The radio source Perseus A is much smaller in size than the optical galaxy. NGC 1275 belongs to a class of galaxies known as Seyfert galaxies, which have small, very bright nuclei with strong, broad optical emission with velocities up to 3,000 km sec^{-1} indicated. Burbidge and Burbidge (1964) suggest that Seyfert galaxies may represent a class of objects undergoing a violent event, such as seems to have occurred in most of the powerful radio galaxies. Possibly the event in the Seyfert galaxy is at an earlier stage, or it is occurring in a different type of galaxy. Only two of the seven galaxies classified as Seyfert type have been found to be radio sources, but Burbidge and Burbidge (1964) feel that the small nuclei and large velocities indicated by the broad emission lines point to a relation between the Seyfert galaxies and the quasi-stellar radio sources rather than to the Cygnus A type of radio galaxy.

Virgo A (M 87) is at a distance of about 35 million light-years and has a radio power output of 10^{35} watts. It is unusual in that the central core of the galaxy, shown in the short-exposure photograph of Fig. 8-57, has a bright, blue jet about 20 sec of arc, or 3,000 light-years, long extending from the nucleus. In long-exposure photographs this feature is completely masked by the outer parts of this giant elliptical galaxy, which has a diameter of about 100,000 light-years. In the radio spectrum a central double source is observed to coincide with the jet. This is surrounded by a larger source of approximately the same extent as the optical galaxy. The jet has a marked degree of linear polarization in the optical spectrum but does not appear to be polarized in the radio spectrum, although the galaxy

Fig. 8-57. Short-exposure photograph showing nucleus of Virgo A with blue jet 3,000 light-years long. (*Official U.S. Navy photograph; courtesy of Dr. K. Aa. Strand; photograph by J. B. Priser with U.S. Navy 61-in. astrometric reflector, Flagstaff, Ariz., with* 103aD + GG14 *plate.*)

as a whole shows a very small degree of polarization (Morris, Radhakrishnan, and Seielstad, 1964). It is thought that the jet consists of matter ejected from the nucleus of the galaxy, with high-energy particles moving in the jet's magnetic field producing the radio emission by the synchrotron mechanism.

Centaurus A is a large, complex radio source with a radio power output of 10^{35} watts. It has a north-south angular extent of about 10° and is double. There is also a radio nucleus, which is itself double and coincides in position with the peculiar elliptical galaxy NGC 5128 at a distance of about 15 million light-years. A composite of the optical and radio appearance is presented in Fig. 8-58. The radio galaxy is seen to be at least 50 times larger than the optical object. In linear dimensions the optical galaxy is about 50,000 light-years in diameter, but the radio object is 2 to 3 million light-years in extent.

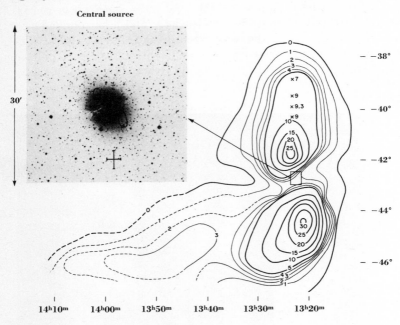

Fig. 8-58. Radio contours and optical appearance of Centaurus A. (*Matthews, Morgan, and Schmidt, 1964; courtesy of Dr. T. A. Matthews; the photograph is by R. Minkowski with the 48-in. Mount Palomar Schmidt telescope on a red plate. The radio-contour map is from Bolton and Clark, 1960; the position and extent of the double inner source is from Maltby, 1961.*)

The optical galaxy is characterized by a dark band across the middle. The galaxy rotates about an axis perpendicular to this band (Burbidge and Burbidge, 1959). Both the inner and outer double radio sources lie roughly along this axis. It is thought that more than one explosion may have occurred in the nucleus of NGC 5128, resulting in the complex radio structure (Matthews, Morgan, and Schmidt, 1964). The outer radio sources show considerable fine structure under higher resolution (Kerr, 1962). Both inner and outer double sources also show marked linear polarization amounting to 13 percent at 3 cm wavelength, 9 percent at 5 cm, and 7 percent at 9 cm, according to Naval Research Laboratory measurements (Mayer et al., 1964; Hollinger et al., 1964).

Fornax A is a two-component radio source with a power output of 10^{35} watts. The optical galaxy NGC 1316 is situated midway between the two sources (Wade, 1961). The galaxy is at a distance of about 55 million light-years. It has prominent optical extensions, which may indicate the ejection of matter in a prior epoch.

Hercules A is a strong radio source which has been identified with a faint optical galaxy at a distance of about 1,500 million light-years.† The Hercules A radio source, like Cygnus A, is double, with two widely separated components. The separation amounts to about 2 min of arc, which corresponds to a distance of 800,000 light-years. The radio power output is about 10^{37} watts.

The *radio source* 3C 273 is an example of a quasi-stellar radio source (Greenstein 1963; Greenstein and Schmidt, 1964). The structure of the radio object, as determined by Hazard, Mackey, and Shimmins (1963), consists of two components separated by about 20 sec of arc. The optical object also has two components, one a 13th magnitude "star" near the center of radio component B and the other a narrow jet about 10 sec of arc long, whose far end is near the center of radio component A. The star has broad emission lines with a red shift of 0.16, which may be interpreted to mean that the object is at a distance of about 1,500 million light-years.‡ At this distance the 20 sec of arc separation of the two radio components amounts to about 150,000 light-years. The radio and optical structure of 3C 273 is depicted in Fig. 8-59, along with the structure of a number of other radio sources, all to approximately the same scale.¶ 3C 273 is one of the radio galaxies in which the radio and optical components coincide in position. Virgo A (M 87) is another example.

If the stellar nucleus of 3C 273 is actually at a distance of some 1,500 million light-years, it cannot be a star in the usual sense but must represent a galaxy of small size or the very bright nucleus of a somewhat larger galaxy, whose full extent is not visible on optical photographs. Much of the radio power output of 10^{37} watts from 3C 273 comes from component A, which coincides with the optical jet. However, the optical power output of 10^{39} watts, corresponding to an absolute photographic magnitude of -26.5, comes almost entirely from the stellar nucleus. This high optical luminosity is about 100 times that of normal galaxies. Radio measurements at several frequencies show distinctly different spectra for the two radio components. Component A, coinciding with the jet, has an index of about 0.9, typical of extragalactic sources and suggesting that the synchrotron mechanism may

† Luminosity, or Hubble distance (see Table 8-9).

‡ This assumes that the red shift results from a cosmological velocity of recession (see Sec. 8-11). Alternative explanations of the red shift might involve a gravitational effect or high collapse velocity in the object.

¶ The source structure in Fig. 8-59 is from data given by Greenstein (1963); Kraus (1964); Matthews, Morgan, and Schmidt (1964); Moffet (1964); and Wade (1961).

be responsible. The spectral index for component B, coinciding with the stellar nucleus, is about 0.0. This is an unusual spectrum for an extragalactic object, and its significance is not yet apparent (Hazard, Mackey, and Shimmins, 1963).

One of the most perplexing problems associated with 3C 273 and other quasi-stellar radio sources is that of very large total energies in a relatively small volume of space. Assuming a lifetime of a million years, a steady light power output of 10^{39} watts would require an energy source in excess

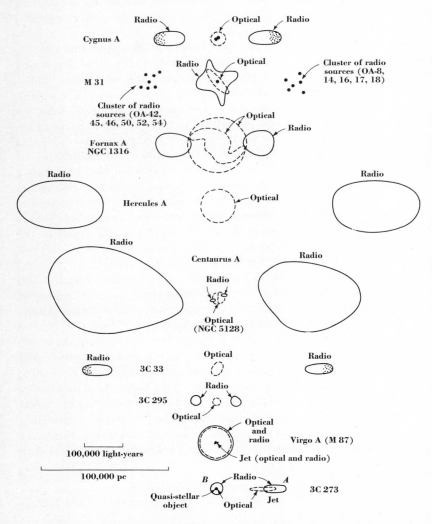

Fig. 8-59. Radio and optical structure of a number of radio sources to approximately the same scale.

of 10^{52} joules.† The conversion of the entire mass of a typical star, such as the sun, into energy at 100 percent efficiency by Einstein's relation ($\mathcal{E} = mc^2$) yields 10^{47} joules. Since the mass-to-energy conversion would not be 100 percent efficient, millions or even billions of solar masses must be involved in nuclear or gravitational mass-energy reactions in the source. One hypothesis (Hoyle and Fowler, 1965), involves the formation and collapse of extremely massive superstars of millions of solar masses with gravitational energy output becoming dominant in the later stages of the collapse as the size of the superstar approaches the Swartzschild radius (about 1 AU for 10^8 solar masses) (Burbidge and Burbidge, 1964).

Measurements of photographic plates back to 1887 show variations in the light output of 3C 273 of the order of 0.5 magnitude with an apparent 10-year periodicity (Smith and Hoffleit, 1963). Also photoelectric measurements during 1963 showed a 0.2-magnitude decrease in a 10-month period (Sandage, 1964). If such short-term light variations are real, it may be inferred that the object involved has dimensions that are much smaller than the distance light would propagate during one cycle of light variation. This would be consistent with the hypothesis that the source involves an ultramassive superstar. An object of ordinary galactic dimensions would seem to be ruled out.

Although the light emission of 3C 273 may come from a small volume, the radio emission would appear to originate over large regions of space, with synchrotron emission from energetic particles moving in magnetic fields being responsible. Recently, however, in a series of measurements over about 3 years, Dent (1965) has detected a 40 percent increase in the radio emission.

Haddock and Sciama (1965) have suggested that if some variable radio sources are detected, it may be possible to measure their distance from the observed difference in the propagation time of the fluctuations at two wavelengths.

Cygnus A is the second strongest radio source (other than the sun) in the sky. The identification of this source in 1954 with a remote galaxy now placed at a distance of about 600 million light-years, provided the first indication that some radio sources were extremely powerful objects. The power is 10^{38} watts in the case of Cygnus A. A photograph of the galaxy is shown in Fig. 8-60. There are two faint condensations about 2 sec of arc apart embedded in a halo of elliptical outline (Baade and Minkowski, 1954). By means of interferometer measurements Jennison and Das Gupta (1953) found that the radio emission does not come from the optical galaxy but from two regions of less than 1 min of arc diameter situated with centers about 1.5 min of arc, or 250,000 light-years, apart. The radio-source posi-

† An energy content of 10^{55} joules is indicated for the strongest radio sources. This value is a consequence of the large energy reservoir needed to produce synchrotron radiation and is independent of the estimated lifetime.

tions are shown by the contours in Fig. 8-60.   Clearly these are far outside the optical galaxy.   Many other radio galaxies are also of this well-separated double type, of which Cygnus A may be regarded as the prototype.   Other examples are Hercules A and 3C 33.   Recent, higher-resolution observations of Cygnus A reveal a distribution which, although basically double, is more complex in its details (Swarup, Thompson, and Bracewell, 1963).

Cygnus A shows a significant degree of linear polarization (about 8 percent at 3 cm) which gives support to the hypothesis that the synchrotron mechanism is responsible for the radio emission.   The degree of linear polarization decreases rapidly with wavelength to about 1.5 percent at 5 cm and even less at longer wavelengths (Hollinger, Mayer, and Mennella, 1964).   If this rapid depolarization is due to differences in the Faraday

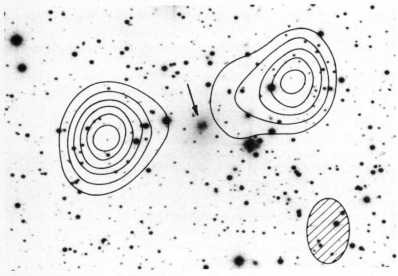

Fig. 8-60.   Cygnus A galaxy (arrow) with map of continuum radio emission at 1400 Mc made with the Cambridge University radio telescope (after Ryle, Elsmore and Neville, 1965).   The antenna beam (cross-hatched) is about 1/2 by 1/3 min of arc.   The contour interval is 52,000°K.   The radio contours are superposed on a photograph by Dr. W. Baade made with the 200-in. Mt. Palomar telescope using a yellow plate (Composite figure courtesy Dr. Alan T. Moffet).

rotation of the emission from the two radio regions or different parts of each region, a $\sin \phi/\phi$ variation in the degree of polarization would be expected, where ϕ is the total rotation of the plane of polarization between the source and the observer (Woltjer, 1962).   Evidence has been obtained for more than one plane of linear polarization in the source (Morris, Radhakrishnan, and Seielstad, 1963).

It is difficult to explain the large power output from Cygnus A coming from the two regions so far outside the optical object.   The evidence for synchrotron emission suggests that the regions contain high-energy particles and magnetic fields. It is thought that these regions may have been ejected from the nucleus of the optical object in an earlier epoch during a titanic explosion. An age of a million years, measured from the explosion of the nucleus, seems indicated. If the radio power of 10^{38} watts is assumed constant for this length of time, an energy of 10^{51} joules is involved. As mentioned in the case of 3C 273, mass-to-energy reactions involving millions of solar masses are required to furnish such enormous energies.

Another quasi-stellar radio source with a larger red shift than 3C 273 is 3C 48. It is placed at a distance of 3,500 million light-years and has a radio power output of 10^{38} watts. Like 3C 273, it has a very faint wisp of nebulosity associated with a bright stellar object in the photographs. It is also like 3C 273 in that photometry of 3C 48 over a year shows the stellar nucleus to be variable by about 0.5 magnitude, while the radio flux appears to be constant (Matthews and Sandage, 1963).

A number of other quasi-stellar radio sources have been discovered including 3C 9, 47, 93, 147, 186, 196, 208, 216, 228, 245, 286, 287, and 345 (Ryle and Sandage, 1964; Sandage and Wyndham, 1965). Of these 3C 147 has the largest red shift measured to date for any object, amounting to 0.545; the corresponding distance is some 5,000 million light-years (Schmidt and Matthews, 1964).† It now appears that a significant fraction of the strongest extragalactic radio objects may be quasi-stellar radio sources, so that such objects should play an important part in studies of the most distant parts of the universe. Their high optical power output also extends the range of optical telescopes for such studies to greater distances than had previously been thought possible. Quasi-stellar radio sources tend to be blue and to show an ultraviolet excess.

It has been reported that some quasi-stellar radio sources fluctuate randomly in flux density at a rapid rate (order of seconds of time) (Hewish, Scott, and Wills, 1964). It is believed that this fluctuation is a scintillation effect resulting from irregularities in the interplanetary medium of the solar system. The effect is observed only for objects, such as some quasi-stellar radio sources, which have very small angular diameters.

The extragalactic radio objects show a wide variety of characteristics. Many, however, provide evidence of cataclysmic events at some phase of their history. Perhaps the different properties we observe are characteristic of different stages in the development of all galaxies. There is some evidence that even so-called normal galaxies, like M 31, may have undergone such events (Kraus, 1964; Kraus and Dixon, 1965). The complex structure of our own galaxy also points to a violent past. On the other hand, we may be dealing with objects which are fundamentally dissimilar.

---

† *Note added in proof:* Red shifts of 2 or greater have been measured recently for some quasi-stellar radio sources, such as 3C 9.

8-11   Cosmological Considerations   Cosmology is the science which treats of the universe as a whole.   The astronomer in dealing with stars or galaxies, has many such objects to study; however, he has only one universe, and only a part of this universe, the part nearest us, is readily accessible for investigation.   For a theory of the universe to be correct it should explain the dynamics of the entire universe, but with only the nearest part within observational range, tests of any theory cannot be complete.   It is with such limitations that a cosmologist must work.

An early problem of cosmologists was that posed by Olbers, who theorized that if stars were uniformly distributed throughout an infinite space the night sky should be uniformly bright, because if we look far enough in any direction we should see the disk of a star.   However, we observe the night sky to be dark, which implies that the universe contains only a finite number of stars.   Clustering of the stars and a red shift of their light are also factors.   The light of distant stars (or integrated light of many stars in distant galaxies) is reddened.   The fainter or more distant the object, the greater the reddening or *red shift*.   This is interpreted as indicating that the objects are receding from us, with the velocity of recession increasing with distance.

The basic evidence for an expanding universe was obtained by Hubble (1937) through measurements of the luminosity and red shift of galaxies. The relation between the luminosity or flux density and the magnitude of a celestial object may be expressed

$$S = K2.512^{-m} \tag{8-70}$$

where S = luminosity or flux density, watts m^{-2} cps^{-1}

$\quad K$ = const

$\quad m$ = magnitude

If the object is at a standard distance, the flux density S_0 is related to the absolute magnitude M by

$$S_0 = K2.512^{-M} \tag{8-71}$$

Assuming that the flux density varies inversely as the square of the distance R, we have from (8-70) and (8-71) that

$$\frac{S_0}{S} = 2.512^{m-M} = \left(\frac{R}{R_0}\right)^2 \tag{8-72}$$

where R = distance of source of flux density S

$\quad R_0$ = standard distance (source flux density = S_0 at this distance)

Knowing S_0 and R_0, the distance R may be obtained by measuring S.   The distance deduced in this way is called a *luminosity distance*.   Hubble found that this distance was related to the measured red shift z by

$$R = \frac{c}{H_0} z \tag{8-73}$$

where R = distance, Mpc

$z = \Delta\lambda/\lambda_0$ = red shift, dimensionless

c = velocity of light ($= 3 \times 10^8$ m sec^{-1})

H_0 = const, m sec^{-1} Mpc^{-1}

$\Delta\lambda$ = wavelength difference, or shift in wavelength

λ_0 = original, or unshifted, wavelength

The correct value for H_0, called *Hubble's constant*, is uncertain. A nominal value in wide present-day usage is 10^5 m sec^{-1} Mpc^{-1} ($= 100$ km sec^{-1} Mpc^{-1}). Values between 180 and 75 km sec^{-1} Mpc^{-1} have been used during the past decade.

Substituting R from (8-73) in (8-72) gives

$$m - M = 5 \log cz - 5 \log H_0 - 5 \log R_0 \qquad (8\text{-}74)$$

where $m - M$ is the modulus of distance.

The red shifts studied by Hubble were relatively small ($z < 0.15$), and the Doppler relation

$$z = \frac{\Delta\lambda}{\lambda_0} = \frac{v}{c} \qquad (8\text{-}75)$$

could be used since v was small compared to the velocity of light. Introducing (8-75) into (8-73) yields

$$R = \frac{v}{H_0} \qquad (8\text{-}76)$$

When v is not small compared to the velocity of light, (8-75) is no longer valid. The relativistic Doppler relation for this case is

$$z = \frac{\Delta\lambda}{\lambda_0} = \frac{1 + v/c}{\sqrt{1 - (v/c)^2}} - 1 \qquad (8\text{-}77)$$

A graph of this relation is presented by the curve in Fig. 8-61. The red shift becomes infinite as v approaches c. When $v \ll c$, (8-77) reduces to (8-75). Using (8-75) for all values of v yields the straight line in Fig. 8-61 with $v = c$ for $z = 1$.

For a uniformly expanding universe the distance-velocity relation of (8-76) is appropriate, with v related to z as in (8-77). For $v = c$ and $H_0 = 100$ km sec^{-1} Mpc^{-1}, the distance $R = 3{,}000$ Mpc (9.8×10^9 light-years). This may be regarded as a *horizon distance*, or radius of the observable universe for the case of uniform expansion. The bottom scales in Fig. 8-61 give the relativistic distance for a uniformly expanding universe in billions of parsecs and light-years when related to z through the upper curve.

Depending on the assumed relation between v or z and R, different distances will be obtained, but, at present, none can be designated as the "correct" distance.   The relations given above by (8-73) and (8-76) are very elementary and do not take into account acceleration, deceleration, or other factors.   For discussions of *world models*, or theories of the universe, that do, the reader is referred to the many excellent texts on cosmology, e.g., McVittie (1956).

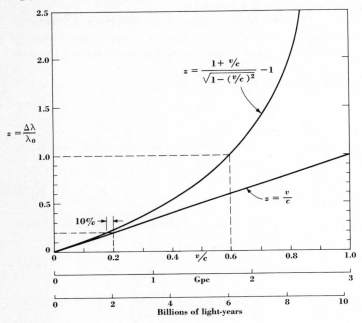

Fig. 8-61.   Red shift z vs. relative velocity of recession for relativistic (upper curve) and nonrelativistic case (straight line).   The lower abscissa scales give the corresponding distance for a uniformly expanding universe in billions of light years and billions of parsecs (Gpc).   A Hubble constant of 100 km sec^{-1} Mpc^{-1} is assumed.

For small red shifts all the expanding-universe models become alike, with distance given by (8-73).   For larger values of z the distance is a function of both z and the model, so that the distance which, in general, is measured along a geodesic curve, loses its simple Euclidean significance. However, the physically observable red shift z is always a significant quantity.   Distances, if used for appreciable red shifts, need to be explicitly labeled with the model involved.

If (8-73) is used for all values of z, instead of being restricted in its use to small values of z, the distance R becomes infinite for infinite red shift. In spite of this, it is often convenient to use this relation, and the distance R so obtained is called the *luminosity or Hubble distance*.   This nonrelativistic distance has the advantage that it is a linear function of z, is independent of any model, and depends numerically only on the value used for H_0. The luminosity distance in billions of parsecs and light years is given by the bottom scales in Fig. 8-61 when related to z by the straight line. For $z = 0.2$ the velocity or distance by the Hubble relation is about 10 percent greater than by the relativistic relation (upper curve), as suggested in the figure.

A method for comparing cosmological theory with observation involves a study of the number of celestial objects as a function of their magnitude or flux density. Assuming that the number N of objects within a radius R is proportional to R^3 (sources uniformly distributed),

$$N = K_1 R^3 \tag{8-78}$$

Assuming further that the flux density S of the sources varies inversely as the square of the distance or

$$S = K_2 R^{-2} \tag{8-79}$$

we have, on eliminating R, that

$$N = K_3 S^{-1.5} \tag{8-80}$$

where K_1, K_2, and K_3 are different constants. Equation (8-80) applies to a static situation. In general we can write that

$$N = K_3 S^{\alpha - 1.5} \tag{8-81}$$

where α is a number-flux index which would be different for different models of the universe. For a static universe with uniformly distributed sources $\alpha = 0$, as in (8-80). Taking this number as N_0, we have, on dividing (8-81) by (8-80), that

$$\frac{N}{N_0} = S^\alpha \tag{8-82}$$

This ratio, or relative source count, as a function of S yields the horizontal straight line in Fig. 8-62 for $\alpha = 0$. If there are more weak sources (or more sources at large distances) than for the $\alpha = 0$ case, a curve with index less than 0 would be obtained, as suggested in Fig. 8-62. On the other hand, if there are fewer weak sources (or fewer sources at large distances) than for the $\alpha = 0$ case, an index greater than 0 would result. The spread between the curves for cases with plus or minus index increases rapidly with decreasing flux density (increasing number) so that with sufficiently sensitive radio telescopes it should be possible to select the theory or theories with indices agreeing most closely with observation. At present, available source counts do not go to sufficiently low flux densities or large numbers to warrant any definite conclusions. Surveys now in progress or ones made with radio telescopes now in design or under construction may prove adequate.

However, there are other complications. For example, if the energy output of a typical galaxy varies with age, a correction for this effect would be required, but it is not known how much this variation might be. Another method for comparing cosmological theory with observation involves a study of the number of celestial objects as a function of their angular diameter. The possibility of measuring very small angular diameters by the method of lunar occultation makes this method one of potential value to radio astronomers (von Hoerner, 1964a, 1964b).

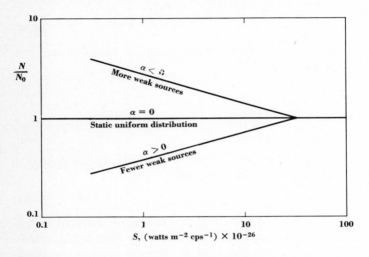

Fig. 8-62. Relative number of radio sources as a function of the flux density for different number-flux indices.

Sandage (1961) has shown that the 200-in. telescope on Mount Palomar is not sensitive enough to be used for deciding between theories or world models on the basis of optical galaxy counts and that other optical tests which might be performed are either very difficult or of marginal significance. If sufficient numbers of quasi-stellar radio sources are discovered, however, this situation may be modified.

In searching for quasi-stellar radio sources, Sandage (1965) has found some objects having the blue starlike optical characteristics of these sources but which have no detectable radio emission. It may be that these blue quasi-stellar objects (or galaxies) and the quasi-stellar radio sources represent the same class of object but at different stages in an evolutionary development, being radio quiet in one phase and radio noisy in another. However, no confirmation of this has been forthcoming.

AAS-NASA Symposium on the Physics of Solar Flares, Goddard Space Flight Center, Greenbelt, Md., Oct. 28–30, 1963.

ALFVÉN, H., and N. HERLOFSON: Cosmic Radiation and Radio Stars, *Phys. Rev.*, vol. 78, p. 616, 1950.

ALLEN, C. W.: Interpretation of Electron Densities from Corona Brightness, *Monthly Notices Roy. Astron. Soc.*, vol. 107, p. 426, 1947.

Annual Reviews of Astronomy and Astrophysics, Annual Reviews, Inc., Palo Alto, California. Yearly review since 1963.

ARGYLE, E.: A Spectrometer Survey of Atomic Hydrogen in the Andromeda Nebula, *Astrophys. J.*, vol. 141, pp. 750–758, 1965.

BAADE, W., and R. MINKOWSKI: Identification of the Radio Sources in Cassiopeia, Cygnus A, and Puppis A, *Astrophys. J.*, vol. 119, pp. 206–214, January, 1954.

BARRETT, A. H.: Microwave Absorption and Emission in the Atmosphere of Venus, *Astrophys. J.*, vol. 133, pp. 281–293, January, 1961.

BARRETT, A. H.: The Detection of the OH and Other Molecular Lines in the Radio Spectrum of the Interstellar Medium, *IEEE Trans. Military Electron.*, vol. MIL-8, pp. 156–164, July–October, 1964.

BENNETT, A. S.: The Revised 3C Catalogue of Radio Sources, *Mem. Roy. Astron. Soc.*, vol. 68, pp. 163–172, 1962.

BIGG, E. K.: Influence of the Satellite Io on Jupiter's Decametric Radiation, *Nature*, vol. 203, pp. 1008–1010, Sept. 5, 1964.

BLAAUW, A., and M. SCHMIDT (eds.): "Galactic Structure," vol. 5 of Stars and Stellar Systems, G. P. Kuiper and B. M. Middlehurst (eds.), University of Chicago Press, 1965.

BOISCHOT, A.: Les Émissions de type IV, in R. N. Bracewell (ed.), "Paris Symposium on Radio Astronomy," Stanford University Press, Stanford, Calif., 1959.

BOK, B. J., H. GOLLNOW, J. V. HINDMAN, and M. MOWAT: Radial Velocities Associated with Selected Emission Nebulae in the Small Magellanic Cloud, *Australian J. Phys.*, vol. 17, pp. 404–408, September, 1964.

BOLTON, J. G.: Discrete Sources of Galactic Radio-frequency Noise, *Nature*, vol. 162, p. 141, 1948.

BOLTON, J. G., and B. G. CLARK: A Study of Centaurus A at 31 cm, *Publ. Astron. Soc. Pacific*, vol. 72, pp. 29–35, 1960.

BRACEWELL, R. N.: The Sunspot Number Series, *Nature*, vol. 171, pp. 649–651, Apr. 11, 1953.

BROWN, R. H., and C. HAZARD: Radio Emission from the Andromeda Nebula, *Monthly Notices Roy. Astron. Soc.*, vol. 3, pp. 357–367, 1951.

BROWN, R. H., and C. HAZARD: Extra-galactic Radio-frequency Radiation, *Phil. Mag.*, vol. 43, pp. 137–152, 1952.

BRUNDAGE, W. D., and J. D. KRAUS: Preliminary Results of a Hydrogen-line Survey of M31 with the O.S.U. Radio Telescope, *Astron. J.*, vol. 70, p. 669, November, 1965.

BURBIDGE, E. M., and G. R. BURBIDGE: Rotation and Internal Motions in NGC 5128, *Astrophys. J.*, vol. 129, pp. 271–281, March, 1959.

BURBIDGE, E. M., and G. R. BURBIDGE: Theories of the Origin of Radio Sources, *IEEE Trans. Military Electron.*, vol. MIL-8, pp. 165–172, July–October, 1964.

BURKE, B. F., and K. L. FRANKLIN: Observations of a Variable Radio Source Associated with the Planet Jupiter, *J. Geophys. Res.*, vol. 60, pp. 213–217, June, 1955.

CARR, T. D., A. G. SMITH, H. BOLLHAGEN, N. F. SIX, and N. E. CHATTERTON: Recent Decameter Wavelength Observations of Jupiter, Saturn, and Venus, *Astrophys. J.*, vol. 134, pp. 105–125, July, 1961.

CARR, T. D., G. W. BROWN, A. G. SMITH, C. S. HIGGINS, H. BOLLHAGEN, J. MAY, and J. LEVY: Spectral Distribution of the Decametric Radiation from Jupiter in 1961, *Astrophys. J.*, vol. 140, pp. 778–795, Aug. 15, 1964.

CHRISTIANSEN, W. N., D. S. MATHEWSON, and J. L. PAWSEY: Radio Pictures of the Sun, *Nature*, vol. 180, pp. 944–945, Nov. 9, 1957.

CHRISTIANSEN, W. N., and D. S. MATHEWSON: Scanning the Sun with a Highly Directional Array, *Proc. IRE*, vol. 46, pp. 127–131, January, 1958.

CLARK, B. G.: An Interferometer Investigation of the 21 cm Hydrogen Line Absorption, *Calif. Inst. Tech. Radio Obs. Rept.* 6, 1965.

CLEETON, C. E., and N. H. WILLIAMS: Electromagnetic Waves of 1.1 cm Wavelength and the Absorption Spectrum of Ammonia, *Phys. Rev.*, vol. 45, pp. 234–237, Feb. 15, 1934.

COATES, R. J.: Lunar Brightness Variations with Phase at 4.3 mm Wavelength, *Astrophys. J.*, vol. 133, pp. 723–725, March, 1961.

CONWAY, R. G., K. I. KELLERMANN, and R. J. LONG: The Radio Frequency Spectra of Discrete Radio Sources, *Monthly Notices Roy. Astron. Soc.*, vol. 125, pp. 261–284, 1963.

COOK, J. J., L. G. CROSS, M. E. BAIR, and C. B. ARNOLD: Radio Detection of the Planet Saturn, *Nature*, vol. 188, pp. 393–394, October, 1960.

COVINGTON, A. E.: Solar Radio Astronomy, *J. Roy. Astron. Soc. Can.*, vol. 51, pp. 298–307, October, 1957.

DENISSE, J. F.: Report of the URSI 12th General Assembly, Boulder, Colo., p. 127, 1957.

DENT, W. A., and F. T. HADDOCK: A New Class of Radio Spectra, *Astron. J.*, vol. 70, pp. 136–137, March, 1965.

DENT, W. A.: Quasi-stellar Sources: Variation in the Radio Emission of 3C 273, *Science*, vol. 148, pp. 1458–1460, 1965.

DICKE, R. H., and R. BERINGER: Microwave Radiation from the Sun and Moon, *Astrophys. J.*, vol. 103, p. 375, 1946.

DIXON, R. S., S. Y. MENG, and J. D. KRAUS: Maps of the Perseus Region at 600 and 1415 Megacycles per Second, *Nature*, vol. 205, pp. 755–758, Feb. 20, 1965.

DODSON, H. W.: Studies at the McMath-Hulbert Observatory of Radio Frequency Radiation at the Time of Solar Flares, *Proc. IRE*, vol. 46, pp. 149–169, January, 1958.

DOUGLAS, J. N., and H. J. SMITH: Decametric Radiation from Jupiter, *Astron. J.*, vol. 68, pp. 163–180, April, 1963*a*.

DOUGLAS, J. N., and H. J. SMITH: Change in Rotation Period of Jupiter's Decameter Radio Source, *Nature*, vol. 199, pp. 1080–1081, September, 1963*b*.

DOUGLAS, J. N.: Decameter Radiation from Jupiter, *IEEE Trans. Military Electron.*, vol. MIL-8, pp. 173–187, July–October, 1964.

DRAKE, F. D., and H. I. EWEN: A Broad-band Microwave Source Comparison Radiometer for Advanced Research in Radio Astronomy, *Proc. IRE*, vol. 46, pp. 53–60, January, 1958.

DRAKE, F. D., and H. HVATUM: Non-thermal Microwave Radiation from Jupiter, *Astron. J.*, vol. 64, pp. 329–330, October, 1959.

DRAKE, F. D.: 10 cm Observations of Venus in 1961, *Publ. Natl. Radio Astron. Obs.*, vol. 1, no. 11, pp. 165–178, February, 1962.

DRÖGE, F., and W. PRIESTER: Durchmusterung der allgemeinen Radiofrequenz-Strahlung bei 200 Mhz, *Z. Astrophys.*, vol. 40, pp. 236–248, 1956.

ELDER, F. R., R. V. LANGMUIR, and H. C. POLLACK: Radiation from Electrons Accelerated in a Synchrotron, *Phys. Rev.*, vol. 74, pp. 52–56, 1948.

ELLIS, G. R. A.: Radiation from Jupiter at 4.8 Mc, *Nature*, vol. 194, pp. 667–668, May, 1962.

ELSMORE, B.: Radio Observations of the Lunar Atmosphere, *Phil. Mag.*, vol. 2, pp. 1040–1046, August, 1957.

EPSTEIN, E. E.: Anomalous Continuum Radiation from Jupiter, *Nature*, vol. 184, p. 52, July, 1959.

EVANS, J. V.: chap. 12 in Z. Kopal (ed.), "Physics and Astronomy of the Moon," Academic Press Inc., New York, 1961.

EWEN, H. I., and E. M. PURCELL: Radiation from Galactic Hydrogen at 1,420 Mc/s, *Nature*, vol. 168, pp. 356–357, Sept. 1, 1951.

FINDLAY, J. W., H. HVATUM, and W. B. WALTMAN: An Absolute Flux-density Measurement of Cassiopeia A at 1,440 Mhz, *Astrophys. J.*, vol. 141, pp. 873–884, Apr. 1, 1965.

FLÜGGE, S. (ed.): Stellar Systems, "Handbuch der Physik," vol. 53, Springer-Verlag OHG, Berlin, 1959.

GALLET, R. M.: Radio Observations of Jupiter, chap. 14, in G. P. Kuiper and B. M. Middlehurst (eds.), "Planets and Satellites," University of Chicago Press, Chicago, 1961.

GARDNER, F. F., B. J. ROBINSON, J. G. BOLTON, and K. J. VAN DAMME: Detection of the Interstellar OH Lines at 1,612 and 1,720 Mc/sec, *Phys. Rev. Letters*, vol. 13, pp. 3–5, July, 1964.

GIBSON, J. E.: Lunar Thermal Radiation at 35 KMC, *Proc. IRE*, vol. 46, pp. 280–286, January, 1958.

GIBSON, J. E., and R. J. McEWAN: Observations of Venus at 8.6 mm Wavelength, in R. N. Bracewell (ed.), "Paris Symposium on Radio Astronomy," pp. 50–52, Stanford University Press, Stanford, Calif., 1959.

GIORDMAINE, J. A., L. E. ALSOP, C. H. TOWNES, and C. H. MAYER: Observations of Jupiter and Mars at 3 cm Wavelength, *Astron. J.*, vol. 64, pp. 332–333, October, 1959.

GOLDSTEIN, R. M., and R. L. CARPENTER: Rotation of Venus: Period Estimated from Radar Measurements, *Science*, vol. 139, pp. 910–911, March, 1963.

GOLDSTEIN, R. M.: Radar Investigations of the Planets, *IEEE Trans. Military Electron.*, vol. MIL-8, pp. 199–206, July-October, 1964.

GREENSTEIN, J. L., and R. MINKOWSKI: The Crab Nebula as a Radio Source, *Astrophys. J.*, vol. 118, pp. 1–15, July, 1953.

GREENSTEIN, J. L.: Quasi-stellar Radio Sources, *Sci. Am.*, vol. 209, pp. 54–62, December, 1963.

GREENSTEIN, J. L., and M. SCHMIDT: The Quasi-stellar Radio Sources 3C 48 and 3C 273, *Astrophys. J.*, vol. 140, pp. 1–34, July 1, 1964.

HADDOCK, F. T., C. H. MAYER, and R. M. SLOANAKER: Radio Emission from the Orion Nebula and Other Sources at 9.4 cm, *Astrophys. J.*, vol. 199, pp. 456–459, March, 1954.

HADDOCK, F. T.: Introduction to Radio Astronomy, *Proc. IRE*, vol. 46, pp. 3–12, January, 1958.

HADDOCK, F. T., and D. W. SCIAMA: A Proposal for the Detection of Dispersion in Radio-wave Propagation through Intergalactic Space, *Phys. Rev. Letters*, vol. 14, p. 1007, June 21, 1965.

HAGEN, J. P., and E. F. McCLAIN: Galactic Absorption of Radio Waves, *Astrophys. J.*, vol. 120, pp. 368–370, Sept., 1954.

HAGEN, J. P., A. E. LILLEY, and E. F. McCLAIN: Absorption of 21-cm Radiation by Interstellar Hydrogen, *Astrophys. J.*, vol. 122, pp. 361–375, November, 1955.

HAZARD, C., M. B. MACKEY, and A. J. SHIMMINS: Investigation of the Radio Source 3C 273 by the Method of Lunar Occultations, *Nature*, vol. 197, pp. 1037–1039, Mar. 16, 1963.

HECKMANN, O., and E. SCHUCKING: Newtonsche and Einsteinsche Kosmologie, in S. Flügge (ed.), "Handbuch der Physik," vol. 53, pp. 489–519, Springer-Verlag OHG, Berlin, 1959.

HECKMANN, O., and E. SCHUCKING: Andere kosmologische Theorien, in S. Flügge (ed.), "Handbuch der Physik," vol. 53, pp. 520–537, Springer-Verlag OHG, Berlin, 1959.

HEESCHEN, D. S.: A Color Absolute Magnitude Diagram for Extra Galactic Radio Sources, *Publ. Astron. Soc. Pacific*, vol. 72, pp. 368–376, October, 1960.

HEESCHEN, D. S.: Radio Galaxies, *Sci. Am.*, vol. 206, pp. 41–49, March, 1962.

HEWISH, A.: A Report on the URSI 12th General Assembly, Boulder, Colo., p. 125, 1957.

HEWISH, A., P. F. SCOTT, and D. WILLS: Interplanetary Scintillation of Small Diameter Radio Sources, *Nature*, vol. 203, pp. 1214–1217, Sept. 19, 1964.

HEY, J. S.: Solar Radiations in the 4 to 6 Meter Radio Wavelength Band, *Nature*, vol. 157, p. 47, 1946.

HOERNER, S. VON: Requirements for Cosmological Studies in Radio Astronomy, *IEEE Trans. Military Electron.*, vol. MIL-8, pp. 282–288, July–October, 1964a.

HOERNER, S. VON: Lunar Occultations of Radio Sources, *Astrophys. J.*, vol. 140, pp. 65–79, July 1, 1964b.

HÖGBOM, J. A., and J. R. SHAKESHAFT: Secular Variation of the Flux Density of the Radio Source Cassiopeia A, *Nature*, vol. 189, pp. 561–562, Feb. 18, 1961.

HOGLUND, B., and P. G. MEZGER: The Detection of the Hydrogen Emission Line n_{110} to n_{109} at the Frequency 5,009 Mhz in Galactic HII Regions, *Science*, vol. 150, pp. 339–348, 1965.

HOLLINGER, J. P., C. H. MAYER, and R. A. MENNELLA: Polarization of Cygnus A and Other Sources at 5 cm, *Astrophys. J.*, vol. 140, pp. 656–665, Aug. 15, 1964.

HOWARD, W. E., A. H. BARRETT, and F. T. HADDOCK: Measurements of Microwave Radiation from the Planet Mercury, *Astrophys. J.*, vol. 136, pp. 995–1004, 1962.

HOYLE, F., and W. A. FOWLER: "Quasi-stellar Sources and Gravitational Collapse," University of Chicago Press, Chicago, 1965.

HUBBLE, E. P.: "The Observational Approach to Cosmology," Clarendon Press, Oxford, 1937.

IAU: *Intern. Astron. Union Inform. Bull.* 8, p. 4, March, 1962.

JAMES, J. C.: Radar Echoes from the Sun, *IEEE Trans. Military Electron.*, vol. MIL-8, pp. 210–225, July–October, 1964.

JENNISON, R. C., and M. K. DAS GUPTA: Fine Structure of the Extra-terrestrial Radio Source Cygnus I, *Nature*, vol. 172, p. 996, Nov. 28, 1953.

KARDASHEV, N. S.: On the Possibility of Detection of Allowed Lines of Atomic Hydrogen in the Radio-frequency Spectrum, *Soviet Astron. A.J. English Transl.*, vol. 3, pp. 813–820, 1959–1960.

KELLERMANN, K. I.: The Spectra of Non-thermal Radio Sources, *Astrophys. J.*, vol. 140, pp. 969–991, Oct. 1, 1964.

KERR, F. J., J. V. HINDMAN, and B. J. ROBINSON: Observations of the 21 cm Line from the Magellanic Clouds, *Australian J. Phys.*, vol. 7, pp. 297–314, 1954.

KERR, F. J.: 210-foot Radio Telescope's First Results, *Sky & Telescope*, vol. 24, pp. 254–260, November, 1962.

KERR, F. J., and G. WESTERHOUT: Distribution of Hydrogen in the Galaxy, in A. Blaauw and M. Schmidt (eds.), "Galactic Structure," vol. 5, chap. 8, University of Chicago Press, Chicago, 1964.

KIEPENHEUER, K. O.: Cosmic Rays as the Source of the General Galactic Radio Emission, *Phys. Rev.*, vol. 79, pp. 738–739, 1950.

KIEPENHEUER, K. O.: Solar Activity, in G. P. Kuiper (ed.), "The Sun," pp. 322–465, University of Chicago Press, 1953.

Ko, H. C., and J. D. KRAUS: Radio-frequency Radiation from the Rosette Nebula, *Nature*, vol. 176, p. 221, July 30, 1955.

Ko, H. C., and J. D. KRAUS: A Radio Map of the Sky at 1.2 Meters, *Sky & Telescope*, vol. 16, pp. 160–161, February, 1957.

KRAUS, J. D.: Some Observations of the Impulsive Radio Signals from Jupiter, *Astron. J.*, vol. 61, pp. 182–183, May, 1956.

KRAUS, J. D.: Planetary and Solar Radio Emission at 11 Meters Wavelength, *Proc. IRE*, vol. 46, pp. 266–274, January, 1958.

KRAUS, J. D., R. C. HIGGY, and W. R. CRONE: The Satellite Ionization Phenomenon, *Proc. IRE*, vol. 48, pp. 672–678, April, 1960.

KRAUS, J. D., and M. E. TIURI: Observations of Satellite-related Ionization Effects between 1958 and 1960, *Proc. IRE*, vol. 50, pp. 2076–2081, October, 1962.

KRAUS, J. D.: Maps of M 31 and Surroundings at 600 and 1415 Megacycles per Second, *Nature*, vol. 202, pp. 269–272, Apr. 18, 1964.

KRAUS, J. D., and R. S. DIXON: A High Sensitivity Survey of M 31 and Surroundings at 1,415 Mc/sec, *Nature*, vol. 207, pp. 587–589, Aug. 7, 1965.

KRAUS, J. D., R. S. DIXON, and R. O. FISHER: A New High Sensitivity Study of the M 31 Region at 1,415 Mc, *Astrophys. J.*, vol. 144, May, 1966.

KUIPER, G. P.: "The Atmospheres of the Earth and Planets," University of Chicago Press, Chicago, 1952.

KUIPER, G. P., and B. M. MIDDLEHURST (eds.): "Planets and Satellites," University of Chicago Press, Chicago, 1961.

KUNDU, M. R.: Solar Radio Astronomy, *Univ. Mich. Radio Astron. Obs., Rept.* 64-4, March, 1964.

LESLIE, P. R. R., and B. ELSMORE: Radio Emission from the Perseus Cluster, *Observatory*, vol. 81, pp. 14–16, February, 1961.

LODGE, O. J.: "Signalling across Space without Wires," "The Electrician" Printing and Publishing Company, Limited, London, 1900.

LOVELL, B., F. L. WHIPPLE, and L. H. SOLOMON: Radio Emission from Flare Stars, *Nature*, vol. 198, pp. 228–230, April, 1963.

LOVELL, B.: Radio Emitting Flare Stars, *Sci. Am.*, vol. 211, pp. 13–19, August, 1964.

LYNDS, C. R., and A. R. SANDAGE: Evidence for an Explosion in the Center of the Galaxy M 82, *Astrophys. J.*, vol. 137, pp. 1005–1021, May 15, 1963.

MACHIN, K. E., and F. G. SMITH: The Occultation of a Radio Star by the Solar Corona, *Nature*, vol. 170, p. 319, 1952.

MALTBY, P.: Central Component of the Radio Source Centaurus A, *Nature*, vol. 191, pp. 793–794, 1961.

MARAN, S. P., and A. G. W. CAMERON (eds.): "Physics of Non-thermal Radio Sources," Goddard Institute for Space Studies, NASA, June, 1964.

MATTHEWS, T. A., and A. R. SANDAGE: Optical Identification of 3C 48, 3C 196, 3C 286 with Stellar Objects, *Astrophys. J.*, vol. 138, pp. 30–56, July 1, 1963.

MATTHEWS, T. A., W. W. MORGAN, and M. SCHMIDT: A Discussion of Galaxies Identified with Radio Sources, *Astrophys. J.*, vol. 140, pp. 35–49, July 1, 1964.

MATHEWSON, D. S., J. R. HEALEY, and B. E. WESTERLUND: A Supernova Remnant in the Large Magellanic Cloud, *Nature*, vol. 199, p. 681, Aug. 17, 1963.

MATHEWSON, D. S., and D. K. MILNE: A Pattern in the Large-scale Distribution of Galactic Polarized Radio Emission, *Nature*, vol. 203, pp. 1273–1274, Sept. 19, 1964.

MAYER, C. H., T. P. McCULLOUGH, and R. M. SLOANAKER: Evidence for Polarized Radio Radiation from the Crab Nebula, *Astrophys. J.*, vol. 126, pp. 468–470, September, 1957.

MAYER, C. H., T. P. McCULLOUGH, and R. M. SLOANAKER: Observations of Venus at 3.15 cm Wavelength, *Astrophys. J.*, vol. 127, pp. 1–10, January, 1958a.

MAYER, C. H., T. P. McCULLOUGH, and R. M. SLOANAKER: Observations of Mars and Jupiter at a Wavelength of 3.15 cm, *Astrophys. J.*, vol. 127, pp. 11–16, January, 1958b.

MAYER, C. H., T. P. McCULLOUGH, and R. M. SLOANAKER: Chap. 12 of G. P. Kuiper and B. M. Middlehurst (eds.), "Planets and Satellites," University of Chicago Press, Chicago, 1961.

MAYER, C. H., T. P. McCULLOUGH, and R. M. SLOANAKER: 3.15 cm Observations of Venus in 1961, *Mem. Soc. Roy. Sci. Liege*, vol. 7, pp. 357–363, 1963.

MAYER, C. H., T. P. McCULLOUGH, and R. M. SLOANAKER: Linear Polarization of the Centimeter Radiation of Discrete Sources, *Astrophys. J.*, vol. 139, pp. 248–268, Jan. 1, 1964.

MAYER, C. H.: Thermal Radio Radiation from the Moon and Planets, *IEEE Trans. Military Electron.*, vol. MIL-8, pp. 236–247, July–October, 1964.

McCLAIN, E. F.: A Test for Non-thermal Radiation from Jupiter at a Wavelength of 21 cm, *Astron. J.*, vol. 64, pp. 339–340, 1959.

McCLAIN, E. F., J. H. NICHOLS, and J. A. WAAK: Investigation of Variations in the Decimeter-wave Emission from Jupiter, *Astron. J.*, vol. 67, pp. 724–727, December, 1962.

McVITTIE, G. C.: "General Relativity and Cosmology," Chapman & Hall, Ltd., London, 1956.

McVITTIE, G. C.: Distance and Time in Cosmology: The Observational Data, in S. Flügge (ed.), "Handbuch der Physik," vol. 53, pp. 445–488, Springer-Verlag OHG, Berlin, 1959.

MENON, T. K.: A Study of the Rosette Nebula, *Astrophys. J.*, vol. 135, pp. 394–407, March, 1962.

MENON, T. K.: Thermal Galactic Sources, *IEEE Trans. Military Electron.*, vol. MIL-8, pp. 247–252, July–October, 1964.

MINKOWSKI, R.: Optical Observations of Non-thermal Galactic Radio Sources, in R. N. Bracewell (ed.), "Paris Symposium on Radio Astronomy," Stanford University Press, Stanford, Calif., 1959.

MOFFET, A. T.: Component Shapes in Double Radio Galaxies, *Owens Valley Radio Obs., Rept.* 8, 1964.

MORRIS, D., and G. L. BERGE: Measurements of the Polarization and Extent of the Decimeter Radiation from Jupiter, *Calif. Inst. Tech. Radio Obs. Pub.* 7, 1961.

MORRIS, D., V. RADHAKRISHNAN, and G. A. SEIELSTAD: Preliminary Measurements on the Distribution of Linear Polarization over Eight Radio Sources, *Astrophys. J.*, vol. 139, pp. 560–569, Feb. 15, 1964.

MULLER, C. A., and J. H. OORT: The Interstellar Hydrogen Line at 1,420 Mc/sec and an Estimate of Galactic Rotation, *Nature*, vol. 168, pp. 357–358, 1951.

MULLER, C. A., 21-cm Absorption Effects in the Spectra of Two Strong Radio Sources, *Astrophys. J.*, vol. 125, pp. 830–834, 1957.

MULLER, C. A., E. M. BERKHUIJSEN, W. N. BROUW, and J. TINBERGEN: Galactic Background Polarization at 610 Mc/s, *Nature*, vol. 200, pp. 155–156, Oct. 12, 1963.

OORT, J. H., F. J. KERR, and G. WESTERHOUT: The Galactic System as a Spiral Nebula, *Monthly Notices Roy. Astron. Soc.*, vol. 118, pp. 379–389, 1958.

OORT, J. H., and T. WALRAVEN: Polarization and Composition of the Crab Nebula, *Bull. Astron. Inst. Neth.*, vol. 12, no. 462, pp. 285–308, May 5, 1956.

PETTIT, E., and S. B. NICHOLSON: Lunar Radiation and Temperatures, *Astrophys. J.*, vol. 71, pp. 102–135, March, 1930.

PETTIT, E.: Lunar Radiation as Related to Phase, *Astrophys. J.*, vol. 81, pp. 17–36, January, 1935.

PETTIT, E.: Radiation Measurements on the Eclipsed Moon, *Astrophys. J.*, vol. 91, pp. 408–420, May, 1940.

PIDDINGTON, J. H., and H. C. MINNETT: Microwave Thermal Radiation from the Moon, *Australian J. Sci. Res.*, vol. A2, p. 63, March, 1949.

PURCELL, E. M., and H. I. EWEN: Radiation from Galactic Hydrogen at 1,420 Mc/sec, *Nature*, vol. 168, p. 356, 1951.

RADHAKRISHNAN, V., and J. A. ROBERTS: Polarization and Angular Extent of the 960 Mc/sec Radiation from Jupiter, *Phys. Rev. Letters*, vol. 4, pp. 493–494, May, 1961.

REBER, G.: Cosmic Static, *Astrophys. J.*, vol. 100, pp. 279–287, November, 1944.

REBER, G.: Hectometer Cosmic Static, *IEEE Trans. Military Electron.*, vol. MIL-8, pp. 257–263, July–October, 1964.

RIIHIMAA, J. J.: High-resolution Spectral Observations of Jupiter's Decametric Radio Emission, *Nature*, vol. 202, June, 1964.

ROBERTS, J. A., and G. J. STANLEY: Radio Emission from Jupiter at a Wavelength of 31 cm, *Publ. Astron. Soc. Pacific*, vol. 71, pp. 485–496, December, 1959.

ROBERTS, M. S.: The Neutral Hydrogen Content of Late-type Spiral Galaxies, *Astrophys. J.*, vol. 67, pp. 437–446, September, 1962.

ROBERTS, W. O.: Corpuscles from the Sun, *Sci. Am.*, vol. 192, pp. 40–45, February, 1955.

ROBINSON, B. J.: The Galaxy and the Magellanic Clouds, *Nature*, vol. 199, pp. 322–325, July 27, 1963.

ROBINSON, B. J., F. F. GARDNER, K. J. VAN DAMME, and J. G. BOLTON: An Intense Concentration of OH near the Galactic Center, *Nature*, vol. 202, pp. 989–991, 1964.

ROSE, W. K., J. M. BOLOGNA, and R. M. SLOANAKER: Linear Polarization of the 3,200 Mc/sec Radiation from Saturn, *Phys. Rev. Letters*, vol. 10, pp. 123–125, February, 1963.

ROUGOOR, G. W., and J. H. OORT: *Proc. Natl. Acad. Sci. U.S.*, vol. 46, 1960.

RYLE, M.: Radio Astronomy and Cosmology, *Nature*, vol. 190, pp. 852–854, June 3, 1961.

RYLE, M.: The New Cambridge Radio Telescope, *Nature*, vol. 194, pp. 517–518, May 12, 1962.

RYLE, M., and A. SANDAGE: The Optical Identification of Three New Radio Objects of the 3C 48 Class, *Astrophys. J.*, vol. 139, pp. 419–421, Jan. 1, 1964.

RYLE, M., B. ELSMORE, and A. C. NEVILLE: High Resolution Observations of the Radio Sources in Cygnus and Cassiopeia, *Nature*, vol. 205, pp. 1259–1262, 1965.

SALOMONOVICH, A. E., and B. YA. LOSOVSKII: Radio Brightness Distribution on the Lunar Disc at 0.8 cm, *Soviet Astron. AJ English Transl.*, vol. 6, pp. 833–839, May–June, 1963.

SANDAGE, A.: The Ability of the 200-inch Telescope to Discriminate between Selected World Models, *Astrophys. J.*, vol. 133, pp. 355–392, March, 1961.

SANDAGE, A.: Intensity Variations of 3C 48, 3C 196, and 3C 273 in Optical Wavelengths, *Astrophys. J.*, vol. 139, pp. 416–419, Jan. 1, 1964.

SANDAGE, A., and J. D. WYNDHAM: On the Optical Identification of Eleven New Quasi-stellar Radio Sources, *Astrophys. J.*, vol. 141, pp. 328–331, Jan. 1, 1965.

SANDAGE, A.: The Existence of a Major New Constituent of the Universe: the Quasi-stellar Galaxies, *Astrophys. J.*, vol. 141, pp. 1560–1578, 1965.

SCHMIDT, M., and T. A. MATTHEWS: Redshifts of the Quasi-stellar Radio Sources 3C 47 and 3C 147, *Astrophys. J.*, vol. 139, pp. 781–785, Feb. 15, 1964.

SCHOTT, E.: 175-Mc Solar Radiation, *Phys. Blätter*, vol. 3, pp. 159–160, 1947.

SCHOTT, G. A.: Electromagnetic Radiation and the Mechanical Reactions Arising from It, Cambridge University Press, Cambridge, 1912.

SCHWINGER, J.: On the Classical Radiation from Accelerated Electrons, *Phys. Rev.*, vol. 75, pp. 1912–1925, 1949.

SCOTT, P. F., M. RYLE, and A. HEWISH: First Results of Radio Star Observations Using the Method of Aperture Synthesis, *Monthly Notices Roy. Astron. Soc.*, vol 122, pp. 95–111, 1961.

SEEGER, C. L., G. WESTERHOUT, and R. G. CONWAY: Observations of Discrete Sources, the Coma Cluster, the Moon, and the Andromeda Nebula at a Wavelength of 75 cm, *Astrophys. J.*, vol. 126, pp. 585–587, November, 1957.

SHAIN, C. A.: 18.3 Mc/s Radiation from Jupiter, *Australian J. Phys.*, vol. 9, pp. 61–73, March, 1956.

SHAIN, C. A.: Galactic Absorption of 19.7 Mc/s Radiation, *Australian J. Phys.*, vol. 10, pp. 195–203, 1957.

SHKLOVSKY, I. S.: "Cosmic Radio Waves," translated by R. B. Rodman and C. M Varsavsky, Harvard University Press, Cambridge, Mass., 1960.

SIMON, M.: Solar Observations at 3.2 mm, *Astrophys. J.*, vol. 141, pp. 1513–1522, 1965.

SLEE, O. B.: Observations of the Solar Corona Out to 100 Solar Radii, *Monthly Notices Roy. Astron. Soc.*, vol. 123, pp. 223–231, 1961.

SLEE, O. B., and C. S. HIGGINS: Long Baseline Interferometry of Jovian Radio Bursts, *Nature*, vol. 197, pp. 781–783, February, 1963.

SLEE, O. B., L. H. SOLOMON, and G. E. PATSTON: Radio Emission from Flare Star V 371 Orionis, *Nature*, vol. 199, pp. 991–993, Sept. 7, 1963.

SLOANAKER, R. M.: Apparent Temperature of Jupiter at a Wavelength of 10 cm, *Astron. J.*, vol. 64, p. 346, October, 1959.

SMERD, S. F.: Radio Frequency Radiation from the Quiet Sun, *Australian J. Sci. Res.*, vol. 3A, pp. 34–59, 1950.

SMITH, A. G.: University of Florida Radio Observatory, *Astron. J.*, vol. 68, pp. 627–629, November, 1963.

SMITH, A. G., T. D. CARR, and N. F. SIX: Nonthermal Radiation from Jupiter, in "Third Symposium on Engineering Aspects of Magnetohydrodynamics," Gordon and Breach Science Publishers, Inc., 1963.

SMITH, H. J., and D. HOFFLEIT: Light Variability and Nature of 3C 273, *Astron. J.*, vol. 68, pp. 292–293, June, 1963.

SOUTHWORTH, G. C.: Microwave Radiation from the Sun, *J. Franklin Inst.*, vol. 239, p. 285, 1945.

STRÖMGREN, B.: On the Density Distribution and Chemical Composition of the Interstellar Gas, *Astrophys. J.*, vol. 108, pp. 242–275, 1948.

STRUVE, O.: Flare Stars, *Sky & Telescope*, vol. 18. pp. 612–613, September, 1959.

SWARUP, G., A. R. THOMPSON, and R. N. BRACEWELL: The Structure of Cygnus A, *Astrophys. J.*, vol. 138, pp. 305–309, July 1, 1963.

TIURI, M. E., and J. D. KRAUS: Ionospheric Disturbances Associated with Echo I as Studied with 19 Mc/sec Radar, *J. Geophys. Res.*, vol. 68, pp. 5371–5385, 1963.

TIURI, M. E., and J. D. KRAUS: Is the Satellite Ionization Phenomenon Responsible for the Decametric Radiation from Jupiter?, *Astron. J.*, vol. 70, p. 695, November, 1965.

TOLBERT, C. W.: Millimeter Wavelength Spectra of the Crab and Orion Nebulae, *Nature*, vol. 206, p. 1304, 1965.

TWISS, R. Q.: On the Nature of the Discrete Radio Sources, *Phil. Mag.*, vol. 45, pp. 249–258, March, 1954.

VAN DE HULST, H. C.: Radio Waves from Space, *Ned. Tijdschr. Natuurk.*, vol. 11, pp. 201–221, 1945.

VAN DE HULST, H. C., C. A. MULLER, and J. H. OORT: The Spiral Structure of the Outer Part of the Galactic System Derived from Hydrogen Emission at 21 cm Wavelength, *Bull. Astron. Inst. Neth.*, vol. 12, no. 452, pp. 117–149, May 14, 1954.

VITKEVITCH, V. V.: Report on the URSI 12th General Assembly, Boulder, Colo., p. 125, 1957.

WADE, C. M.: The Structure of Fornax A, *Publ. Natl. Radio Astron. Obs.*, vol. 1, no. 6, August, 1961.

WARWICK, J. W.: Relation of Jupiter's Radio Emission at Long Wavelengths to Solar Activity, *Science*, vol. 132, pp. 1250–1252, October, 1960.

WARWICK, J. W.: Dynamic Spectra of Jupiter's Decametric Emission, 1961, *Astrophys. J.*, vol. 137, pp. 41–60, January, 1963.

WEINREB, S., A. H. BARRETT, M. L. MEEKS, and J. C. HENRY: Radio Observations of OH in the Interstellar Medium, *Nature*, vol. 200, pp. 829–831, Nov. 30, 1963.

WESTERHOUT, G.: The Structure of the Galaxy from Radio Observations, *IEEE Trans. Military Electron.*, vol. MIL-8, pp. 288–297, July–October, 1964.

WESTERHOUT, G., C. L. SEEGER, W. N. BROUW, and J. TINBERGEN: Polarization of the Galactic 75-cm Radiation, *Bull. Astron. Inst. Neth.*, vol. 16, no. 518, pp. 187–212, July 6, 1962.

WESTERLUND, B. E., and D. S. MATHEWSON: Supernova Remnants in the Large Magellanic Cloud, *Monthly Notices Roy. Astron. Soc.*, vol. 130, 1966.

WILD, J. P.: The Radio Frequency Line Spectrum of Atomic Hydrogen and Its Application in Astronomy, *Astrophys. J.*, vol. 115, pp. 206–221, 1952.

WILD, J. P.: The Radio Emission of the Sun, in H. P. Palmer, R. D. Davies, and M. I. Large (eds.), "Radio Astronomy Today," University of Manchester Press, Manchester, 1963.

WILD, J. P., S. F. SMERD, and A. A. WEISS: Solar Bursts, *Ann. Rev. Astron. Astrophys.*, vol. 1, pp. 291–366, 1963.

WILLIAMS, P. J. S.: Absorption in Radio Sources of High Brightness Temperature, *Nature*, vol. 200, pp. 56–57, Oct. 5, 1963.

WOLTJER, L.: The Polarization of Radio Sources, *Astrophys. J.*, vol. 136, pp. 1152–1154, November, 1962.

Problems

8-1. For an ionized-hydrogen cloud with an electron temperature of 10,000°K show that the ratio of the optical depths τ_1 and τ_2 at two frequencies ν_1 and ν_2 is given by

$$\frac{\tau_1}{\tau_2} = \left(\frac{\nu_1}{\nu_2}\right)^{-2} \frac{\ln(4 \times 10^{14}/\nu_1)}{\ln(4 \times 10^{14}/\nu_2)}$$

8-2. An H I absorption feature peaks at a frequency 100 Kc higher than the standard of rest frequency. Calculate its velocity with respect to the standard of rest.

8-3. An H I emission feature has an average optical depth of unity and temperature of 125°K. If the bandwidth of the feature is 70 Kc, calculate the total number of neutral hydrogen atoms associated with the feature in a column of 1 m² cross section.

8-4. (*a*) A radio source has flux densities of 12.1 and 8.3 flux units at 600 and 1,415 Mc, respectively. Calculate its spectral index. (*b*) Do the same for a source with flux densities of 3.7 and 2.2 flux units at the two frequencies. (*c*) Are the spectra thermal or nonthermal? *Ans.* (*a*) 0.44, (*b*) 0.61, (*c*) nonthermal

8-5. An ionized-hydrogen cloud has a brightness temperature of 150°K at 600 Mc. Calculate the emission measure and the average free-electron density if the Strömgren sphere is 25 pc in diameter.

8-6. What electron-energy spectral index is required for a nonthermal radio spectrum of index 1.0? It is assumed that relativistic electrons radiating in the synchrotron mode are responsible for the radiation.

8-7. (*a*) Calculate the frequency of maximum synchrotron radiation of a 500-Mev electron in a magnetic field of 10^{-8} weber m^{-2}. (*b*) What is the velocity of the electron in terms of the velocity of light?

8-8. Find the cone angle θ of synchrotron radiation for an electron with 500 Mev energy.

8-9. Why is it believed that lunar radio emission comes from deeper surface layers than optical emission?

8-10. What does the "Jupiter rocking effect" refer to?

8-11. Describe the radio and optical characteristics of UV Ceti.

8-12. What does the "galactic spur" refer to?

8-13. Describe the sequence of events and associated radio effects that usually follow a large solar flare.

8-14. Discuss the differences between normal and radio galaxies.

8-15. Is the spectral index of nonthermal radio sources a function of galactic latitude? Explain.

8-16. An object has a red shift of 0.5. What is its luminosity distance and its relativistic distance assuming uniform expansion and a Hubble constant of 100 km sec^{-1} Mpc^{-1}?

8-17. Name as many characteristics of quasi-stellar radio sources, such as small optical size, large optical and radio-power output, ultraviolet excess, and red shift, as you can, and discuss each characteristic quantitatively.

8-18. What are the reasons for the bent spectra of ionized-hydrogen regions?

8-19. A radio source has a flux density S_1 at a frequency ν_1 and a flux density S_2 at a frequency ν_2. Show that its spectral index $n = [\log (S_1/S_2)]/[\log (\nu_2/\nu_1)]$. A straight spectrum is assumed.

8-20. Determine the spectral indices of 3C 84, 111, 231, and 380 in the 500 to 2,800 Mc range. (See Appendix 3b for flux densities.)   *Ans.* 0.7 for 3C 111.

8-21. Take the spectral index of the brightness temperature of a nonthermal source as 2.7, of an optically thin thermal source as 2.0, and of an optically thick thermal source as 0.0. Show that the antenna temperature T_A and flux density S spectral indices are as tabulated:

	Point source		Extended source	
	T_A	S	T_A	S
Nonthermal	0.7	0.7	≤ 2.7	0.7
Thermal (thin)	0.0	0.0	≤ 2.0	0.0
Thermal (thick)	-2.0	-2.0	≤ 0.0	-2.0

A_e is constant and the extended source fills the main beam at the longest wavelength.

appendix 1

List of Physical Quantities and Units

Name	Symbol	Unit	Equivalent units
Absorption coefficient	K	$m^2\ kg^{-1}$	
Acceleration	a	$m\ sec^{-2}$	
Angle	θ, ϕ	rad	$57°296$
Area, aperture	A	m^2	
Attenuation constant	α	nepers m^{-1}	
Bandwidth	$\Delta\nu$	cps	
Beam area	Ω	rad^2	$3,283\ deg^2$
Brightness	B	watts $m^{-2}\ cps^{-1}\ rad^{-2}$	
Brightness, total	B'	watts $m^{-2}\ rad^{-2}$	
Charge	e	coul	6.25×10^{18} electron charges
			$=$ amp sec
			$= 3 \times 10^9$ cgs esu
			$= 0.1$ cgs emu
Conductivity	σ	mhos m^{-1}	(ohm-meter)$^{-1}$
Current	I	amp	coul sec^{-1}
Current density	J	amp m^{-2}	
Density (charge)	ρ	coul m^{-3}	
Density (mass)	ρ	kg m^{-3}	
D vector	D	coul m^{-2}	
Electric field intensity	E	volt m^{-1}	newton $coul^{-1}$
			$= \frac{1}{3} \times 10^{-4}$ cgs esu
			$= 10^6$ cgs emu
Electron density	N	number m^{-3}	
Electron volt (energy)	eV	ev	1.602×10^{-19} joule
Emf	υ	volt	joule $coul^{-1}$
Emission coefficient	j	watt $kg^{-1}\ cps^{-1}$	
Energy or work	$\mathcal{E}$	joule	newton-m $=$ watt sec
			$=$ volt-coul $= 10^7$ ergs
			$= 10^7$ dyne-cm
			$= 6.25 \times 10^{18}$ ev
Energy density	$\mathcal{E}_d$	joules $m^{-3}\ cps^{-1}$	
Flux density	S	watt $m^{-2}\ cps^{-1}$	jansky $= 10^{26}$ flux units
Flux density, total	S'	watt m^{-2}	
Force	F	newtons	kg-m $sec^{-2} =$ joule $m^{-1} = 10^5$ dyne

List of physical quantities and units (Continued)

Name	Symbol	Unit	Equivalent units
Frequency	ν	cps	hertz (1 Kc $= 10^3$ cps; 1 Mc $= 10^6$ cps; 1 Gc $= 10^9$ cps)
H vector	H	amp m^{-1}	$4\pi \times 10^{-3}$ oersted $= 400\pi$ gammas
Impedance	Z	ohms	volt amp^{-1}
Intensity	I	watts m^{-2} cps^{-1} rad^{-2}	
Length	L, r	m	100 cm
Luminosity	l	watts m^{-2}	
Magnetic flux density	B	webers m^{-2}	10^4 gauss
Mass	M, m	kg	1,000 g
Momentum	mv	newton-sec	kg m sec^{-1} = joule-sec m^{-1}
Noise figure	F	dimensionless	
Optical depth	τ	nepers	
Period	t	sec	
Permeability	μ	henrys m^{-1}	
Permittivity	ϵ	farads m^{-1}	
Power	W	watts	joule sec^{-1} = newton m sec^{-1} = kg m^2 sec^{-3}
Power pattern, antenna	P	watts	
Power pattern, antenna, normalized	P_n	dimensionless	
Poynting vector	S	watts m^{-2}	
Pressure	p	kg m^{-1} sec^{-2}	
Propagation constant	β	rad m^{-1}	$= 2\pi/\lambda$
Radian frequency	ω	rad sec^{-1}	$= 2\pi\nu$
Reactance	X	ohms	volts amp^{-1}
Resistance	R	ohms	volts amp^{-1}
Solid angle	Ω	rad^2	steradian $= 3,283$ deg^2
Spectral power	w	watts cps^{-1}	
Temperature	T	°K	
Time	t	sec	1/60 min = 1/3,600 hr = 1/86,400 day
Vector potential	α	webers m^{-1}	
Velocity	v	m sec^{-1}	
Wavelength	λ	m	

List of Constants[†]

Astronomical unit (AU) = 1.496×10^8 km
Boltzmann's constant (k) = 1.3805×10^{-23} joule °K^{-1}
Cosmic year (rotation time of our galaxy) = 3×10^8 years
$e = 2.71828$
$1/e = 0.36788$
Earth, mass = 5.98×10^{24} kg
Electron, charge = -1.602×10^{-19} coul
Electron, mass = 9.1066×10^{-31} kg
Electron, ratio charge to mass = -1.76×10^{11} coul kg^{-1}
Hubble's constant (H_0) = 100 km sec^{-1} Mpc^{-1}
Hydrogen atom, mass = 1.673×10^{-27} kg
Hydrogen-line frequency = 1,420.405 Mc
Galaxy (our), mass of, = 10^{41} kg = 5×10^{10} solar masses
Light, velocity (c) = 2.99776×10^8 m sec^{-1}
Light-year (LY) = 9.46×10^{12} km = 5.9×10^{12} miles
$\log x = \log_{10} x$ (common logarithm)
$\ln x = \log_e x$ (natural logarithm)
$\ln x = 2.3026 \log x$
$\log x = 0.4343 \ln x$
Parsec (pc) = 3.086×10^{13} km = 3.26 light-years = 2.06×10^5 AU
$\pi = 3.1415927$
$\pi^2 = 9.8696044$
Permeability of vacuum (μ_0) = $4\pi \times 10^{-7}$ henry m^{-1}
Permittivity of vacuum (ϵ_0) = 8.85×10^{-12} farad m^{-1}
Planck's constant (h) = 6.6254×10^{-34} joule sec
Radian = 57°.2958
Square degrees in a sphere = 41,253
Space, impedance of = 376.7 ohms
Stefan-Boltzmann constant = 5.67×10^{-8} watt m^{-2}°K^{-4}
Steradian (square radian) = 3,283 deg^2
Sun, mass of = 1.99×10^{30} kg

[†] For more accurate values and other constants see C. W. Allen, "Astrophysical Quantities," University of London Press, London, 1963.

appendix 3

Lists of
Radio Sources

Six lists of radio sources are tabulated in this section. Five of these are the Third Cambridge (3C) revised list (178 Mc), the Sydney (MSH) list (86 Mc), the California Institute of Technology CTA and CTB lists (960 Mc), and the Ohio State University (OA) list (600 and 1,415 Mc). These are preceded by a list of a few of the better-known radio sources arranged alphabetically according to name. One function of this list is to serve as a finder for the other lists. Thus, from the name of a source its position can be determined. This can then assist in finding a source in the other lists where the sources are arranged in order of increasing right ascension.

All positions are epoch 1950.0. No errors for positions or fluxes are included. For these reference should be made to the original lists. Some positions and fluxes have been rounded off from the more accurate values given in the original lists.

Abbreviations used are: SNR for supernova remnant, EMN for emission nebula (H II region), HB for Hanbury Brown, MSH for Mills, Slee, and Hill, and quasar for quasi-stellar radio source.

Recent, more accurate positions are given, when available†, instead of the positions in the original list, with references for the improved positions indicated by superscripts as follows:

> [a] Tyler, W. C., D. E. Hogg, and C. M. Wade: First results with the National Radio Astronomy Observatory Interferometer, *Astron. J.*, vol. 70, p. 332, 1965. Measurements at 2,700 Mc.
>
> [b] Read, R. B.: Accurate Measurement of the Declinations of Radio Sources, *Astrophys. J.*, vol. 138, pp. 1–29, 1963; Wyndham, J. D., and R. B. Read: Further Accurate Declinations of Radio Sources, *Owens Valley Radio Obs. Rept.* 9, 1964; Formalont, E. B., T. A. Matthews, D. Morris, and J. D. Wyndham: Accurate Right Ascensions for 227 Radio Sources, *Owens Valley Radio Obs. Rept.* 5, 1964. Measurements at 958 and 1,390 Mc.
>
> [d] Bolton, J. G., F. F. Gardner, and M. B. Mackey: The Parkes Catalogue of Radio Sources, Declination Zone −20° to −60°, *Australian J. Phys.*, vol. 17, pp. 340–372, 1964; Price, R. M., and D. K. Milne: The Parkes Catalogue of Radio Sources, Declination Zone −60° to −90°, *Australian J. Phys.*, vol. 18, pp. 329–347, 1965. Position measurements at 1,410 Mc.

Where reliable flux densities are available at a number of frequencies, these are given in parentheses following the flux density of the original survey with the source of the supplementary data indicated by superscripts as follows:

> [c] Kellermann, K. I.: Measurements of the Flux Density of Discrete Radio Sources at Decimeter Wavelengths, *Astron. J.*, vol. 69, pp. 205–214, 1964. Flux densities

† Except for List a.

are given for 475, 710, 958, 1,420, and 2,841 Mc. For example, the flux densities listed for 3C 20 in Appendix 3*b* are 41 flux units at the original survey frequency of 178 Mc and 23, 19, 15, 11, and 7 flux units at 475, 710, 958, 1,420, and 2,841 Mc, respectively. When a flux density is not available at a given frequency, a dash is used to indicate a gap in the sequence.

d Same as (*d*) above for positions. Flux densities are given for 408, 1,410, and 2,650 Mc.

e Conway, R. G., K. I. Kellermann, and R. J. Long: The Radio Frequency Spectra of Discrete Radio Sources, *Monthly Notices Roy. Astron. Soc.*, vol. 125, pp. 261–284, 1963. Flux densities are given for 38, 178, 240, 408, 412, 710, 958, 1,420 and 3,200 Mc.

The number of digits used for positions or flux densities may generally† be considered as an indication of the presumed accuracy, the values being uncertain in the last digit. For detailed information regarding the method and accuracy of measurements the original references should be consulted.

A catalog of 1,292 sources with detailed data on each has been compiled by W. E. Howard and S. P. Maran: General Catalog of Discrete Radio Sources, *Astrophys. J.*, supp. no. 98, vol. 10, pp. 1–330, 1965.

a. A List of a Few Well-known Radio Sources

Abbreviations: SNR = supernova remnant
             EMN = emission nebula

Object	Position (1950.0)				
	RA			Dec	
	h	*m*	*s*	*Deg*	*Min*
Andromeda nebula (M 31)	00	40	00	+41	00
Auriga A, SNR	04	57		+46	30
California Institute of Technology objects					
CTA 21	03	16	22	+16	19
CTA 102	22	29	53	+11	28
Cambridge objects					
3C 47, Quasar	01	33	40	+20	42
3C 48, Quasar	01	34	49	+32	53
3C 147, Quasar	05	38	43	+49	50
3C 273, Quasar	12	26	33	+02	20
3C 295, Distant galaxy	14	09	34	+52	26

† Except for List a.

A list of a few well-known radio sources (Continued)

Object	RA h	RA m	RA S	Dec Deg	Dec Min
				Position (1950.0)	
Cassiopeia A, SNR	23	21	11	+58	33
Centaurus A, NGC 5128	13	22	28	−42	46
Crab Nebula, M 1, SNR	05	31	30	+21	58
Cygnus A, distant galaxy	19	57	45	+40	36
Cygnus loop, SNR	20	49	30	+29	50
Cygnus X, extended EMN	20	27		+27	
Fornax A, NGC 1316	03	20	25	−37	22
HB 21 (Hanbury Brown) SNR	20	44		+50	20
Hercules A	16	48	43	+05	06
Horse's head nebula, EMN	05	38	24	−01	54
Horseshoe or Omega nebula (M 17), EMN	18	17	48	−16	09
Hydra A	09	15	43	−11	52
IC 443, SNR	06	14	36	+22	43
Kepler's SNR (1604)	17	27	43	−21	28
Lagoon nebula (M 8) EMN	18	01	00	−24	22
Magellanic cloud, small (SMC)	01	15		−73	
Magellanic cloud, large (LMC)	05	40		−69	
M 33, Sc galaxy	01	31	00	+30	24
M 77, Seyfert galaxy	02	40	12	−00	13
M 82, exploding galaxy	09	51	28	+69	56
M 101, Sc galaxy	14	01	24	+54	36
North America nebula, EMN	20	54	24	+43	52
Omega nebula (M 17), EMN	18	17	48	−16	09
Orion nebula (M 42), EMN	05	32	48	−05	27
Perseus A, NGC 1275	03	16	27	+41	21
Puppis A, SNR	08	20	18	−42	48
Rosette nebula, EMN	06	29	18	+04	57
Sagittarius A, galactic nucleus	17	42	30	−28	55
Taurus A, Crab nebula	05	31	30	+21	58
Tycho's SNR (1572)	00	22	28	+63	52
Virgo A, M 87	12	28	18	+12	40

b. The Third Cambridge (3C) List (Revised)*

178 Mc; 328 Sources

Reference: A. S. Bennett: The Revised 3C Catalog of Radio Sources, *Mem. Roy. Astron. Soc.*, vol. 68, pp. 163–172, 1962. See (page 412) for meaning of superscripts *a*, *b*, *c*, etc.

		Position (1950.0)					
		RA		Dec			
Number	h	m	s	Deg	Min	Flux density†	Remarks
3C 2^b	00	03	48.3	−00	21.3	15	
6.1	00	14	12	+79	24	16	
9	00	17	50	15	23	15	Quasar
10	00	22	28	63	51.9	134 (84, 68, 55, 43, —)^c	Tycho's SNR
11.1	00	27	06	63	24	12	
13	00	31	35	39	03	11	
14	00	33	31	18	21	10	
14.1	00	33	36	59	30	14	
15^b	00	34	29.8	−01	25.6	15	
16	00	35	06	+12	50	12	
17^b	00	35	46	−02	23.9	21 (13, 11, 8, 5, —)^c	
18^b	00	38	12.4	+09	46.8	23	
19^b	00	38	12.4	32	53.6	9	
20^b	00	40	17.8	51	47.2	51 (23, 19, 15, 11, 7)^c	
21.1	00	42	30	67	48	9	
22	00	48	06	50	57	15	
27	00	52	42	68	13	24 (16, 14, 10, 7, 3)^c	
28	00	53	10	26	08	14	
29	00	55	03	−01	35	15	
31	01	04	42	+32	08	16	
33	01	06	14	13	04	49 (29, 23, 18, 12, 7)^c	
33.1	01	06	24	72	54	12	
33.2	01	06	54	69	06	9	
34	01	07	32	31	30	11	
35	01	09	07	49	10	13	
36	01	15	14	45	23	10	
40	01	23	35	−01	32	24 (15, 11, 9, 6, —)^c	
41	01	23	55	+32	56	10	
42	01	25	44	28	48	12	
43^b	01	27	13.8	23	22.7	12	
44	01	28	46	06	10	11	
46^b	01	32	32	37	38.5	10	
47^b	01	33	39.5	20	41.9	20	Quasar
48^b	01	34	49.2	32	53	47 (34, 26, 21, 16, 8)^c	Quasar
49	01	38	32	13	48	12	

* A 4C list of several thousand sources to a limit of 2 flux units at 178 Mc is in preparation. The limit of the 3C list is 9 flux units.

† At 178 Mc in units of 10^{-26} watt m^{-2} cps^{-1}.

The Third Cambridge (3C) list (Revised) (Continued)

Number	Position (1950.0) RA h	m	s	Dec Deg	Min	Flux density†	Remarks
3C 52	01	45	18	53	21	10	
54	01	52	28	43	33	10	
55[b]	01	54	18.3	28	36.4	21	
58	02	01	41	64	38	24 (37, 37, 32, 34, —)[c]	
61.1	02	12	00	86	00	29	
63[b]	02	18	22.1	−02	10.5	15	
65	02	20	38	+39	48	27	
66[b]	02	19	56	42	45.8	33 (27, 15, 13, 9, —)[c]	
67	02	21	18	27	37	10	
68.1	02	29	29	34	30	17	
68.2	02	31	24	31	20	12	
69	02	34	20	58	59	23	
71[b]	02	40	07.3	−00	13.5	14 (10, 8, 6, 5, 4)[c]	
75[b]	02	55	04.4	+05	50.7	23 (14, 10, 8, 6, —)[c]	
76.1	03	00	28	16	15	10	
78[b]	03	05	49.0	03	55.2	15	
79[b]	03	07	12	16	54.6	24	
83.1	03	14	55	41	44	28	NGC 1265
84[a]	03	16	29.61	41	19.85	58 (26, 20, 18, 13, 7)[c]	NGC 1275, Perseus A
86	03	23	31	55	08	21	
88	03	25	19	02	23.1	16	
89	03	31	46	−01	21	19	
91[a]	03	34	02.8	+50	36.10	13	
93	03	40	53	04	48	10	Quasar
93.1	03	45	36	33	44	10	
98[b]	03	56	11.7	10	17.6	41 (23, 17, 14, 10, 7)[c]	
99	03	58	30	00	17	10	
103[b]	04	04	35.7	42	52.3	25	
105[b]	04	04	45.0	03	33.3	15	
107	04	09	55	−01	01	11	
109[b]	04	10	55.7	+11	04.5	20	
111	04	15	02	37	53	57 (34, 26, 20, 15, 10)[c]	
114	04	17	30	17	47	10	
119[a]	04	29	07.8	41	32.13	15 (17, 12, 10, 8, 5)[c]	
123	04	33	56	29	34	175 (110, 79, 61, 46, 24)[c]	
124	04	39	24	01	38	10	
125	04	42	50	39	37	10	
129	04	45	25	44	57	21	
129.1	04	46	42	44	48	11	
130	04	48	50	51	46	13	

The Third Cambridge (3C) list (Revised) (Continued)

Number	h	m	s	Deg	Min	Flux density†	Remarks
			Position (1950.0)				
		RA		Dec			
3C 131[b]	04	50	11.1	31	24.4	14	
132	04	53	43	22	44	13	
133	04	59	55	25	12	20	
134	05	01	18	38	01	66 (31, 20, 15, 10, 4)[c]	
135[b]	05	11	33	00	53.1	16	
136.1	05	12	54	24	54	13	
137	05	15	38	50	53	9	
138[a]	05	-18	16.4	16	35.4	19	
139.1	05	19	21	34	00	20	
139.2	05	21	19	28	11	11	
141	05	23	26	32	47	13	
142.1	05	28	49	06	30	18	
144	05	31	30	21	58	1,420	Taurus A; Crab nebula; M 1
147[b]	05	38	43.2	49	49.6	58 (43, 35, 29, 21, 12)[c]	Quasar
147.1	05	39	03	−01	48	15	
152	06	01	31	+20	21	11	
153[b]	06	05	43.4	48	04.9	18	
153.1	06	06	42	21	12	19	
154[b]	06	10	42.8	26	05.4	19	
157	06	14	36	22	43	210 (500, 230, —, —, —, —, 185, —, —)[e]	IC 443
158[a]	06	18	50.1	14	33.7	14	
165	06	40	04	23	22	11	
166[b]	06	42	24.7	21	25.0	13	
169.1	06	47	42	45	06	10	
171[b]	06	51	12	54	12.8	23	
173	06	58	57	38	04	10	
172	06	59	05	25	18	14	
173.1	07	03	00	75	12	12	
175	07	10	15	11	52	16	
175.1	07	11	27	15	36	12	
177	07	21	45	15	35	10	
180[b]	07	24	33.2	−01	58.5	14	
181	07	25	20	+14	45	13	
184	07	34	34	70	20	11	
184.1	07	34	54	80	30	12	

The Third Cambridge (3C) list (Revised) (Continued)

Number	\multicolumn{5}{c}{Position (1950.0)}	Flux density†	Remarks
	\multicolumn{3}{c}{RA}	\multicolumn{2}{c}{Dec}		
	h	m	s	Deg	Min		
3C 186	07	41	04	38	05	14	Quasar
187	07	42	26	01	54	11	
190	07	58	52	14	27	11	
191^b	08	02	02	10	23.7	11	
192	08	02	36	24	19	20 (11, 8, 7, 5, —)^c	
194	08	06	39	42	37	9	
196^b	08	10	00.1	48	22.1	59 (36, 26, 20, 14, 7)^c	Quasar
196.1	08	13	21	−02	48	15	
197.1	08	18	21	+46	42	10	
198	08	19	52	06	07	17	
200	08	24	20	29	28	13	
204	08	32	59	65	32	10	
205	08	35	28	57	50	13	
207	08	38	12	12	48	10	
208	08	50	24	14	03	16	Quasar
208.1	08	52	00	14	42	13	
210	08	55	13	28	04	11	
212^b	08	55	55	14	21.4	16	
213.1	08	58	07	29	13	11	
215	09	03	44	16	58	10	
216^a	09	06	17.33	43	05.99	19 (10, 7, 5, 4, —)^c	Quasar
217	09	05	55	38	04	12	
219	09	17	52	45	52	44 (24, 15, 12, 8, 5)^c	
220.1	09	26	42	80	12	12	
220.2	09	27	51	36	06	10	
220.3	09	32	00	83	48	15	
222	09	33	55	04	39	9	
223	09	36	53	36	08	15	
223.1	09	38	20	39	56	10	
225^b	09	39	31.1	14	00.9	25	
226	09	41	41	10	05	14	
227^b	09	45	09	07	39.3	28 (18, 13, 10, 7, 4)^c	
228^b	09	47	28.3	14	34.1	16	Quasar
230^b	09	49	24	00	13.0	21	
231^b	09	51	45	69	55.0	13 (13, 11, 10, 9, 6)^c	
234	09	58	57	29	02	29	
236	10	03	06	35	03	9	
237^a	10	05	22.04	07	44.95	24 (14, 11, 9, 6, 4)^c	
238^b	10	08	25	06	39.6	13	
239	10	08	48	46	40	12	

The Third Cambridge (3C) list (Revised) (Continued)

Number	Position (1950.0)					Flux density†	Remarks
	RA			Dec			
	h	m	s	Deg	Min		
3C 241	10	19	10	22	12	10	
244.1	10	30	30	58	42	19	
245ᵃ	10	40	06.09	12	19.74	10	Quasar
247	10	56	10	43	18	16	
249	10	59	37	−01	00	20	
249.1	11	00	12	77	30	12	
250	11	06	12	25	15	11	
252	11	08	58	35	54	12	
254	11	11	54	40	54	19	
255	11	17	02	−02	31	17	
256	11	18	05	23	46	10	
257	11	20	43	05	52	9	
258	11	22	12	19	39	9	
263	11	36	55	66	07	13	
263.1	11	40	50	22	25	18	
264	11	42	32	19	52	24	
265ᵇ	11	42	53.4	31	50.6	18	
266	11	43	06	50	02	9	
267	11	47	26	13	10	14	
268.1	11	57	54	73	24	20	
268.2	11	58	25	31	48	9	
268.3	12	04	12	64	48	10	
268.4	12	06	45	43	57	9	
270ᵇ	12	16	49.9	06	06.5	44 (37, 31, 28, 21, —)ᶜ	
270.1	12	18	05	33	56	12	
272	12	22	03	42	23	10	
272.1ᵇ	12	22	33.0	13	09.5	18	
273Bᵃ	12	26	33.0	02	19.8	67	Quasar
274	12	28	18	12	40	970 (3.2, 1.1, 0.8, —, 0.5, —, 0.3, 0.2, 0.1 × 10³)ᵉ	Virgo A; M 87
274.1	12	32	57	21	36	15	
275ᵇ	12	39	45.0	−04	41.6	10	
275.1	12	41	28	+16	43	16	
277	12	49	29	50	52	13	
277.1	12	50	18	57	12	12	
277.2	12	51	03	15	48	10	
277.3	12	51	48	27	53	12	
280ᵇ	12	54	40	47	36.2	20 (12, 10, 7, 5, —)ᶜ	
280.1	12	58	14	40	23	11	
284	13	08	43	27	42	11	
285	13	19	08	42	52	11	

The Third Cambridge (3C) list (Revised) (Continued)

Number	RA h	RA m	RA s	Dec Deg	Dec Min	Flux density†	Remarks
3C 286*a*	13	28	49.64	30	45.98	21 (24, 21, 19, 16, 11)*c*	Quasar
287*a*	13	28	16.1	25	24.67	14 (12, 9, 8, 7, 4)*c*	Quasar
287.1	13	30	21	02	19	12	
288	13	36	38	39	04	15	
288.1	13	40	48	60	12	10	
289	13	43	29	50	01	10	
293	13	50	03	31	39	12	
293.1	13	52	21	16	42	10	
294	14	04	35	34	32	9	
295*a*	14	09	33.5	52	26.2	73 (52, 36, 31, 22, 11)*c*	Red shift = 0.5
296	14	14	23	08	31	13	
297	14	14	45	−03	58	10	
298*a*	14	16	38.72	+06	42.15	14 (23, 13, 9, 6, 2)*c*	
299	14	19	08	41	59	11	
300	14	20	40	19	51	16	
300.1*b*	14	25	58.3	−01	11.1	11	
303	14	14	21	+52	18	13	
303.1	14	44	06	77	12	12	
305	14	48	09	63	33	14	
305.1	14	48	24	76	54	12	
306.1	14	52	45	−03	48	13	
309.1	14	59	42	+72	42	17	
310*b*	15	02	48.5	26	12.6	51 (25, 16, 12, 7, 4)*c*	
313	15	08	33	08	01	28	
314.1	15	10	48	71	18	9	
315	15	11	31	26	18	18	
317	15	14	18	07	11	43	
318*a*	15	17	50.6	20	26.12	10	
318.1	15	19	24	07	50	11	
319*b*	15	22	45	54	39.1	17	
320	15	29	37	35	53	12	
321	15	29	40	24	13	13	
322	15	34	08	56	05	10	
323	15	40	57	60	18	10	
323.1	15	45	33	21	03	9	
324	15	47	39	21	36	12	
325	15	49	08	62	50	11	
326	15	50	14	20	16	10	
326.1	15	53	58	20	14	9	
327*b*	15	59	56	02	06.2	40 (25, 13, 12, 8, —)*c*	

The Third Cambridge (3C) list (Revised) (Continued)

Number	Position (1950.0)					Flux density†	Remarks
	RA			Dec			
	h	m	s	Deg	Min		
3C 327.1	16	02	14	01	24	24 (15, 7, 6, 4, —)^c	
330^b	16	09	15	66	04.8	24	
332	16	15	47	32	28	11	
334	16	18	07	17	32	10	
336	16	22	33	23	52	14	
337	16	27	16	44	20	17	
338^b	16	26	54.6	39	39.5	41 (17, 10, 6, 4, 2)^c	
340	16	27	30	23	30	11	
341	16	26	05	27	49	11	
343	16	34	18	62	48	10	
343.1	16	38	06	62	48	9	
345^a	16	41	17.58	39	54.17	10 (10, 9, 8, 7, 6)^c	Quasar
346	16	41	35	17	20	11	
348^b	16	48	41.2	05	04.4	325 (130, 89, 68, 47, 23)^c	Hercules A
349	16	58	10	47	08	12	
351	17	04	03	60	48	11	
352	17	09	15	46	02	11	
353	17	17	56	−00	57	203 (118, 89, 75, 57, 29)^c	
356	17	23	06	+51	01	15	
357	17	26	27	31	47	9	
363.1	17	50	30	05	36	90	
368	18	02	42	10	57	14	
371	18	07	04	69	55	10	
372.1	18	12	42	−03	36	41	
379.1	18	26	24	+74	24	10	
380^a	18	28	13.46	48	42.67	57 (30, 24, 18, 14, 9)^c	
381	18	32	28	47	24	13	
382	18	33	12	32	39	20	
386	18	36	12	17	07	24 (15, 11, 9, 7, 4)^c	
388	18	42	37	45	32	22 (15, 11, 8, 6, 4)^c	
390^b	18	43	15.2	09	50.3	19	
389	18	43	41	−03	23	21	
390.1	18	44	00	+06	36	51	
390.2	18	45	12	−03	12	110	
390.3	18	46	12	+80	00	44 (25, 18, 13, 10, 6)^c	
391	18	46	47	−00	59	24 (39, 23, 20, 10, —)^c	
392	18	53	34	+01	14	400	
394^b	18	57	05	12	54.8	12	
396	19	01	36	05	24	22	
396.1	19	04	30	−03	12	30	

The Third Cambridge (3C) list (Revised) (Continued)

Number	RA			Dec		Flux density†	Remarks
	h	m	s	Deg	Min		
397	19	05	07	+07	05	28	
398	19	08	44	09	01	55	
399.1	19	13	59	30	12	13	
399.2	19	15	42	10	36	19	
400	19	21	30	14	24	540	
400.1	19	22	48	35	54	17	
400.2	19	36	30	17	24	22	
401	19	39	37	60	35	19 (12, 8, 7, 5, —)^c	
402^b	19	40	22.5	50	29.5	13	
403	19	49	44	02	29	26	
403.1	19	49	54	−02	00	13	
403.2	19	52	00	+31	54	50	
405	19	57	45	40	36	8,100 (22, 8.7, —, 4.7, —, 2.8, 2.1, 1.5, 0.6 × 10³)^e	Cygnus A
409^a	20	12	18.02	23	25.70	73 (37, 26, 19, 13, 5)^c	
409.1	20	18	00	45	12	230	Part of Cygnus X
410	20	18	03	29	32	32 (23, 16, 13, 11, 6)^c	
410.1	20	19	36	40	06	410	Part of Cygnus X
411	20	19	47	09	59	14	
415.1	20	31	06	43	18	24	
415.2	20	31	21	53	35	10	
416.1	20	33	24	46	52	60	
416.2	20	34	00	41	33	160	
418	20	37	07	51	07	13	
419.1	20	39	06	42	06	74	
424	20	45	44	06	53	13	
427	21	05	12	76	36	23	
428	21	06	45	49	22	19	
430^b	21	17	03	60	35	29 (15, 12, 9, 7, 4)^c	
431^b	21	17	09	49	22	21 (12, 6, 5, 3, —)^c	
432	21	20	26	16	52	12	
433	21	21	32	24	53	52 (32, 22, 16, 12, 6)^c	
434	21	21	45	15	49	11	
434.1	21	23	06	51	48	23	
435	21	26	39	07	27	10	
435.1	21	35	12	84	24	12	

The Third Cambridge (3C) list (Revised) (Continued)

Number	\-\-h	RA\n m	s	Dec\n Deg	Min	Flux density†	Remarks
436	21	41	59	27	56	15	
437	21	45	00	15	03	11	
437.1	21	48	24	13	42	15	
438	21	53	46	37	46	37 (20, 14, 10, 7, 3)*c*	
441	22	03	50	29	17	13	
442*b*	22	12	20	13	35.5	20	
445*b*	22	21	15.4	−02	21.7	23 (12, 11, 7, 7, —)*c*	
449	22	29	09	+38	57	14	
452	22	43	34	39	25	49 (29, 20, 15, 11, 6)*c*	
454	22	49	10	18	34	9	
454.1	22	49	36	71	36	10	
454.2	22	50	24	64	48	11	
454.3	22	51	42	15	00	15	
455*b*	22	52	36	12	57.1	13	
456	23	10	04	09	06	13	
458	23	10	20	05	06	10	
459*a*	23	14	02.2	03	49.1	22	
460	23	19	00	23	32	10	
461*b*	23	21	07	58	33.8	11,000 (36, 12, 8.5, 5.5, 5.3, 4.0, 2.9, 2.4, 1.3×10^3)*e*	Cassiopeia A
465*b*	23	35	57	26	45.0	35 (20, 15, 11, 8, 3)*c*	
468.1	23	48	36	64	24	34	
469.1	23	53	30	79	48	12	
470	23	56	04	43	50	9	

Position (1950.0)

c. **California Institute of Technology List A (CTA)**

960 Mc; 106 Sources

Reference: D. E. Harris, and J. A. Roberts, Radio Source Measurements at 960 Mc/s, *Publ. Astron. Soc. Pacific*, vol. 72, pp. 237–255, August, 1960. See page 412 for meaning of superscripts *a*, *b*, *c*, etc.

	Position (1950.0)						
	RA			Dec		Flux	
Number	h	m	s	Deg	Min	density†	Remarks
CTA 1	00	02	35	+72	05	75	
2	00	22	46	+63	51	57	
3–5							3C 15, 17, 20
6	00	50	05	+56	20	18	NGC 281
7, 8							3C 27, 28
9	01	05	47	−16	20	7	
10	01	06	21	+13	06	18	3C 33
11	01	18	05	−15	44	8	
12–20							3C 40, 41, 47, 48, 58, 63, 66, 75, 79
21[a]	03	16	09.16	+16	17.66	9 (10, 9, 9, 8, 4)[c]	
22	03	16	27	+41	21	21	NGC 1275, Perseus A
23	03	20	25	−37	22	150	Fornax A
24, 25							3C 86, 89
26	03	36	54	−01	55	4	
27–32							3C 98, 103, 109, 111, 123, 129
33	04	57	30	+46	26	160	HB 9
34	05	01	12	+38	00	15	3C 134
35	05	18	19	−45	52	84	Pictor A
36	05	31	30	+21	59	1,030	Taurus A, Crab nebula
37	05	32	49	−05	25	360	Orion nebula, M 42
38	05	36		+28		13	
39, 40							3C 147, 154
41	06	14	16	+22	36	129	IC 443
42	06	24	37	−05	51	24	
43	06	29	24	+04	53	105	Rosette nebula
44, 45							3C 171, 196
46	08	21	20	−42	52	102	Puppis A
47	09	15	43	−11	52	67	Hydra A
48–53							3C 227, 234, 237, 264, 270, 273
54	12	28	18	+12	40	300	M 87
55	12	52	00	−12	25	10	
56	12	53	37	−05	41	7	
57	12	54	42	+47	35	7	3C 280
58	13	08	50	−22	11	10	

† At 960 Mc in units of 10^{-26} watt m^{-2} cps^{-1}.

California Institute of Technology list A (CTA) (Continued)

Number	RA h	m	s	Dec Deg	Min	Flux density†	Remarks
CTA 59	13	22	28	−42	46	2,010	Centaurus A, NGC 5128
60–72							3C 286, 287, 295, 298, 310, 313, 315, 317, 318, 324, 327, 330, 338
73	16	36	56	+62	42	11	
74	16	41	57	+39	54	8	3C 345
75	16	48	43	+05	06	74	Hercules A
76	17	17	59	−00	56	84	3C 353
77	17	19	24	−18	45	36	MSH 17-16
78	17	27	47	−21	16	20	
79	18	28	17	+48	41	18	3C 380
80	18	33	21	+32	41	7	
81–87							3C 386, 388, 392, 398, 401, 402, 403
88	19	57	45	+40	36	2,160	Cygnus A
89, 90							3C 409, 410
91	20	45		+50	30	180	HB 21
92	20	45	43	+06	50	6	3C 424
93	20	50		+30		252	Cygnus loop
94–96							3C 430, 433, 436
97	21	51	37	+46	50	5	
98	21	53	54	+37	43	11	3C 438
99	22	11	33	−17	14	14	
100	22	21	17	−02	27	8	3C 445
101	22	23	03	−05	15	8	
102ᵃ	22	30	07.5	+11	28.40	7	
103, 104							3C 452, 459
105	23	21	11	+58	32	3,120	Cassiopeia A
106	23	35	53	+26	46	12	3C 465

d.   California Institute of Technology List B (CTB)

960 Mc; 110 galactic plane objects

Reference: R. W. Wilson and J. G. Bolton: A Survey of Galactic Radiation at 960 Mc/s, *Publ. Astron. Soc. Pacific*, vol. 72, pp. 331–347, October, 1960.

Number	RA h	RA m	Dec Deg	Dec Min	Flux density†	Remarks
CTB 1	00	00.3	+62	11	56	
2	00	00.5	67	06	220	NGC 7822, EMN¶
3	00	10.4	57	47	5	
4	00	22.8	63	51	57	HB1, SNR¶ of 1572
5	00	25.5	55	21	48	
6	00	39.9	51	47	13	CTA 5, 3C 20
7	00	50.1	56	20	24	NGC 281, EMN
8	02	01.7	64	36	33	CTA 16, 3C 58
9	02	23.7	61	43	400	IC 1795/1805, EMN
10	02	37.4	58	59	25	
11	02	48.5	60	15	175	IC 1848, EMN
12	04	00.8	51	12	33	IC 1491, EMN
13	04	24	47	00	225	
14	04	59	46	25	150	HB 9, SNR
15	05	01.2	38	00	15	CTA 34, 3C 134
16	05	12.4	33	48	40	IC 405, EMN
17	05	18.7	33	26	60	IC 410, EMN
18	05	31.5	21	59	1,030	M 1, Crab nebula, SNR
19	05	37.1	36	00	20	Sharpless 232, EMN
20	06	14.3	22	36	195	IC 443, SNR
21	06	29.4	04	53	342	Rosette nebula, EMN
22	07	03	−10	45	125	EMN
23	07	48	−26	00	13	NGC 2467, EMN
24	08	15	−36	00	8	NGC 2568
25	08	21.4	−42	58	187	Puppis A, SNR
26	08	32.8	−45	37	1,400	Vela X
27	08	37.6	−40	34	33	Stromlo 14, EMN
28	08	43	−43	31	120	Vela Y
29	08	45.7	−45	03	40	Vela Z
30	08	48.1	−42	15	33	Stromlo 17, EMN
31	08	57.4	−47	16	250	Stromlo 23
32	08	57.8	−43	34	59	Stromlo 20
33	16	31.3	−47	42	274	
34	16	36.8	−46	36	300	
35	16	56.7	−40	13	66	EMN

† At 960 Mc in units of 10^{-26} watt m^{-2} cps^{-1}.

¶ EMN, emission nebula; SNR, supernova remnant.

California Institute of Technology list B (CTB) (Continued)

Number	RA h	RA m	Dec Deg	Dec Min	Flux density†	Remarks
CTB 36	17	01	−44	20	30	
37	17	12.8	−38	03	100	
38	17	15.7	−39	00	37	
39	17	17.7	−35	58	180	NGC 6334, EMN
40	17	23	−34	21	540	NGC 6357, EMN
41	17	28.7	−21	36	13	SNR of 1604
42	17	42.9	−28	50	1,800	Sagittarius A, galactic nucleus
43	17	48	−17	15	35	MSH 17-110
44	17	57	−32	15	15	
45	17	57.8	−23	30	240	M 20, NGC 6514, EMN
46	18	00.7	−24	21	106	M 8, NGC 6523, EMN
47	18	01.7	−21	39	210	
48	18	04.8	+00	40	30	
49	18	06	−20	00	165	
50	18	15.3	−12	00	350	NGC 6604, EMN
51	18	16.2	−13	50	106	NGC 6611, EMN
52	18	17.5	−16	18	500	M 17, NGC 6618, EMN
53	18	23.5	−12	30	200	
54	18	27.3	+00	30	67	
55	18	28.6	−02	12	26	
56	18	31.2	−08	45	135	
57	18	33.8	−07	15	65	
58	18	34	−12	12	17	
59	18	44.7	−02	06	345	
60	18	53.6	+01	18	240	MSH 18+011, 3C 392
61	18	54.6	08	14	22	
62	18	55	09	15	6	
63	18	56.4	15	37	60	
64	18	57.7	04	04	87	
65	19	01.0	05	32	27	3C 396
66	19	01.4	11	14	25	
67	19	04.6	07	12	27	3C 397
68	19	07.9	09	02	110	3C 398
69	19	08.4	05	04	31	MSH 19+01
70	19	13.8	11	36	44	
71	19	15.7	02	05	12	
72	19	15.9	06	00	23	
73	19	20.7	14	04	370	
74	19	22.8	16	00	60	
75	19	39	02	30	70	

California Institute of Technology list B (CTB) (Continued)

Number	RA h	RA m	Dec Deg	Dec Min	Flux density†	Remarks
CTB 76	19	40	26	06	15	NGC 6823, EMN
77	19	44	24	45	20	
78	19	44.8	28	00	15	
79	19	46	09	30	320	
80	19	51.8	33		130	
81	19	57.7	40	36	2,190	Cygnus A
82	19	59.2	33	24	34	
83	19	59.4	35		16	
84	20	05	34	06	18	
85	20	12	36	05	80	
86	20	14.3	33	55	50	
87	20	14.3	37	03	10	
88	20	16.6	45	00	200	
89	20	19.5	40	52	190	
90	20	19.7	37	02	120	
91	20	21	40		800	HB 20, γ Cygni nebula
92	20	27	41	10	31	
93	20	30.4	43	48	90	
94	20	32	39	30	220	
95	20	32.7	46	58	100	
96	20	37.5	41	55	800	
97	20	43.5	50	12	186	HB 21, SNR
98	20	45.2	48	03	6	
99	20	48.2	29	30	252	Cygnus loop, SNR
100	20	52.5	44	09	350	NGC 7000, IC 5068, IC 5070
101	20	05.9	49	48	13	
102	21	10.6	52	16	60	
103	21	20	44	03	80	Sharpless 119, EMN
104	21	30.4	50	38	66	
105	21	35.4	57	30	210	HB 22, IC 1396, EMN
106	22	18.5	55	58	190	
107	22	22.6	63	17	100	
108	22	56.3	62	20	60	
109	23	00.3	58	45	75	
110	23	21.2	58	32	3,120	Cassiopeia A, SNR

e.   Sydney (MSH) List of Radio Sources

86 Mc; 340 Sources

References: B. Y. Mills, O. B. Slee, and E. R. Hill: A Catalogue of Radio Sources between Declinations +10° and −20°, *Australian J. Phys.*, vol. 11, pp. 360–377, 1958; B. Y. Mills, O. B. Slee, and E. R. Hill: A Catalogue of Radio Sources between Declinations −20° and −50°, *Australian J. Phys.*, vol. 13, pp. 676–699, 1960; B. Y. Mills, O. B. Slee, and E. R. Hill: A Catalogue of Radio Sources between Declinations −50° and −80°, *Australian J. Phys.*, vol. 14, pp. 497–507, 1961; E. R. Hill and B. Y. Mills: Source Corrections to the Sydney Radio Source Survey Catalogue, *Australian J. Phys.*, vol. 15, pp. 437–440, 1962.   See page 412 for meaning of superscripts *a*, *b*, *c*, etc.

	Position (1950.0)				
	RA		Dec		Flux
Number	h	m	Deg	Min	density†
00−01	00	03.3	−00	56	35
00+01	00	04.9	+06	05	35
00−43[d]	00	07.96	−44	40.0	60 (6, 2, 1)[d]
00−16	00	12.4	−15	07	34
00−17	00	15.9	−13	02	52
00−29[b]	00	22.00	−29	44.7	33 (8, 3, 2)[d]
00+06	00	30.0	+01	40	68
00−313	00	36.2	−39	24	30
00−09	00	36.4	−02	50	120
00+011	00	37.4	+09	30	37
00−011	00	39.2	−09	43	56
00−410[d]	00	39.77	−44	30.8	35 (12, 4, 2)[d]
00−411[d]	00	43.87	−42	24.3	52
00−71[d]	00	55.0	−73	00	550 (110, 67, —)[d]
00−017	00	54.5	−01	39	90
01−21	01	00.1	−22	11	35
01−41[d]	01	03.10	−45	22.0	41 (8, 3, 2)[d]
01−12	01	05.9	−16	15	53
01−35	01	07.9	−35	01	30
01−45	01	14.0	−47	34	34
01+03	01	17.3	+03	20	33
01−19	01	18.0	−15	34	45
01−05	01	23.5	−01	35	88
01−111	01	25.1	−14	13	30
01−48	01	25.2	−41	17	33

† At 86 Mc in units of 10^{-26} watt m^{-2} cps^{-1}; only sources with flux densities of 30 or more units are included in the list here.   The original lists include many more (weaker) sources.

Sydney (MSH) list of radio sources (Continued)

Number	RA h	m	Dec Deg	Min	Flux density†
01−311^d	01	31.70	−36	44.6	56 (16, 7, 3)^d
01−217‡	01	48.7	−29	44	63
01+013	01	52.1	+03	32	49
02−13	02	08.0	−11	18	30
02−15	02	13.2	−13	19	42
02−43^d	02	14.88	−48	03.4	65 (10, 2, 1)^d
02−07	02	18.6	−02	11	74
02−71	02	21.3	−70	50	30
02−110^b	02	35.41	−19	44.8	44 (13, 4, 2)^d
02−014	02	40.0	−00	09	35
02−53^d	02	41.87	−51	22.7	37 (12, 3, 1)^d
02−54^d	02	45.45	−55	54.2	48 (12, 2, 1)^d
02−65	02	51.7	−67	42	34
02+010	02	55.1	+05	53	51
03+03	03	05.4	+03	50	34
03−31¶	03	20.6	−37	23	950
03+05	03	25.2	+02	25	41
03−03	03	31.7	−01	25	64
03+08	03	40.5	+04	55	35
03−17	03	44.1	−11	13	34
03−36^d	03	44.67	−34	31.5	33 (9, 3, 2)^d
03−19^b	03	49.16	−14	38.1	44
03−212^b	03	49.55	−27	53.2	53 (15, 10, 8, 6, —)^c
04+03	04	04.7	+03	45	37
04−12^b	04	05.5	−12	19.5	31 (9, 5, 4, 3, —)^c
04−62^d	04	08.10	−65	49	36 (45, 15, —)^d
04−71^d	04	10.18	−75	15.3	87
04−06	04	09.6	−01	02	35
04−07	04	15.5	−05	35	36
04−63^d	04	20.5	−62	30	38 (9, 3, —)^d
04−36^d	04	27.87	−36	37.5	35 (7, 2, 1)^d
04−54^d	04	27.85	−53	56.1	50 (15, 6, 3)^d
04−64^d	04	28.80	−61	34	35 (7, 2, —)^d
04−112^b	04	31.85	−13	30.6	38
04+010	04	41.8	+02	15	30

‡ Complex.

¶ Fornax A.

Sydney (MSH) list of radio sources (Continued)

Number	RA h	RA m	Dec Deg	Dec Min	Flux density†
Position (1950.0)					
04−218[b]	04	42.63	−28	14.7	82 (22, 7, 4)[d]
04−017	04	47.3	−04	33	46
04−314[d]	04	53.35	−30	11.3	43 (13, 3, 2)[d]
04−221[d]	04	54.08	−22	03.7	21 (4, 2, 1)[d]
05−22	05	03.6	−26	41	60
05−13	05	08.5	−18	42	41
05+02	05	10.9	+01	02	38
05−42[d]	05	11.58	−48	28.0	41 (13, 4, 2)[d]
05−63[d]‡	05	22.0	−69	00	4,000 (1,100, 620, —)[d]
05−43¶	05	18.40	−45	49.8	570
05−36[d]	05	21.23	−36	30.0	66 (37, 19, 11)[d]
05−64[d]	05	26.0	−66	08	34
05+05	05	28.9	+06	35	30
05−011§	05	32.5	−05	24	83
05−012	05	38.0	−02	20	88
05−310	05	38.1	−33	40	30
05−66	05	39.8	−69	20	108
05−410[d]	05	47.80	−40	51.9	31 (9, 3, 1)[d]
06+03	06	05.4	+08	08	109
06−15	06	17.8	−16	36	63
06+06	06	20.3	+09	00	153
06−53[d]	06	20.60	−52	39.5	30 (9, 3, 2)[d]
06−04	06	25.0	−05	56	120
06−55[d]	06	25.30	−53	39.3	172 (26, 7, 4)[d]
06+08*	06	29.6	+05	01	250
06−210[d]	06	34.42	−20	34.3	85 (19, 10, 10, 7, —)[c]
06+010	06	34.8	+07	15	72
06−07	06	39.0	−08	01	50
06−111	06	42.2	−10	19	84
06−08	06	45.0	−02	06	33
06−113	06	49.7	−12	43	55
06−216[d]	06	56.53	−24	12.7	59 (13, 3, 1)[d]
07−11	07	03.2	−11	02	55
07−22	07	06.9	−29	14	45
07−23[d]	07	09.65	−20	37.3	33 (9, 2, 1)[d]

‡ Centroid of LMC.
¶ Pictor A.
§ Orion nebula, M 42.
* Rosette nebula.

Sydney (MSH) list of radio sources (Continued)

	Position (1950.0)				
	RA		Dec		Flux
Number	h	m	Deg	Min	density†
07−04	07	22.3	−09	49	36
07−05	07	23.1	−06	10	94
07+04	07	41.9	+02	05	36
07−117	07	45.5	−19	00	52
08−11	08	00.3	−14	40	30
08−31	08	07.6	−38	50	38
08−05	08	13.4	−02	53	35
08+03	08	19.8	+06	07	125
08−44‡	08	20.9	−42	52	690
08−34	08	28.5	−32	51	30
08−45	08	33.7	−45	38	1,100
08−71[d]	08	42.20	−75	29.5	58 (11, 4, 2)[d]
08−53	08	44.7	−57	20	30
08−219[b]	08	58.6	−25	43.2	54 (16, 6, 3)[d]
09+02	09	15.2	+09	35	40
09−14¶	09	15.7	−11	53	690
09−52[d]	09	16.10	−54	42.7	36 (5, 3, 2)[d]
09−09	09	42.0	−09	39	100
09−212[d]	09	55.83	−28	50.2	30 (6, 1, 1)[d]
10−21[d]	10	02.87	−21	33.3	48 (6, 5, 3, 2, —)[c]
10−51	10	05.6	−56	53	68
10+01	10	05.7	+07	54	30
10−12	10	07.6	−11	47	32
10+02	10	08.6	+06	32	39
10−53	10	17.2	−58	50	71
10−44§	10	17.9	−42	26	51
10−54	10	20.3	−57	34	151
10+06	10	24.0	+06	41	35
10−57	10	43.5	−59	30	339
11−61	11	00.0	−60	36	162
11−11	11	00.6	−15	01	56
11−62	11	10.2	−60	38	42
11−08	11	16.9	−02	46	31
11−54	11	22.2	−59	10	39
11−44	11	23.0	−48	13	53

‡ Puppis A.
¶ Hydra A.
§ Complex.

Sydney (MSH) list of radio sources (Continued)

Number	RA h	RA m	Dec Deg	Dec Min	Flux density†
11−55	11	24.1	−56	04	30
11−16	11	30.9	−19	22	32
11−63	11	34.0	−62	26	100
11−64[d]	11	36.28	−67	55.2	34 (7, 3, 2)[d]
11−18	11	36.5	−13	41	44
11−65	11	37.0	−63	02	38
12−12	12	02.4	−17	39	48
12−13	12	04.0	−12	53	56
12−51	12	08.1	−52	20	182
12−61	12	10.5	−62	25	140
12+04	12	14.8	+04	00	30
12+05	12	16.7	+05	59	100
12−63	12	17.4	−63	40	30
12+08	12	26.6	+02	17	167
12−110	12	28.4	−16	59	38
12−012	12	37.1	−07	19	52
12−45[d]	12	45.90	−41	01.7	45 (10, 4, 2)[d]
12−212	12	51.7	−29	03	47
12−118	12	52.3	−12	19	53
13−23[d]	13	09.00	−22	00.9	61 (22, 5, 3)[d]
13−05	13	12.8	−08	05	45
13+06	13	18.7	+01	00	50
13−42‡	13	22.4	−42	41	8,700
13−33¶	13	33.9	−33	42	70
13−25[d]	13	34.18	−29	36.3	36 (9, 3, 1)[d]
13−011	13	35.7	−06	21	35
13−013	13	43.0	−07	48	53
13−62	13	43.4	−60	07	795
13−44	13	47.1	−45	36	100
13+011	13	50.0	+06	19	54
13−45[d]	13	55.93	−41	38.0	35 (13, 5, 3)[d]
13−52	13	59.6	−57	56	71
14−32[d]	14	00.97	−33	48.0	57 (10, 1, 0.3)[d]
14−52	14	05.7	−51	20	39
14−53	14	05.7	−59	08	44

‡ Centaurus A.

¶ Complex.

Sydney (MSH) list of radio sources (Continued)

Number	Position (1950.0)				Flux density†
	RA		Dec		
	h	m	Deg	Min	
14−61	14	06.3	−61	24	196
14−62	14	09.9	−65	05	35
14−42	14	12.6	−43	23	38
14−14	14	16.0	−15	47	34
14+05	14	16.7	+06	43	114
14−55	14	19.6	−55	53	36
14−28^d	14	20.10	−27	14.2	40 (9, 3, 1)^d
14−45	14	25.2	−43	07	90
14+010^b	14	34.5	+03	37.1	47
14−57	14	37.8	−59	44	167
14−63	14	38.9	−62	24	110
14−49	14	45.0	−46	57	33
14−118	14	46.9	−15	53	42
14−58	14	50.0	−59	03	71
14−64	14	51.9	−60	50	147
14−121^b	14	53.21	−10	56.8	41
14−415	14	59.9	−41	42	55
15−23	15	05.8	−29	23	35
15+02	15	08.2	+08	09	42
15−16	15	10.6	−19	23	49
15−52	15	11.5	−58	55	252
15+05	15	14.2	+07	11	140
15−53	15	15.6	−57	30	55
15+07	15	19.3	+07	55	50
15−43	15	27.4	−42	21	100 (13, 9, 7, 4, —)^c
15−61	15	28.0	−60	50	81
15−54	15	28.3	−56	08	50
15−012	15	38.1	−01	54	37
15−46	15	41.2	−48	04	37
15−55	15	44.0	−54	30	70
15−56	15	48.6	−55	56	323
15−57	15	50.8	−53	14	146
15−48	15	55.3	−46	21	50
15−213^d	15	56.13	−21	32.2	36 (9, 3, 1)^d
15−58	15	58.0	−52	07	91

Sydney (MSH) list of radio sources (Continued)

Number	Position (1950.0) RA h	m	Dec Deg	Min	Flux density†
16+01	16	00.0	+02	13	100
16−21^d	16	02.12	−28	50.4	35 (8, 3, 1)^d
16+02	16	02.8	+01	05	45
16+03^b	16	04.7	+00	07.9	35
16−61^d	16	10.78	−60	48.2	1,190 (—, 56, —)^d
16−32	16	10.9	−39	24	56
16−43	16	19.9	−49	28	76
16−71	16	22.6	−75	11	38
16−44	16	24.9	−48	40	64
16−45	16	25.1	−42	10	30
16−62	16	33.8	−61	47	30
16−47	16	36.3	−46	31	330
16−117	16	40.4	−15	19	30
16+09	16	44.7	+01	43	33
16−119	16	45.4	−10	48	37
16+010‡	16	48.8	+05	04	890
16−012	16	49.3	−00	18	80
16−48	16	51.1	−44	00	55
16−49	16	52.1	−42	25	32
16−014	16	52.6	−02	17	60
16−410	16	57.6	−46	22	30
16−411	16	59.0	−41	18	90
17−31	17	02.1	−31	54	32
17+01	17	03.2	+09	16	78
17−12	17	05.4	−17	13	60
17−32	17	06.6	−37	10	100
17−23	17	09.5	−23	18	42
17−42	17	10.4	−45	36	35
17−25	17	10.8	−25	00	47
17−33	17	11.2	−38	23	500
17−26	17	13.2	−29	50	38
17−61	17	13.4	−62	10	35
17−05	17	16.9	−04	25	31
17−27	17	17.9	−28	56	32
17−06	17	18.1	−00	55	475

‡ Hercules A.

Sydney (MSH) list of radio sources (Continued)

Number	RA h	RA m	Dec Deg	Dec Min	Flux density†
17−16	17	19.4	−18	45	150
17−34	17	19.8	−35	40	75
17−71	17	21.2	−70	01	46
17−35	17	21.8	−38	14	75
17+02	17	22.3	+05	44	36
17−36	17	26.2	−34	45	60
17−37	17	26.7	−34	02	120
17−211‡	17	27.8	−21	29	110
17−38	17	28.3	−32	43	95
17−39	17	36.6	−30	50	165 (52, 34, 27, 22, —)^c
17−011	17	37.7	−01	18	39
17−310	17	39.9	−36	01	31
17−213¶	17	43.7	−28	41	4,500
17−012	17	48.1	−02	06	53
17−110	17	48.7	−17	28	30
17−214	17	52.2	−21	30	194
17−215	17	52.4	−23	05	82
17−014	17	54.4	−05	34	55
17−48	17	54.4	−41	33	30
17−015	17	55.7	−01	24	50
17+03	17	56.1	+02	46	48
17−216	17	57.4	−23	26	900
17−217	17	59.9	−28	00	35
18−11	18	00.1	−17	49	40
18−21	18	02.7	−21	23	253
18+01	18	03.9	+00	12	33
18−23	18	04.0	−20	10	70
18−41	18	04.6	−45	28	77
18−02	18	05.2	−00	59	62
18−24	18	05.8	−27	00	34
18−13^b	18	11.28	−17	12.9	160
18−03	18	12.4	−05	59	82
18−25	18	12.8	−24	09	114
18−04	18	14.9	−07	03	30
18−15	18	14.9	−10	57	35

‡ Kepler's SNR of 1604.
¶ Sagittarius A, galactic nucleus.

Sydney (MSH) list of radio sources (Continued)

Number	RA h	RA m	Dec Deg	Dec Min	Flux density†
18−61	18	16.1	−64	00	71
18−27	18	16.2	−25	45	30
18−05	18	17.6	−09	32	50
18−33	18	17.8	−39	11	41
18−06	18	20.6	−01	34	76
18−17	18	21.5	−13	50	40
18−18	18	21.8	−12	24	150
18−28	18	22.5	−20	05	70
18−71	18	22.6	−75	11	38
18−29	18	23.8	−24	03	31
18−19	18	25.0	−11	17	50
18−07	18	25.3	−04	38	40
18+05	18	26.5	+00	25	50
18−111	18	27.5	−12	46	40
18−112	18	28.7	−14	36	30
18+06	18	29.6	+09	40	45
18−113	18	30.1	−10	01	230
18−08	18	31.6	−08	42	160
18−211	18	33.6	−23	59	39
18−43	18	39.8	−48	36	41
18−010	18	41.7	−03	51	180
18−114	18	42.1	−19	40	56
18+08	18	42.7	+09	30	54
18−014	18	53.0	−02	42	150
18+011	18	53.7	+01	29	550
18−015	18	57.9	−04	13	34
19−21	19	00.1	−23	33	35
19−01	19	04.2	−03	06	53
19+01	19	09.0	+05	05	59
19−23	19	12.8	−26	58	53
19−24	19	13.6	−24	49	35
19−28	19	29.2	−26	40	42
19+05	19	32.4	+09	43	30
19−46[d]	19	32.33	−46	28.2	141 (35, 23, 16, 12, —)[c]
19−110	19	32.2	−10	55	75

Sydney (MSH) list of radio sources (Continued)

Number	RA h	m	Dec Deg	Min	Flux density†
19+07	19	37.4	+04	13	34
19−111ᵇ	19	38.4	−15	31.5	38
19−410	19	40.60	−40	37.8	38 (6, 2, 0.4)ᵈ
19−011	19	40.8	−07	29	33
19+010	19	49.8	+02	26	64
19−413ᵈ	19	53.67	−42	30.9	31 (10, 4, 2)ᵈ
19−57ᵈ	19	54.35	−55	17.8	54
19−35ᵈ	19	55.72	−35	43.3	45
20−52ᵈ	20	06.52	−56	37.8	81 (12, 2, 0.4)ᵈ
20−55ᵈ	20	20.47	−57	33.9	36 (9, 4, 2)ᵈ
20−37ᵈ	20	32.58	−35	05.1	41 (13, 10, 7, 5, —)ᶜ
20−61ᵈ	20	41.32	−60	29.7	55
20+012	20	55.4	+00	42	104
20−215ᵈ	20	58.63	−28	13.3	59 (16, 7, 3)ᵈ
21−21ᵈ	21	04.38	−25	39.0	100 (27, 18, 15, 11, 4)ᶜ
21−19	21	20.2	−16	49	30
21+05	21	27.1	+01	06	67
21−115	21	34.7	−14	39	33
21−47ᵈ	21	40.38	−43	27.1	27 (10, 4, 2)ᵈ
21−64ᵈ	21	52.97	−69	55.8	253 (80, 32, 18)ᵈ
22+03	22	10.8	+08	48	31
22−17	22	12.0	−17	11	127
22−06	22	16.3	−03	46	33
22−09	22	21.5	−02	18	60
22−010	22	23.1	−05	13	30
22−52ᵈ	22	23.83	−52	49.4	30 (8, 3, 1)ᵈ
22−71	22	44.3	−77	19	43
22−46ᵈ	22	50.12	−41	14.3	42 (13, 5, 3)ᵈ
22−019	22	53.9	−00	18	32
23+05	23	14.0	+03	53	57
23−112ᵇ	23	22.7	−12	23.9	30
23−010	23	24.3	−05	15	35
23−44ᵈ	23	31.75	−41	42.8	50 (16, 6, 3)ᵈ
23−37ᵈ	23	54.45	−35	02.4	39 (6, 1, 0.5)ᵈ
23−64ᵈ	23	56.40	−61	11.7	296 (66, 22, 11)ᵈ

f.   The Ohio (OA) List of Radio Sources

600 and 1,415 Mc; 128 Sources

Detected at the Ohio State University Radio Observatory.   The prefix OA stands for Ohio list A.   *References:* J. D. Kraus: Maps of M 31 and Surroundings at 600 and 1,415 Megacycles per Second, *Nature*, vol. 202, pp. 269–272, Apr. 18, 1964; R. S. Dixon, S. Y. Meng, and J. D. Kraus: Maps of the Perseus Region at 600 and 1,415 Megacycles per Second, *Nature*, vol. 205, pp. 755–758, Feb. 20, 1965; J. D. Kraus and R. S. Dixon: A High Sensitivity Survey of M 31 and Surroundings at 1,415 Mc/sec, *Nature*, vol. 207, pp. 587–589, Aug. 7, 1965; J. D. Kraus, R. S. Dixon, and R. O. Fisher: A New High Sensitivity Study of the M 31 Region at 1,415 Mc/sec, *Astrophys. J.*, vol. 144, May, 1966.

	Position (1950.0)					Flux density†		Remarks
	RA			Dec				
Source	h	m	s	Deg	Min	600 Mc	1,415 Mc	
OA 2	23	42	05	38	56			
4	23	44	42	42	37		0.4	
5	23	49	18	40	56		0.6	
6	23	49	34	45	01		0.9	
8	23	51	27	39	53	2.7	0.7	
10	23	55	55	45	46		2.0	
12	23	56	15	39	11			
14	23	58	23	40	37	2.7	1.0	
16	23	58	35	41	33		0.6	
16.1	00	04	14	41	16		1.0	
17	00	06	33	39	44		0.6	
18	00	10	54	40	33	3.1	2.3	
20	00	12	55	38	41			
22	00	17	30	45	01			
23	00	15	37	42	07		0.2	
24	00	19	01	43	15	2.0	0.6	
25	00	21	53	39	58		0.5	
26	00	22	35	39	05			
27	00	22	51	42	25		0.4	
28	00	27	08	39	45		0.2	
29	00	29	52	39	36		0.3	
30	00	34	58	38	39		1.1	
32	00	35	32	39	18			
33	00	35	41	41	20		0.4	
34	00	39	00	39	17			

† In units of 10^{-26} watt m^{-2} cps^{-1}.

The Ohio (OA) list of radio sources (Continued)

Source		Position (1950.0)				Flux density†		Remarks
		RA		Dec				
	h	m	s	Deg	Min	600 Mc	1,415 Mc	
OA 35	00	39	28	37	19		0.9	
35.1	00	37	35	40	16		0.2	M 31 feature‡
35.3	00	39	54	41	02		0.4	M 31 feature
35.4	00	41	13	41	05		0.2	M 31 feature
35.5	00	42	20	41	30		0.3	M 31 feature
35.6	00	41	57	42	30		0.7	M 31 feature
35.8	00	43	53	42	08		0.2	M 31 feature
36	00	39	32	39	42		0.7	M 31 feature
37	00	45	36	40	03	4.0	0.8	
37.1	00	49	05	42	30		0.3	
38	00	51	33	40	31	2.7	0.8	
38.1	00	52	59	39	26		0.5	
39	00	53	19	41	12		0.4	
40	00	54	05	43	26		0.9	
42	01	00	35	42	19		0.6	
43	01	02	00	39	56		0.2	
44	01	03	15	43	26			
45	01	03	58	42	09		0.7	
46	01	07	01	39	46		0.4	
48	01	08	45	43	22			
50	01	09	23	41	35	2.0	1.0	
52	01	10	32	40	10		0.6	
53	01	12	23	40	17		1.2	
54	01	13	33	39	58		0.5	
56	01	14	23	41	45		0.2	
58	01	15	25	42	45		0.3	
59	01	20	31	40	35		0.7	
60	01	21	14	42	52		0.4	
61	01	28	21	39	45		0.4	
62	01	30	25	43	21			
64	01	36	37	39	41		0.9	
66	01	40	06	44	20			
67	01	44	50	43	16		0.9	
68	01	44	55	44	40			
69	01	46	40	36	16			

‡ The integrated total flux density of M 31 is 14.0 flux units at 600 Mc and 8.0 flux units at 1,415 Mc.

The Ohio (OA) list of radio sources (Continued)

	Position (1950.0)						Flux density†		
	RA			Dec					
Source	h	m	s	Deg	Min		600 Mc	1,415 Mc	Remarks
OA 70	01	47	17	39	42			0.4	
72	01	53	22	41	39			0.5	
74	01	54	20	45	09				
76	01	57	05	40	33		3.4	1.3	
78	01	57	30	44	11			1.4	
80	01	58	00	43	19				
82	02	00	00	39	54				
84	02	01	15	41	56		2.0	0.8	
86	02	01	38	36	16				
88	02	01	58	43	46				
100	02	16	02	42	25		1.8	1.2	
102	02	18	42	39	44			1.4	
104	02	22	42	40	25			1.7	
106	02	28	02	40	02			1.0	
107	02	28	44	41	03			1.2	
108	02	32	34	37	21			1.0	
109	02	38	42	39	44			1.4	
110	02	41	28	39	22			1.9	
112	02	44	22	37	39			1.0	
114	02	47	08	39	25		4.6	2.4	
116	02	48	28	43	05			1.2	
117	03	01	29	44	00			1.0	
118	03	07	18	44	32			1.0	
120	03	07	36	42	37		2.8	1.2	
122	03	09	05	39	05		4.6	2.6	
124	03	09	35	41	23			1.7	
126	03	17	17	42	42			1.9	
128	03	18	25	44	00			1.2	
130	03	25	40	38	03			1.9	
132	03	26	01	39	24		4.6	2.2	
133	03	41	48	39	41			1.4	
134	03	45	30	40	46			1.7	
138	03	50	03	38	25		3.7	1.7	
140	03	54	35	41	45		3.7	1.9	
142	04	02	34	37	53			1.0	

The Ohio (OA) list of radio sources (Continued)

Source	Position (1950.0)					Flux density†		Remarks
	RA			Dec				
	h	m	s	Deg	Min	600 Mc	1,415 Mc	
OA 144	04	05	49	38	44	5.5	2.5	
146	04	07	03	39	22		1.2	
148	04	18	33	37	27		1.2	
149	04	19	51	40	45		2.4	
150	04	19	56	38	02		1.9	
151	04	19	56	36	41		1.2	
152	04	20	32	41	45		3.0	
154	04	20	38	43	22		1.7	
156	04	24	17	41	45		1.4	
158	04	25	55	42	25	1.8	1.2	
160	04	27	00	40	45		1.2	
162	04	27	24	43	22	1.8	1.0	
164	04	28	11	44	33		1.2	
166	04	43	26	38	42		1.4	
168	04	47	05	43	51		1.0	
169	04	50	18	38	42		1.4	
170	04	52	37	42	02		2.2	
172	04	54	24	37	22		1.4	
174	04	55	05	40	45		1.4	
176	04	56	07	43	22	1.8	1.2	
178	05	02	35	37	45		1.4	
180	05	03	24	40	33		1.2	
181	05	09	42	40	42		1.9	
182	05	12	56	38	02		1.9	
184	05	13	23	41	45	12.1	8.3	SNR or EMN?‡
186	05	16	08	39	03		1.4	
188	05	22	22	42	42	3.7	2.2	SNR or EMN?
190	05	23	42	40	35		1.2	
192	05	32	41	38	02	1.8	1.4	
193	05	40	12	45	22		2.2	
194	05	45	25	38	25	2.7	1.0	
196	05	48	13	37	22		1.4	
198	05	51	47	39	42	2.7	1.7	
200	05	57	36	38	02		1.4	

‡ SNR, supernova remnant; EMN, emission nebula.

Messier's List of
Nebulous Objects

A considerable number of these 103 objects have been found to be radio sources. *Reference:* H. Shapley and H. Davis: Catalogue of Nebulae and Clusters, *Observatory*, vol. 41, pp. 318–321, August, 1918. This reference gives positions for epoch 1900.0. These have been precessed to 1950.0. Messier's catalog contains nearly all the brightest and most striking nebulas and clusters north of a declination of −35°. Messier's original list was published in "Connaissance des temps" in 1784.

	Position (1950.0)				
	RA		Dec		
Number	h	m	Deg	Min	Remarks
M 1	05	31.5	+21	59	SNR,† Crab nebula, Taurus A, NGC 1952
2	21	30.9	−01	03	Globular cluster, NGC 7089
3	13	39.9	+28	38	Globular cluster, NGC 5272
4	16	20.6	−26	24	Globular cluster, NGC 6121
5	15	16.0	+02	16	Globular cluster, NGC 5904
6	17	36.8	−32	11	Open cluster, NGC 6405
7	17	50.6	−34	48	Open cluster, NGC 6475
8	18	00.7	−24	23	Irregular nebula, NGC 6523
9	17	16.2	−18	28	Globular cluster, NGC 6333
10	16	54.5	−04	02	Globular culster, NGC 6254
11	18	48.4	−06	20	Open cluster, NGC 6705
12	16	44.6	−01	52	Globular cluster, NGC 6218
13	16	39.9	+36	33	Globular cluster, NGC 6205
14	17	35.0	−03	13	Globular cluster, NGC 6402
15	21	27.6	+11	57	Globular cluster, NGC 7078
16	18	16.0	−13	48	Open cluster, EMN,† NGC 6611
17	18	17.9	−16	12	Horseshoe or Omega nebula, EMN, NGC 6618
18	18	17.0	−17	09	Open cluster, NGC 6613
19	16	59.5	−26	12	Globular cluster, NGC 6273
20	17	59.3	−23	02	Trifid nebula, NGC 6514

† SNR, supernova remmant; EMN, emission nebulosity.

Messier's list of nebulous objects (Continued)

Number	Position (1950.0)				Remarks
	RA		Dec		
	h	m	Deg	Min	
M 21	18	01.6	−22	30	Open cluster, NGC 6531
22	18	33.4	−23	57	Globular cluster, NGC 6656
23	17	53.9	−19	01	Open cluster, NGC 6494
24	18	15.5	−18	26	Open cluster, NGC 6603
25	18	28.7	−19	17	Open cluster, IC 4725
26	18	42.5	−09	27	Open cluster, NGC 6694
27	19	57.5	+22	35	Dumbbell nebula, NGC 6853
28	18	21.5	−24	54	Globular cluster, NGC 6626
29	20	22.1	+38	22	Open cluster, NGC 6913
30	21	37.5	−23	25	Globular cluster, NGC 7099
31	00	40.0	+41	00	Andromeda galaxy, NGC 224
32	00	39.9	+40	35	Andromeda galaxy subsystem, NGC 221
33	01	31.0	+30	24	Spiral galaxy, NGC 598
34	02	38.8	+42	34	Open cluster, NGC 1039
35	06	05.8	+24	21	Open cluster, NGC 2168
36	05	32.8	+34	06	Open cluster, NGC 1960
37	05	49.1	+32	32	Open cluster, NGC 2099
38	05	25.4	+35	48	Open cluster, NGC 1912
39	21	30.4	+48	13	Open cluster, NGC 7092
40	12	19.8	+58	23	Two faint stars
41	06	44.8	−20	41	Open cluster, NGC 2287
42	05	32.9	−05	25	Orion nebula, NGC 1976
43	05	33.1	−05	18	Irregular nebula, NGC 1982
44	08	37.2	+20	10	Praesepe, NGC 2632
45	03	44.5	+23	57	Pleiades (position of Alcyone)
46	07	39.5	−14	42	Open cluster, NGC 2437
47	07	52.5	−15	17	Cluster, NGC 2478
48	08	11.5	−01	48	Very open cluster
49	12	27.2	+08	16	Nebula, NGC 4472
50	07	00.6	−08	16	Open cluster, NGC 2323
51	13	27.8	+47	27	Spiral galaxy, NGC 1594
52	23	22.0	+61	19	Cluster, NGC 7654
53	13	10.5	+18	26	Globular cluster, NGC 5024
54	18	50.9	−30	33	Globular cluster, NGC 6715
55	19	36.9	−31	03	Globular cluster, NGC 6809

Messier's list of nebulous objects (Continued)

Number	RA h	RA m	Dec Deg	Dec Min	Remarks
M 56	19	14.6	+30	05	Globular cluster, NGC 6779
57	18	51.8	+32	58	Ring nebula in Lyra, NGC 6720
58	12	35.2	+12	05	Spiral galaxy, NGC 4579
59	12	39.5	+11	56	Spiral galaxy, NGC 4621
60	12	41.1	+11	50	Nebula, NGC 4649
61	12	19.4	+04	45	Spiral galaxy, NGC 4303
62	16	58.0	−30	03	Globular cluster, NGC 6266
63	13	13.5	+42	18	Spiral galaxy, NGC 5055
64	12	54.3	+21	57	Spiral galaxy, NGC 4826
65	11	18.3	+13	22	Spiral galaxy, NGC 3623
66	11	17.6	+13	16	Spiral galaxy, NGC 3627
67	08	48.5	+12	00	Open cluster, NGC 2682
68	12	36.8	−26	29	Globular cluster, NGC 4590
69	18	28.1	−32	23	Globular cluster, NGC 6637
70	18	40.0	−32	20	Globular cluster, NGC 6681
71	19	51.5	+18	39	Open cluster, NGC 6838
72	20	50.8	−12	44	Globular cluster, NGC 6981
73	20	56.2	−12	50	Open cluster, NGC 6994
74	02	34.0	+15	29	Spiral galaxy, NGC 628
75	20	03.2	−22	04	Globular cluster, NGC 6864
76	01	39.1	+51	19	Nebula, NGC 650
77	02	40.2	−00	13	Spiral galaxy, NGC 1068
78	05	44.2	+00	02	Irregular nebula, NGC 2068
79	05	22.2	−24	34	Globular cluster, NGC 1904
80	16	14.1	−22	52	Globular cluster, NGC 6093
81	09	51.5	+69	18	Spiral galaxy, NGC 3031
82	09	51.7	+69	56	Exploding galaxy, NGC 3034
83	13	34.2	−29	36	Spiral galaxy, NGC 5236
84	12	22.5	+13	09	Nebula, NGC 4374
85	12	22.9	+18	28	Nebula, NGC 4382
86	12	23.6	+13	13	Nebula, NGC 4406
87	12	28.3	+12	40	Elliptical galaxy, Virgo A, NGC 4486
88	12	29.4	+14	41	Spiral galaxy, NGC 4501
89	12	33.1	+12	49	Nebula, NGC 4552
90	12	34.3	+12	26	Spiral galaxy, NGC 4569

Messier's list of nebulous objects (Continued)

	Position (1950.0)				
	RA		Dec		
Number	*h*	*m*	*Deg*	*Min*	*Remarks*
M 91	12	38.5	+13	33	?
92	17	42.6	+43	14	Globular cluster, NGC 6341
93	07	42.5	−23	45	Open cluster, NGC 2447
94	12	48.6	+41	24	Spiral galaxy, NGC 4736
95	10	41.3	+11	58	Spiral galaxy, NGC 3351
96	10	44.1	+12	05	Spiral galaxy, NGC 3368
97	11	11.9	+55	18	Owl nebula, NGC 3587
98	12	11.2	+15	10	Spiral galaxy, NGC 4192
99	12	16.3	+14	41	Spiral galaxy, NGC 4254
100	12	20.4	+16	06	Spiral galaxy, NGC 4321
101	14	01.4	+54	36	Spiral galaxy, NGC 5457
102	15	05.2	+55	57	Spiral galaxy, ?
103	01	29.9	+60	27	Open cluster, NGC 581

appendix 5

Frequencies Allocated for
Radio Astronomy

Reference: "Radio Spectrum Utilization," Joint Technical Advisory Committee, IEEE, New York, 1965.

Interference from radio transmitters constitutes a serious threat to future radio-astronomy observations. The situation is rapidly becoming more critical because the transmitters are becoming more numerous and more powerful while the radio telescopes are becoming more sensitive. In order to afford some measure of protection to radio observatories a few frequency bands have been set aside by international agreement and are reserved partially or exclusively for radio-astronomy observations. One of the most important is the band from 1,400 to 1,427 Mc for the neutral-hydrogen line. Unfortunately, however, *L*-band radars have been assigned frequencies just below 1,400 Mc. These radars operate with peak powers of many megawatts and even a small amount of spillover into the hydrogen-line band can disrupt observations. Suppression in excess of 80 dB may be required to reduce this spurious radiation to a sufficiently low level. However, such suppression would need to be applied to all transmitters, a difficult undertaking. Relocation of such powerful transmitters to frequencies more remote from the hydrogen-line band probably would be more effective. Ultimately both suppression and relocation may be required. In the United States the Federal Communications Commission is responsible for all frequency assignments. The U.S. National Academy of Sciences acts in an advisory capacity regarding the requirements of radio astronomy and other areas of scientific research. Allocations are listed internationally by the International Frequency Registration Board of the International Telecommunications Union, Geneva, Switzerland.

Frequency band		*Use*
2.5 Mc	5 Kc†	Shared with time services
5 Mc	5 Kc†	Shared with time services
10 Mc	5 Kc†	Shared with time services
15 Mc	5 Kc†	Shared with time services
20 Mc	5 Kc†	Shared with time services
25 Mc	5 Mc†	Shared with time services
37.25–	38.25 Mc	Shared with fixed and mobile services
73.0 –	74.6 Mc	
79.75–	80.25 Mc	
150.05–	153 Mc	

† Width of band.

Frequencies allocated for radio astronomy (Continued)

Frequency band			Use
322	–	329	Mc	Deuterium line
404	–	410	Mc	
606	–	614	Mc	
1,400	–1,427	Mc	Hydrogen line
1,664.4	–1,668.4	Mc	OH line
1,660	–1,690	Mc	
2,690	–2,700	Mc	
3,165	–3,195	Mc	
4,800	–4,810	Mc	
4,990	–5,000	Mc	
5,800	–5,815	Mc	
8,680	–8,700	Mc	
10.68–	10.7	Gc	
15.35–	15.4	Gc	
19.3 –	19.4	Gc	
31.3 –	31.5	Gc	
33.0 –	33.4	Gc	
33.4 –	34	Gc	
36.5 –	37.5	Gc	

Relation of Beam Width and Side-Lobe Level to Aperture Distribution

Aperture field distribution			
Rectangular or linear apertures	$E(x)$	Half-power beam width†	Level of first side lobe, dB
(1) Field only at edge		$\dfrac{29°}{L_\lambda}$	0
(2) Uniform except for zero field over central ⅓ zone‡		$\dfrac{41°}{L_\lambda}$	−4
(3) Uniform except for zero field over center ⅑ area‡		$\dfrac{48°}{L_\lambda}$	−10
(4) Uniform except for zero field over central ⅕ zone‡		$\dfrac{45°}{L_\lambda}$	−6
(5) Uniform except for zero field over central 1/25 area‡		$\dfrac{50°}{L_\lambda}$	−12
(6) Uniform		$\dfrac{51°}{L_\lambda}$	−13

† L_λ = length of rectangular or linear aperture in wavelengths. D_λ = diameter of circular aperture in wavelengths. It is assumed that $L \gg \lambda$ and $D \gg \lambda$. Data on some other tapers are given, for example, by S. Silver, "Microwave Antenna Theory and Design," McGraw-Hill Book Company, New York, 1949.

‡ Plan views:

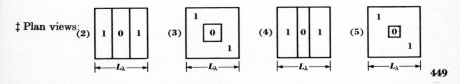

Relation of beam width and side-lobe level to aperture distribution (Continued)

Aperture field distribution		Half-power beam width†	Level of first side lobe, dB
Rectangular or linear apertures	$E(x)$		
(7) Tapered to $\frac{1}{3}$ at edge ($\sim$10 dB down) $E(x) = 1 - 2x^2/3$		$\dfrac{59°}{L_\lambda}$	-19
(8) Tapered to zero at edge $E(x) = 1 - x^2 \simeq \cos(\pi x/2)$		$\dfrac{66°}{L_\lambda}$	-21
(9) Tapered to zero at edge $E(x) = \cos^2(\pi x/2)$		$\dfrac{83°}{L_\lambda}$	-32
Circular apertures	$E(r)$		
(1) Uniform		$\dfrac{58°}{D_\lambda}$	-18
(2) Tapered to $\frac{1}{3}$ at edge ($\sim$10 dB down) $E(r) = 1 - 2r^2/3$		$\dfrac{66°}{D_\lambda}$	-23
(3) Tapered to zero at edge $E(r) = 1 - r^2$		$\dfrac{73°}{D_\lambda}$	-25
(4) Tapered to zero at edge $E(r) = (1 - r^2)^2$		$\dfrac{84°}{D_\lambda}$	-31

appendix **7**

Precession Charts

Figures A-7a and b show the precession in right ascension in seconds of time per year and of declination in seconds of arc per year as a function of position as given by the relations

$$\Delta\alpha = 3\overset{s}{.}07 + 1\overset{s}{.}34 \sin\alpha \tan\delta \qquad \text{per year}$$
$$\Delta\delta = 20'' \cos\alpha \qquad \text{per year}$$

The chart at the right of Fig. A-7b indicates the precession in declination in minutes of arc as a function of the interval in years. For example, to find the precession for an object at RA = 04^h00^m for an 18-year interval one enters Fig. A-7b (left) at 4 hr, finding a precession of 10 sec of arc per year. Then moving horizontally to the right-hand chart to a point above 18 years the precession is found to be 3 min of arc for the 18-year interval.

Figures A-7a and b are useful for approximate precession determinations for intervals between epochs 1800.0 to 2000.0. For more accurate determinations see Sec. 2-9.

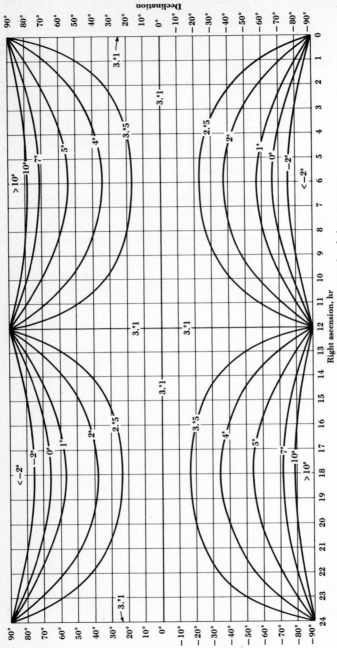

Fig. A-7a. Right-ascension precession in seconds of time per year.

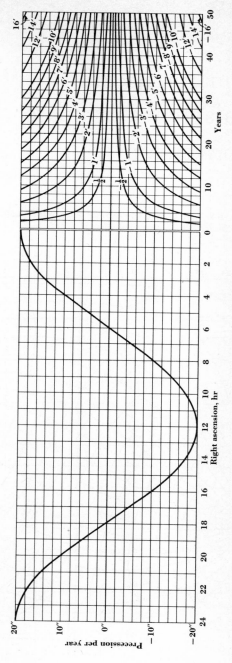

Fig. A-7b. Declination precession in seconds of arc per year (left) and minutes of arc as a function of interval in years (right).

Equatorial- to Galactic-Coordinate Conversion

Fig. A-8. Chart for conversion of equatorial (1950.0) coordinates into new galactic coordinates (l^{II}, b^{II}) or vice versa.

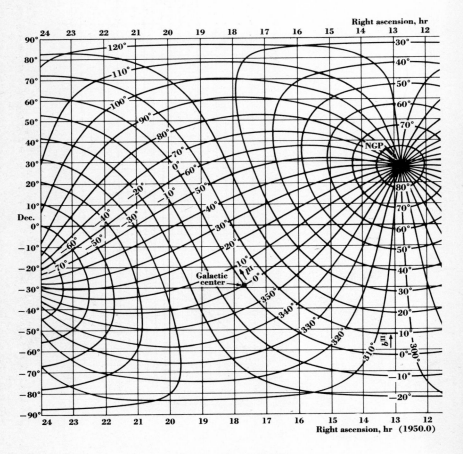

Figure A-8 is useful for approximate conversion of equatorial (1950.0) coordinates into new galactic coordinates (l^{II}, b^{II}) or vice versa. For more precise conversion reference should be made to the following tables: *Annals of the Observatory of Lund*, Nos. 15, 16, and 17, Lund, Sweden, 1961, of which No. 15 is for the conversion of equatorial coordinates (1950.0) into new galactic coordinates (l^{II}, b^{II}), and No. 16 vice versa; No. 17 is for the conversion of the old (l^{I}, b^{I}) to the new (l^{II}, b^{II}) galactic coordinates.

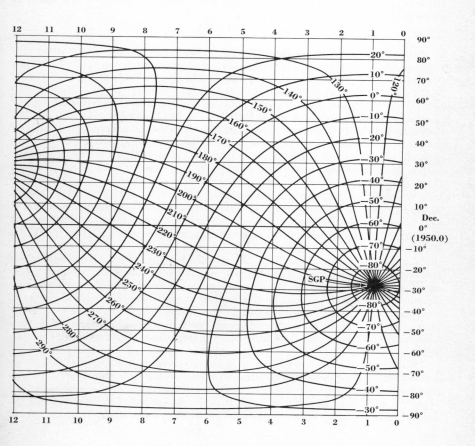

Supplement

This appendix has been added as a supplement to the second printing in order to include material on a few of the new radio astronomy developments of the past year. The subject of quasi-stellar radio sources continues to grow rapidly, and results of recent observations, including a list of 58 quasi-stellar sources with red shifts, are presented. Cosmological versus local interpretations of the red shift are reviewed. Observations of the variability in the power output of quasi-stellar radio sources are summarized and possible evolutionary trends in these sources are discussed. Recent OH line measurements are also mentioned, and a section is included on a wave-to-antenna coupling factor.

a   Quasi-Stellar Radio Sources   The ten quasi-stellar radio sources (quasars) with the largest measured red shifts (z value in parentheses) are as follows: $1116+12$ (2.118), $0106+01$ (2.107), 3C9 (2.012), $1148-00$ (1.982), 3C191 (1.946), 3C432 (1.805), 3C454 (1.757), 3C280.1 (1.659), 4C31.38 (1.557), and 3C270.1 (1.519).   The red shift of 2.118 for the Parkes, Australia, source $1116+12$ corresponds to a recession velocity of about 80 percent of the velocity of light [see Eq. (8-77) and Fig. 8-61].   If this is a cosmological velocity of recession associated with the general expansion of the universe, the relativistic distance of the object is some 8 billion light years, assuming a uniformly expanding universe and a Hubble constant of 100 km sec^{-1} Mpc^{-1}.   The large power output (high luminosity) required of an object at this distance is one of the problems that has led to consideration of alternative explanations for the red shift.   These have been discussed by Burbidge and Hoyle (1966); Arp (1966); and Terrell (1964, 1966). The most likely explanation postulates that the red shifts are Doppler shifts associated with large velocities resulting from the ejection of the quasi-stellar objects from our own and/or other galaxies.   If the objects are ejected from other galaxies, blue shifts should be more common than red shifts, which is contrary to present observations (Setti and Woltjer, 1966; Noerdlinger, Jokipii, and Woltjer, 1966).   If our own galaxy is assumed to be the source of the objects, only red shifts would be expected.   This agrees with observations, but the large power output problem of the cosmologically distant objects is then replaced by the problem of explaining the source of energy in our galaxy that could supply the ejected objects with the required huge amounts of kinetic energy.   Although a red shift could also be produced by an intense gravitational field, this mechanism is not believed to be responsible (Greenstein and Schmidt, 1964).

Aside from the large power requirements involved, it has appeared difficult to account for the rapid variations in radio and optical output if the objects are at cosmological distances.   Time variations of the order of months or even weeks have been observed, which may be taken to imply that the dimensions of the source are of the order of light-months, or light-weeks.   A normal galaxy such as M31 has a diameter of over 100,000 light years and optical and radio outputs of about 10^{37} and 10^{32} watts, respectively. By contrast, a quasi-stellar radio source has perhaps 100 times the optical output and 10^6 or more times the radio output. If these powers are produced in volumes measuring light-weeks in diameter instead of hundreds of thousands of light years, the power emitted per unit volume is of the order of 10^{20} greater than in a normal galaxy. This appears to present difficulties. However, if maser action (or negative absorption) plays an important part in the emission of a quasi-stellar radio source, variations from the entire source are possible on a time scale substantially shorter than the light travel time through the source (McCray, 1966). The total

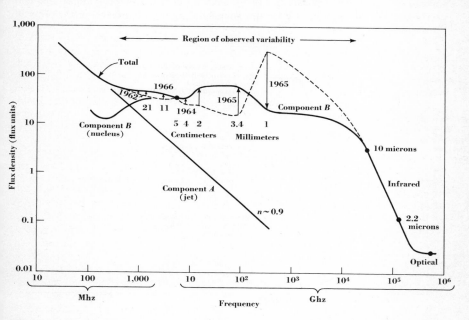

Fig. A-9a. Spectrum of 3C273. The variability of component B is indicated by the vertical arrows. The centimeter data are from Moffet (1966), Pauliny-Toth and Kellermann (1966), and Dent (1966); the millimeter data by Epstein (1965) and Low (1965); and the infra-red and optical data by Low and Johnson (1965).

power and energy requirements are also reduced if the source is anisotropic so that radiation is in a directional beam. It is not clear, however, that such behavior can adequately explain the observed characteristics of the sources. Thus, the problem of the fundamental nature of the quasi-stellar radio sources and their distance remains open.

One of the most interesting and complex radio source spectra is that of the quasi-stellar radio source 3C273, which is presented in Fig. A-9a. This source has two components: one is associated with a jet (component A) and the other with the stellar nucleus (component B). (See pages 389–393 and Fig. 8-59.) At frequencies above 1,000 Mc the spectrum becomes quite flat with the emission due primarily to component B. In the wavelength region between 1 mm and 31 cm significant flux density variations have been observed during the period 1962–1966 as indicated by the vertical arrows. The largest variations were observed at millimeter wavelengths and involved a decrease at 1 mm between January and June, 1965 (Low, 1965) and an increase at 3.4 mm between April and July of the same year (Epstein, 1965).

Variability (see pages 391 and 393) tends to occur in quasi-stellar radio sources which have a relatively flat spectrum at the higher frequencies. One example is 3C273 (see Fig. A-9a). Other examples are 3C sources 47, 48, 279, 345, and 446. A flux density increase of 50 percent was observed in 3C345 between 1962 and 1964 at 11 cm, and a decrease of about the same amount between 1964 and 1966 at the same wavelength (Bartlett, 1966; Moffet, 1966). A 0.4 magnitude increase and subsequent decrease was observed optically for this source during a time interval of the order of

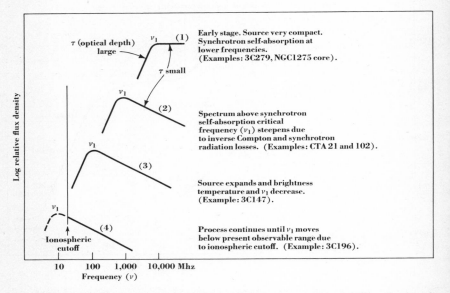

Fig. A-9b. Evolution of quasi-stellar radio source spectra. (*After Dent and Haddock*, 1966.)

one month in 1965 (Goldsmith and Kinman, 1965). The largest observed change was an increase of more than 3 magnitudes in the optical emission of 3C446 between October, 1964, and June, 1966 (Sandage, Westphal, and Strittmatter, 1966). The Seyfert galaxy NGC1275 (see page 386) also has a flat spectrum at the higher frequencies (see Fig. 8-9). It was observed to increase in flux density about 70 percent at 4 cm between 1964 and 1966 (Dent, 1966).

The variability of quasi-stellar radio sources with flat spectra at the higher frequencies suggests that these are very young sources, and according to Dent and Haddock (1966) they may evolve in the manner of the sequence of spectra of Fig. A-9b. After an explosion, synchrotron self-absorption at frequencies below a critical value ν_1 (see page 318) causes the otherwise flat spectrum to have a decrease of flux density at the longer wavelengths (1). Below ν_1 the source is optically thick (optical depth, τ, large) while above ν_1 it is thin (τ small). Next, inverse Compton and synchrotron radiation losses produce a steepening of the spectrum at frequencies above the critical value ν_1 (2). Then as the source expands and its brightness temperature decreases, the critical frequency ν_1 drifts to lower frequencies until ν_1 moves below the observable range (3 and 4).

The spectrum of 3C273 is sufficiently complex to suggest that there are several parts to component B. This may be interpreted to mean that a series of explosions have occurred in the source which have produced the periodic injection of an expanding cloud of energetic particles with gradually decreasing critical frequency (Kellermann, 1966; Pauliny-Toth and Kellermann, 1966). Evidence for a critical frequency of one cloud at a value of about 100 Gc is indicated by the decrease of flux density observed at 1 mm and the concurrent increase at 3.4 mm (see Fig. A-9a). According to Kellermann spectral data of radio galaxies may also be explained by the periodic injection of energetic particles but this would involve much longer intervals between explosions.

Aller and Haddock (1967) have reported polarization variations of the 3C sources 273, 279, and 345 at a wavelength of 4 cm involving both the degree of linear polarization and its position angle. In the case of 273 and 279 the polarized flux densities varied more rapidly than the total flux densities, with the polarized flux densities amounting to 10 percent or more of the total flux densities of the rapidly varying components of the sources.

The radio power output (absolute luminosity) of a source may be calculated from

$$w = 4\pi R^2 S \tag{A-9-1}$$

where w = spectral power (watts cps^{-1})
R = distance of source (m)
S = flux density (jan)
From (3-35) the average brightness of a source is given by

$$B_{avg} = \frac{4S}{\pi\theta^2} \tag{A-9-2}$$

where B_{avg} = average brightness (jan rad^{-2})
S = flux density (jan)
θ = equivalent diameter of source (rad)
Measurements of S, R, and θ have been made for about 50 extragalactic objects, and Heeschen (1966) has used these data to calculate w and B according to (A-9-1) and (A-9-2) with the result shown in Fig. A-9c. The extragalactic objects fall into four groups: quasars, radio galaxies (mostly giant ellipticals), normal galaxies (spirals and irregulars), and cores. The last group is made up of the nuclei of core-halo objects. The central nucleus of the elliptical galaxy M87 (Vigro A) is an example (see page 387 and Figs. 8-57 and 8-59). As a convenience in remembering the radio brightness of the different groups of objects, note from the auxiliary brightness scales in Fig. A-9c that the brightness of normal galaxies lies in the range of 10 to 1000 or more flux units *per square degree*, the brightness of radio galaxies in the range of 1 to 1,000 flux units *per square minute of arc*, and the brightness of quasars in the range of 1 to 1,000 flux units *per square second of arc*.

The long arrows in Fig. A-9c suggest, according to Heeschen, two evolutionary tracks, one along which quasi-stellar radio sources evolve into radio galaxies and the other along which normal galaxies (spirals and irregulars) evolve from objects of much lower luminosity than quasars. It was assumed by Heeschen that the quasi-stellar radio sources are at cosmological distances with R in (A-9-1) related to the observed red shift z as in (8-73). If the quasars are local, their positions would be much lower in the diagram of Fig. A-9c and they would no longer form an extension of the radio galaxy group. However, if the ordinate is replaced by a quantity proportional to Sz^2, the arrangement in Fig. A-9c is unchanged. This might be taken as evidence for a cosmological distance of the quasars unless an explanation can be found for the

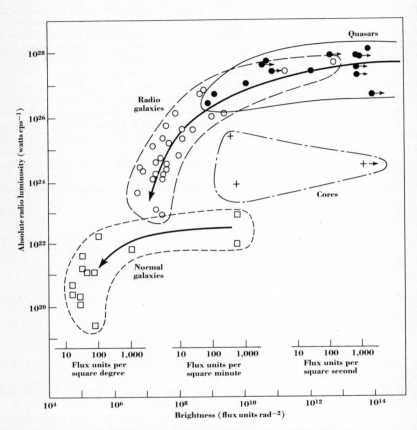

Fig. A-9c. Absolute (spectral) radio luminosity versus brightness of various radio sources. The small arrows indicate that the point represents a lower limit to the brightness. The long arrows suggest two possible evolutionary tracks, one from quasars into radio galaxies and the other from objects of lower luminosity into normal galaxies (*after Heeschen*, 1966).

continuous sequence of quasars and radio galaxies even though the red shift is taken to be cosmological for the radio galaxies but not for the quasars.

From a study of 254 probably extragalactic sources in the revised 3C catalog, which occur in clear fields, Véron (1966) reports that they fall into six groups: (1) 100 certain radio galaxies, (2) 44 possible radio galaxies, (3) 39 certain quasars, (4) 21 possible quasars, (5) 36 empty fields (no visible object near the radio source) and (6) 14 non-identified sources. N versus S graphs, where N is the number of sources of flux density greater than S, are discussed on pages 397 and 398. When N and S are plotted to logarithmic scales an N–S index (equal to $\alpha - 1.5$ in (8-81)) is given directly by the slope of the curve (see Fig. 8-62). For groups (1) and (2) (certain and possible radio galaxies) Véron found an index of -1.55, for groups (3) and (4) (certain and possible quasars) an index of -2.2, and for groups (5) and (6) (empty fields and non-identified sources) an index of -2.2. All groups together had an index of -1.85. For a static, Euclidean universe the index is -1.5. Thus, an excess of weak sources is indicated. Véron concludes from the fact that the index is the same for groups (3) and (4) as for (5) and (6) that the sources in the empty fields and nonidentified groups are quasars. Although Bolton (1966) has obtained similar indices for identified sources in a Parkes, Australia, list he draws a different conclusion. In an Ohio State University survey at 1415 Mc near the north galactic pole, 337 sources were observed at flux densities above 0.3 flux units. These have an index of -1.8 (Scheer and Kraus, 1967).

A list of 58 quasars with their optically determined red shifts is presented in Table A-9-1.

b New OH Observations Emission from the 1665 Mc OH line near the thermal radio source W3 (Westerhout 3) has been observed to have a high degree of circular polarization (Meeks, Ball, Carter, and Ingalls, 1966). The polarization also changes from right to left circular with a radical velocity change as small as about 1 km sec^{-1}. The angular size of the source may be very small, less than 20 sec of arc (Cudaback, Read, and Rougoor, 1966). According to Weaver, Williams, and Dieter (1966) OH emission appears to be associated with intense H II regions with emission by a process which departs radically from thermodynamic equilibrium and involves population inversion among some energy levels. Maser action has been considered (Perkins, Gold, and Salpeter, 1966).

The radio detection of $O^{18}H^1$ at 1639 Mc has been observed in absorption in the galactic nucleus (Sagittarius A) by Rogers and Barrett (1966). An isotopic abundance ratio (O^{16}/O^{18}) was determined which is of the same order as the terrestrial value (500).

c Wave-to-antenna Coupling Factor By rearrangement of (4-86) it follows that

$$w = \left[\frac{1-d}{2} + d \cos^2 \frac{MM_a}{2} \right] A_e S_0 = F A_e S_0 \tag{A-9-3}$$

where w = spectral power (w cps^{-1})
  A_e = effective aperture (m^2)
  S_0 = observed flux density
  d = degree of polarization (dimensionless)
  MM_a = wave-to-antenna matching angle (deg or rad)

The factor F, as given by the expression in the brackets, may be called a *wave-to-antenna coupling factor*. It is a function of the degree of polarization and the angle MM_a, and gives the fraction of the power $A_e S_0$ which is received. It is a dimensionless quantity with values between 0 and 1. When $d = 0, F = \frac{1}{2}$ as in (3-13). When $d = \frac{1}{2}, \frac{1}{4} \leq F \leq \frac{3}{4}$, and when $d = 1, 0 \leq F \leq 1$. At a given d the maximum F (and power received) occurs when the antenna and wave are matched ($MM_a = 0$) and the minimum when the antenna and wave are completely mismatched ($MM_a = 180°$). From (4-82) and (4-90) we can also write $F = [\tilde{a}_i][s_i]/2 = Tr\{[a_{ij}] \times [s_{ij}]\}$. In terms of power, instead of spectral power, we have $W = F A_e S_0'$ where W = power (watts) and S_0' = total flux density (watts m^{-2}). It is assumed that F is constant with respect

Table A-9-1
*A list of 58 quasi-stellar radio sources
with their red shifts z*

Source	z	Ref.	Source	z	Ref.
3C 9	2.012	35	3C345	0.595	6
47	0.425	36	351	0.371	21
48	0.367	36	380	0.691	21, 33
138	0.760	8	432	1.805	35
147	0.545	36	446	1.404	35
181	1.382	35	454	1.757	35
186	1.063	20	4C20.33	0.871	20
191	1.946	9	21.35	0.433	8
196	0.871	20	31.38	1.557	8
204	1.112	35	39.25	0.700	8
208	1.11	5	CTA102	1.037	35
215	0.411	20	CTD141	1.015	35
245	1.028	21, 30	0106+01	2.107	5
249	0.311	35	0859−14	1.327	7
249.1	0.311	35	0922+14	0.896	20
254	0.734	35	0957+00	0.907	20
261	0.614	35	1116+12	2.118	35
263	0.652		1127−14	1.187	8
270.1	1.519	35	1148−00	1.982	8
273	0.158	36	1217+02	0.240	20
275.1	0.557	20	1252+11	0.871	35
277.1	0.320	35	1327−21	0.528	8
279	0.536	21, 33	1354+19	0.720	8
280.1	1.659	20	1454−06	1.249	8
286	0.849	21, 23	1510−08	0.361	8
287	1.055	35	MSH03−19	0.614	35
298	1.439	20	13−011	0.625	8
309.1	0.904	8	14−121	0.942	35
334	0.555	21, 41			
336	0.927	20			

to frequency and position angle over the range of values considered, or, if not, that it is an average.

In the text it is usually assumed that the wave is unpolarized so that $F = \frac{1}{2}$, unless otherwise stated. Thus (3-113) implies unpolarized radiation. To be more general, we could write $S_0 = kT_A/FA_e$. For a point source, or one small compared with the beam size, a complete description of the source would be given by $S = kT_A/FA_e$ and the Stokes parameters s_1, s_2, and s_3.

It is to be noted that an antenna is always (completely) polarized for any given direction but a wave may or not be. Thus, a wave is in general partially polarized with (completely) polarized and (completely) unpolarized cases as limiting conditions. Although the antenna is completely polarized (as given by M_a) for any given direction, the polarization may be a function of position. For example, the main lobe may be linearly polarized in one plane and the minor lobes linearly polarized in another plane. Or the minor lobes may be polarized in various linear or elliptical manners in a random fashion. This randomness does not imply, however, that the antenna's response is unpolarized.

References*

1. ALLER, H. D., and F. T. HADDOCK, *Ap. J.*, **147** (Feb.) 1967.
2. ARP, H., *Science*, **151**, 1214, 1966.
3. BARTLETT, J. F., *A. J.*, **71**, 155, 1966.
4. BOLTON, J. G., *Nature*, **211**, 917, 1966.
5. BURBIDGE, E. M., *Ap. J.*, **143**, 612, 1966.
6. BURBIDGE, E. M., and G. R. BURBIDGE, *Ap. J.*, **143**, 271, 1966.
7. BURBIDGE, G. R., and F. HOYLE, *Ap. J.*, **144**, 534, 1966; *Sci. Am.* **215**, No. 6, 40, 1966.
8. BURBIDGE, E. M., and T. D. KINMAN, *Ap. J.*, **145**, 654, 1966.
9. BURBIDGE, E. M., C. R. LYNDS, and G. R. BURBIDGE, *Ap. J.*, **144**, 447, 1966.
10. CUDABACK, D. D., R. B. READ, and G. W. ROUGOOR, *A. J.*, **71**, 851, 1966.
11. DENT, W. A., *Ap. J.*, **144**, 843, 1966.
12. DENT, W. A., *Science*, **148**, 1458, 1965.
13. DENT, W. A., and F. T. HADDOCK, *Ap. J.*, **144**, 568, 1966.
14. EPSTEIN, E. E., *Ap. J.*, **142**, 1285, 1965.
15. GOLDSMITH, D. W., and T. D. KINMAN, *Ap. J.*, **142**, 1693, 1965.
15a. GREENSTEIN, J. L., and M. SCHMIDT, *Ap. J.*, **140**, 1, 1964.
16. HEESCHEN, D. S., *Ap. J.*, **146**, 517, 1966.
17. KELLERMANN, K. I., *Ap. J.*, **146**, 621, 1966.
18. LOW, F. J., *Ap. J.*, **142**, 1287, 1965.
19. LOW, F. J., and H. L. JOHNSON, *Ap. J.*, **141**, 336, 1965.
20. LYNDS, C. R., S. J. HILL, K. HEERE, and A. N. STOCKTON, *Ap. J.*, **144**, 1244, 1966.
21. LYNDS, C. R., A. N. STOCKTON, and W. C. LIVINGSTON, *Ap. J.*, **142**, 1667, 1965.
22. LUYTEN, W. J., and J. A. SMITH, *Ap. J.*, **145**, 366, 1966.
23. MATTHEWS, T. A., and A. SANDAGE, *Ap. J.*, **138**, 30, 1963.
23a. McCRAY, R., *Science*, **154**, 1320, 1966.
24. MEEKS, M. L., J. A. BALL, J. C. CARTER, and R. P. INGALLS, *Science*, **153**, 978, 1966.
25. MOFFET, A. T., *Owens Valley Repts.* 2 and 6, 1966.
26. NOERDLINGER, P. D., J. R. JOKIPII and L. WOLTJER, *Ap. J.*, **146**, 523, 1966.
27. PAULINY-TOTH, I. K., and K. I. KELLERMAN, *Ap. J.*, **146**, 634, 1966.
28. PERKINS, F., T. GOLD, and E. E. SALPETER, *A. J.*, **71**, 866, 1966.
29. ROGERS, A. E. E., and A. H. BARRETT, *A. J.*, **71**, 868, 1966.
30. RYLE, M., and A. SANDAGE, *Ap. J.*, **139**, 419, 1964.
31. SANDAGE, A., *Ap. J.*, **139**, 416, 1964; **144**, 1234, 1966.
32. SANDAGE, A., J. A. WESTPHAL and R. A. STRITTMATTER, *Ap. J.*, **146**, 322, 1966.
33. SANDAGE, A., and J. D. WYNDHAM, *Ap. J.*, **141**, 328, 1965.
34. SCHEER, D. J., and J. D. KRAUS, *A. J.*, **72**, 1967.
35. SCHMIDT, M., *Ap. J.*, **141**, 1295, 1965; **144**, 443, 1966.
36. SCHMIDT, M., and T. A. MATTHEWS, *Ap. J.*, **139**, 781, 1964.
37. SETTI, G., and L. WOLTJER, *Ap. J.*, **144**, 838, 1966.
38. TERRELL, J., *Science*, **145**, 918, 1964; **154**, 1281, 1966.
39. VÉRON, P., *Nature*, **211**, 724, 1966.
40. WEAVER, H., D. R. WILLIAMS, and N. H. DIETER, *A. J.*, **71**, 184, 1966.
41. WYNDHAM, J. D., *Ap. J.*, **70**, 384, 1965.

* *A. J.: Astronomical Jour.; Ap. J.: Astrophysical Jour.;* volume number in bold face.

Glossary
of Symbols
and Abbreviations

For more information see key equations in text (with symbol lists) and Appendix 1 (List of Physical Quantities and Units) and Appendix 2 (List of Constants) and main index. Italic letters represent scalar quantities; boldface letters represent vector quantities. Abbreviations are always in Roman letters. A bar over a symbol is used to indicate either the Fourier transform as $\bar{B}$ or $\bar{S}$, the mean value as $\overline{\Delta T}$, or a tensor as $\bar{\epsilon}$. Parentheses following a symbol indicate that it is a function of the quantities inside the parentheses. Generally only the simple scalar form is listed in the glossary. Thus, the E given in the glossary represents the scalar electric-field intensity, but possible (unlisted) variants are $\mathbf{E}$ for the vector electric-field intensity, $E(\phi)$ for the electric field as a function of ϕ, $\bar{E}(x_\lambda)$ for the Fourier transform of the electric field as a function of x_λ, etc.

a	acceleration, area, aperture extent	B_i	intrinsic brightness
a_i	antenna polarization parameter	B_g	magnetic flux density in gauss
amp	ampere	$B_{\max}$	maximum brightness
ap	aperture	B_N	output susceptance
atm	atmosphere (of pressure)	B_s	source brightness
avg	average	B_λ	brightness in terms of wavelength
A	area, aperture	B'	total brightness
A_e	effective aperture	BWFN	beam width between first nulls
A_{ep}	effective aperture deduced from pattern	c	velocity of light
$\mathcal{C}$	vector potential	cgs	centimeter-gram-second
a-c	alternating current	cm	centimeters
AGC	automatic gain control	const	constant
AR	axial ratio	coul	coulombs
AU	astronomical unit	cps	cycles per second
b	susceptance (normalized)	cw	clockwise
b_L	tuning susceptance (normalized)	ccw	counterclockwise
b^{I}	galactic latitude (old system)	C	a constant, capacitance
b^{II}	galactic latitude (new system)	C_d	diode capacitance
B	brightness, magnetic flux density	C_p	parallel capacitance
B_{avg}	average brightness	C_0	capacitance (d-c term in Fourier expansion)
B_c	constant brightness	CP	circularly polarized

CSIRO	Commonwealth Scientific and Industrial Research Organization (Australia)	F_R	noise figure of receiver
		F_{R2}	two-channel noise figure
		g	conductance (normalized)
CTA	California Institute of Technology list A	g_G	normalized conductance of generator or signal source
CTB	California Institute of Technology list B	g_L	normalized load conductance
		g_-	negative conductance (normalized)
3C	3d Cambridge catalog		
d	distance, degree of polarization	G	power gain of antenna, exchangeable power gain of amplifier, conductance
d	day		
d_c	degree of circular polarization		
d_i	reactance slope factor of idler circuit	G	special parameter (Eq. 5-89)
		G_A	receiver gain with antenna
d_l	degree of linear polarization	G_C	receiver gain with comparison load
d_o	reactance slope factor of signal circuit	G_e	electronic gain
d-c	direct current	G_G	generator conductance
deg	degree	G_{HF}	high frequency power gain (predetection)
deg^2	square degrees		
dB	decibels	G_{LF}	low frequency power gain (post detection)
ds	dark side		
D	directivity	G_m	gain of mixer
Dec	declination	G_N	output conductance
e	2.71828, charge of particle	Gc	gigacycles per second (10^9 cps)
eff	effective	Gev	billion electron volts (10^9 volts)
emu	electromagnetic unit	Gpc	billion parsecs (10^9 pc)
esu	electrostatic unit	G'_-	negative conductance (unprimed for effective value)
ev	electron volt		
exp	exponent	h	Planck's constant
E	electric field intensity	^{h}	(superscript) hours
E_l	left circularly polarized electric-field intensity	hr	hour
		hz	hertz (1 cps)
E_r	right circularly polarized electric-field intensity	H	magnetic field
		H_0	Hubble constant
ε	energy, energy per unit length	HA	hour angle
ε_d	energy density (per unit bandwidth)	HB	Hanbury Brown
		HF	high frequency
ε_{GV}	energy in billions of electron volts	HP	half power
ε_V	energy in electron volts	HPBW	half-power beam width
EBPA	electron-beam parametric amplifier	H I	neutral hydrogen
		H II	ionized hydrogen
EHF	extremely high frequency (30–300 Gc)	in.	inch
		I	intensity, current, Stokes parameter, radiance.
EM	emission measure	$\mathcal{I}$	short circuit current
EMN	emission nebulosity	$\mathcal{I}_G$	generator current (short circuit)
EST	Eastern Standard Time	$\mathcal{I}_N$	noise current (short circuit)
ET	Ephemeris Time	$\mathcal{I}_s$	signal current (short circuit)
f	noise figure index	IAU	International Astronomical Union
ft	foot		
F	force, noise figure	IC	Index Catalog
F_{IF}	IF amplifier noise figure		

IEE	Institution of Electrical Engineers (London)	max	maximum
IEEE	Institute of Electrical and Electronics Engineers	min	minimum, minute
		mks	meter-kilogram-second
IF	intermediate frequency	mm	millimeters
Im	imaginary part	msec	milliseconds
IRE	Institute of Radio Engineers	M	absolute magnitude, polarization state
ITU	International Telecommunications Union	M_a	polarization state of antenna
I-V	current-voltage	M_p	absolute photographic magnitude
j	emission coefficient, complex operator	M_r	absolute radio magnitude
jan	jansky (1 watt m^{-2} cps^{-1})	M_v	absolute visual magnitude
J	current density	$M_\odot$	mass of sun
k	a constant, Boltzmann's constant	M	Messier
k_a	achievement factor of antenna	Mc	megacycles per second
k_o	ohmic efficiency factor of antenna	Mev	million electron volts
		MHD	magnetohydrodynamic (wave)
k_p	pattern shape factor	Mhz	megahertz (10^6 cps)
k_u	utilization factor of antenna	Mpc	million parsecs
$k_1, k_2, \ldots$	antenna efficiency factors	MSH	Mills, Slee, and Hill
kg	kilogram	MT	meridian transit
km	kilometer	n	a number, an integer, spectral index, population density
kpc	kiloparsec	N	a number, electron density
K	a constant, absorption coefficient	N_d	number of sources detected
K_s	sensitivity constant of receiver	N_r	number of sources resolved
°K	degrees Kelvin	NG	noise generator
Kc	kilocycles per second	NGC	New General Catalog
l	luminosity, length	NCP	north celestial pole
l^I	galactic longitude (old system)	NGP	north galactic pole
l^II	galactic longitude (new system)	NRAO	National Radio Astronomy Observatory, Green Bank, W. Va.
ln	natural logarithm (to base e)		
log	ordinary logarithm (to base 10)	OA	Ohio State University list A
L	length, loss factor	OAO	orbiting astronomical observatory
L_D	coupling loss		
L_m	conversion loss of mixer	OSO	orbiting solar observatory
L_s	series inductance	OSU	Ohio State University
LCP	left circularly polarized	p	pressure, parallax, relative phase velocity
LEP	left elliptically polarized		
LF	low frequency	pc	parsec
LHP	linear horizontal polarization	P	power pattern of antenna, a point, polarization state
LP	linearly polarized		
LVP	linear vertical polarization	P_n	normalized power pattern of antenna
LY	light year		
m	mass, magnitude, an integer	$\tilde{P}$	mirror image of power pattern
m	meters	Q	circuit (decrement) parameter, Stokes parameter
^{m}	(superscript) minutes		
m_0	rest mass	Q_d	Q of diode
m_F	photographic magnitude	Q_{di}	varactor diode Q on idler frequency
m_r	radio magnitude		
m_v	visual magnitude	Q_{ds}	varactor diode Q on signal frequency
ma	milliamperes		

Q_{Li}	loaded Q of idler circuit
Q_m	magnetic quality factor
Q_M	auxiliary quantity
Q_0	intrinsic quality factor
QSG	quasi-stellar galaxy
QSO	quasi-stellar object
QSS	quasi-stellar source
QSRS	quasi-stellar radio source
r	radius, distance, normalized resistance
r_-	normalized resistance of negative resistance device
rad	radian
rad²	square radians (steradians)
rms	root mean square
R	radius, range, resistance
R_d	diode resistance
R_i	idler resistance
R_G	generator resistance
R_N	output resistance
R_0	standard distance, galactic radius
R'_-	negative resistance (unprimed for effective value)
RA	right ascension
RC	resistance-capacitance
RCP	right circularly polarized
Re	real part
REP	right elliptically polarized
RF	radio frequency
s	spacing, standing wave ratio
^{s}	(superscript) seconds
s_i	normalized Stokes parameter
s_λ	spacing in wavelengths
sec	second
ss	sunlit side
sys	system
S	flux density, tension (Eq. 5-98), phase angle (Chap. 7), Poynting vector
S'	total flux density
S_L	left circularly polarized Poynting vector
S_o	observed flux density, flux density of source
S_p	polarized Poynting vector
S_R	right circularly polarized Poynting vector
S_t	Poynting vector of wave from antenna when transmitting
S_u	unpolarized Poynting vector
S_0	flux density at standard distance
SCP	south celestial pole

SGP	south galactic pole
SHF	super high frequency (3-30 Gc)
SNR	supernova remnant
SS	solid state
ST	Sidereal Time
t	time, time constant, integration time
t_{LF}	(low frequency) integration time
t_m	noise temperature ratio
T	temperature, noise temperature, spot noise temperature
T	time (see ST, UT, etc.)
T_A	antenna noise temperature
T_{avg}	average noise temperature
T_b	brightness temperature
T_c	cloud temperature
T_{const}	constant temperature
T_C	comparison temperature
T_d	diode noise temperature, diode physical temperature
T_{dg}	degenerate parametric amplifier noise temperature
T_D	noise temperature of directional coupler
T_G	noise temperature of generator
T_{GC}	temperature of cold load (acting as generator)
T_i	idler noise temperature, idler load physical temperature
T_L	load temperature
T_{LP}	physical temperature of transmission line
T_M	noise temperature of maser
T_{MP}	physical temperature of maser
T_R	receiver noise temperature
T_{RT}	noise temperature of receiver, transmission line combination
T_{R2}	two-channel receiver noise temperature
T_s	source temperature
T_{sys}	system temperature
T_T	noise temperature of transmission line, attenuator
T_0	290°K
T'_-	noise temperature of negative resistance [unprimed for (effective) noise temperature of effective negative resistance]
Tr	trace
TWM	traveling wave maser
TWT	traveling wave tube

U	radiation intensity, Stokes parameter	W_{sys}	system noise power
U_{avg}	average radiation intensity	W_X	mixer noise power
U_{max}	maximum radiation intensity	x	coordinate direction, normalized reactance
UHF	ultra high frequency (0.3–3 Gc)	x_λ	distance in wavelengths
UT	Universal Time	$x_{\lambda 0}$	spatial frequency
v	velocity, voltage, volume	x_{λ_c}	spatial frequency cut off
v_g	group velocity	X	reactance
V	voltage, Stokes parameter	X_i	idler reactance
$\mathcal{U}$	electromotive force (emf), open circuit voltage	X_N	output reactance
		X_s	signal circuit reactance
V_a	cost per aperture	X_+	sum frequency tuning reactance
V_D	detector d-c voltage	y	coordinate direction, normalized admittance
V_{det}	detector output voltage		
V_{eff}	effective voltage	y	year
V_{fp}	forward voltage	yr	year
$\mathcal{U}_G$	generator voltage (open circuit)	y_-	normalized admittance of negative resistance device
V_i	idler voltage		
V_{IF}	IF amplifier output voltage	Y	admittance
V_{out}	output voltage	Y_d	diode admittance
V_p	peak voltage	Y_0	characteristic admittance
V_r	cost of receiver	z	redshift, coordinate direction
V_R	receiver output d-c voltage	z_m	zenith angle
V_s	signal voltage	z_-	normalized impedance of negative resistance device
V_v	valley voltage		
V_0	visibility function	Z	impedance, intrinsic impedance of medium
VHF	very high frequency (30–300Mc)		
w	spectral power (power per unit bandwidth)	Z_0	characteristic impedance
		α	an angle, right ascension, source extent, attenuation constant, number-flux index, receiver frequency index
W	power		
W_d	detector (input) power		
W_D	detector d-c power output		
W_G	generator power	$\propto$	proportionality sign
W_{GN}	generator noise power	β	an angle, phase constant $(= 2\pi/\lambda)$
W_{GNT}	total generator noise power		
W_{GX}	generator noise power through mixer	γ	an angle, polarization parameter, propagation constant
W_{HF}	high frequency power	δ	an angle, declination, phase angle, small quantity, delta function, complex deviation factor
W_L	load power		
W_{LF}	low frequency (output) power		
W_N	noise power		
W_{NC}	noise power from comparison load	Δ	finite increment
		$\Delta\nu$	bandwidth
W_{NT}	total noise power	ΔG	gain variation
W_{NA}	antenna noise power	ΔV	output signal voltage
W_{NR}	noise power due to receiver (including transmission line) referred to antenna terminals	ΔT	change in temperature, signal temperature
		ΔT_{min}	minimum detectable temperature
W_{NRT}	total receiver noise power		
W_{NS}	signal noise power	$\Delta\nu_{HF}$	high frequency bandwidth (predetection)
W_{NT}	total output noise power		

$\Delta\nu_{IF}$	intermediate frequency bandwidth	ν_s	signal frequency
		π	3.1415927
$\Delta\nu_{LF}$	low frequency bandwidth (post-detection)	Π	product sign
		ρ	charge density, mass density, reflection coefficient
$\Delta\nu_N$	noise equivalent bandwidth		
$\Delta\nu_s$	3 dB bandwidth of a single amplifier stage	ρ_G	reflection coefficient towards generator
ϵ	permittivity of medium, polarization parameter, transmission line efficiency coefficient	ρ_L	reflection coefficient towards load
		ρ_-	voltage reflection coefficient of negative resistance device
ϵ_{ap}	aperture efficiency		
ϵ_f	feed efficiency	σ	conductivity
ϵ_m	stray factor	Σ	summation sign
ϵ_M	main beam efficiency	τ	optical depth, tilt angle
ϵ_0	permittivity of vacuum	ϕ	an angle
ζ	damping constant	ϕ_{FN}	angle between first nulls in ϕ direction
η	index of refraction		
θ	an angle	ϕ_G	grating lobe angle
θ_{FN}	angle between first nulls in θ direction	ϕ_{HP}	half-power angle in ϕ direction
		ψ	an angle
θ_{HP}	half power angle in θ direction	ω	radian frequency ($= 2\pi\nu$)
λ	wavelength, longitude	ω_g	radian gyro frequency
μ	permeability of medium	ω_i	radian frequency of idler
μsec	microseconds	ω_p	radian frequency of pump
ν	frequency	ω_s	radian frequency of signal
ν_{CV}	cutoff frequency at bias voltage V	ω_0	critical or plasma (radian) frequency
ν_d	diode resonant frequency	ω_+	radian sum frequency
ν_i	image frequency, idler frequency	Ω	solid angle
ν_{IF}	IF signal frequency	Ω_A	beam solid angle of antenna
ν_M	switch frequency	Ω_m	minor-lobe solid angle
ν_{max}	frequency of maximum radiation	Ω_M	main-beam solid angle
ν_o	oscillator frequency	Ω_M'	main-beam plus near side lobe solid angle
ν_p	pump frequency		
ν_{RF}	RF signal frequency	Ω_s	source solid angle
ν_{ro}	resistive cutoff frequency		

Name Index

Subject Index